IET HEALTHCARE TECHNOLOGIES SERIES 15

Value-based Learning Healthcare Systems

IET Book Series on e-Health Technologies—Call for Authors

Book Series Editor: Professor Joel P. C. Rodrigues, the National Institute of Telecommunications (Inatel), Brazil and Instituto de Telecomunicações, Portugal
While the demographic shifts in populations display significant socioeconomic challenges, they trigger opportunities for innovations in e-Health, m-Health, precision and personalized medicine, robotics, sensing, the Internet of Things, cloud computing, Big Data, Software Defined Networks, and network function virtualization. Their integration is however associated with many technological, ethical, legal, social, and security issues. This new Book Series aims to disseminate recent advances for e-Health Technologies to improve healthcare and people's wellbeing.
Topics considered include Intelligent e-Health systems, electronic health records, ICT-enabled personal health systems, mobile and cloud computing for eHealth, health monitoring, precision and personalized health, robotics for e-Health, security and privacy in e-Health, ambient assisted living, telemedicine, Big Data and IoT for e-Health, and more.
Proposals for coherently integrated International multi-authored edited or co-authored handbooks and research monographs will be considered for this Book Series. Each proposal will be reviewed by the Book Series Editor with additional external reviews from independent reviewers. Please email your book proposal for the IET Book Series on e-Health Technologies to: Professor Joel Rodrigues at joeljr@ieee.org or joeljr@inatel.br.

Value-based Learning Healthcare Systems

Integrative Modeling and Simulation

Bernard P. Zeigler, Mamadou Kaba Traore,
Grégory Zacharewicz and Raphaël Duboz

The Institution of Engineering and Technology

Published by The Institution of Engineering and Technology, London, United Kingdom

The Institution of Engineering and Technology is registered as a Charity in England & Wales (no. 211014) and Scotland (no. SC038698).

The Institution of Engineering and Technology
Michael Faraday House
Six Hills Way, Stevenage
Herts, SG1 2AY, United Kingdom

www.theiet.org

British Library Cataloguing in Publication Data
A catalogue record for this product is available from the British Library

ISBN 978-1-78561-326-5 (hardback)
ISBN 978-1-78561-327-2 (PDF)

Typeset in India by MPS Limited
Printed in the UK by CPI Group (UK) Ltd, Croydon

Contents

About the authors

Bernard P. Zeigler is Professor Emeritus of Electrical and Computer Engineering at the University of Arizona, USA. He received a Ph.D. in Computer and Communication Sciences from the University of Michigan (1968), an M.S. in Electrical Engineering from the Massachusetts Institute of Technology (1964), and a B.S. in Engineering Physics from McGill University in Montreal, Canada (1962). Zeigler is currently affiliated with the C4I Center at George Mason University and serves as the Chief Scientist at RTSync Corp. He is a Fellow of the Society for Modeling and Simulation International and a member of its Hall of Fame. He made a Fellow of the IEEE in 1985 for invention of the DEVS formalism. In addition to leading research in Health Information Technology, he continues to develop the Theory of Modeling and Simulation, now in its third edition since 1976 as well software environments to support its use.

Mamadou Kaba Traore received his M.Sc. (1989) and Ph.D. (1992) in Computer Science from Blaise Pascal University, France, where he is currently heading the Web & Mobile Software Engineering M.Sc. degree. He is also Adjunct Professor at the African University of Science and Technology, Nigeria, where he heads the Computer Science department. His current research is on formal specifications, symbolic manipulation, and automated code synthesis of simulation models.

Grégory Zacharewicz received his Habilitation Thesis (HDR, 2014) from the University of Bordeaux, his Ph.D. (2006), and Engineer Degree (2002) from AMU University, France. He is currently Full Professor at the National Institute of Mines and Telecom (IMT), and Ecole des Mines d'Ales, France. His research topics cover Modeling and Interoperability of Enterprise Modeling & Distributed Simulation; Synchronization Algorithms; High Level Architecture (HLA), Workflow modeling, DEVS, G-DEVS, Simulation, Multi Agents; Domain Ontology, Ontology Ephemeral, and Model transformation; Model Matching. https://orcid.org/0000-0001-7726-1725.

Raphaël Duboz received his Ph.D. in Computer Science (CS) in 2004 after two master degrees, one in Computer Science, and the other one in Marine Environmental Modeling. He is currently working in the field of epidemiology at Cirad, a French research institute working in partnership with the developing countries. He has been seconded in Thailand and Cambodia at the Asian Institute of Technology and Institut Pasteur du Cambodge, respectively. His interests are dynamical systems, multi-modeling and hybrid simulation, discrete event systems, decision making, qualitative modeling, participatory modeling, and networks analysis. He has been involved in different European and French research and development programs which aim at promoting and developing participatory modeling techniques in holistic approaches called "One Health" and "EcoHealth." Dr Duboz is coleading the RED network for the development and promotion of DEVS in France and francophone countries.

Glossary

Activity-based credit assignment (ACA)   a form of system of systems (SoS) learning in which credit is assigned to component systems based on correlation between their activity and the behavior achieved at the SoS composition level allowing alternative versions to be selected.

Agent Based Modeling (ABM)   is a form of modeling that includes autonomous decision making by humans or other entities with agency.

Alternative payment models (APM)   for services that move in steps toward value-based payment schemes.

Business Process Modeling Notation (BPMN)   a modeling formalism to represent business processes that can be used for healthcare processes as well.

Care delivery value chain (CDVC)   all the services needed for the target condition are organized into an *end-to-end interaction* with the patient called a full cycle of care.

Clients   people in need of health and social services referred to the Pathways Community HUB including, e.g., patients at risk of low birth weight pregnancies.

Community care coordination   coordination of care outside the healthcare setting addressing both health and social services needs provided by *care coordinators* such as *community health workers (CHWs)*.

Complex adaptive system (CAS)   a subsystem component inside a SoS that displays intelligent, adaptive, or autonomous or semiautonomous behavior.

Core pathways   the pathways that are approved by the National Pathways Community HUB Certification Group, of which there are currently 20 in number.

Data architecture   a model of the interactions among data systems of an information system setting forth data processing requirements and standards.

Data validity   refers to the level of completeness (i.e., the amount of missing data for a data element), accuracy (i.e., the extent to which the data reflects the underlying state or process of interest), and granularity (i.e., clinical specificity) of a data set.

Decision support system   a computer-based information system that supports business or organizational decision-making activities.

Discrete event system specification (DEVS)   a formalism for developing simulation models based on discrete events and phrased within systems theory.

Electronic health record (EHR)   defines a broader repository and record of client information than an electronic "Medical" record, which is generally defined as only containing clinical data.

Experimental frame characterizes the conditions under which a model is to be simulated and hence the set of experiments for which the model is expected to be apply and be valid.

Formalization providing a definite structure for an informal concept; in this book, through providing a logical, mathematical, and/or systemic definition that supports computer manipulation.

Healthcare system also referred to as health system, healthcare system or healthcare delivery SoS, is an organization of people, institutions, and resources that deliver health care services to meet the health needs of target populations.

Health system capability maturity metrics for assessing how mature a health system is in achieving levels of coordination needed to achieve value-based healthcare.

Health Information Technology (HIT) refers to computer and information technology dedicated to health care.

High-level language for systems specification a visual specification language for constructing multi-analysis system models.

Health system performance metrics for assessing how well a healthcare system is achieving the iron triangle (value-based) objectives.

Holistic simulation a technique to integrate in a common simulation various models of the same system, which have been developed from different perspectives.

Implementation generally, the process of putting a decision or plan into effect; commonly employed in information technology as the process of developing software to the specifications of a data architecture.

Integrated approaches to health (IAH) calls for transdisciplinary efforts integrating scientists, citizens, government, and private sectors to collaborate in designing and implementing actions to enhance sustainable health management across human, animal, and ecosystem interfaces.

Integrated practice unit (IPU) reform of health care that requires that physicians re-organize themselves into such units, which tackle a disease or disorder in end-to-end fashion.

Iron triangle the objectives of value-based healthcare can be broadly stated as simultaneously achieving low cost, high quality, and wide accessibility of services.

Integration process by which different models can be brought together to simulate a target system.

Legal traces sequences of steps that are prescribed by a pathway model (as opposed to illegal sequences that do not conform to the pathway specification).

Level of abstraction the amount of complexity at which a system is represented.

Machine learning (ML) algorithms for supervised and unsupervised discovery of patterns from data including data mining and clustering.

Metadata data that serves to provide context or additional information about other data, in this paper about the data in a database.

Modeling and simulation knowledge the knowledge specific to the behavior and structure of the system under study.

Multi-formalism modeling the use of a mixture of appropriate formalisms to model the different components of a system.

Multi-paradigm modeling a research field focused on breaking the complexity of complex systems into different levels of abstraction and views, each expressed in an appropriate modeling formalism.

Multi-perspective modeling framework a framework for developing simulation models from multiple perspectives simultaneously.

Ontology a formalization of concepts and relationships underlying knowledge about a domain such as medicine; prescribes semantics of terms used in the domain and can be extended to include pragmatics (see definitions below).

Ontology for healthcare systems simulation (O4HCS) an ontology meant to capture and share a common understanding of the abstractions necessary/used for the simulation of the entire healthcare domain (beyond unit specific and facility specific modeling).

Parameter a value introduced in a model to stand for some of the model's inherent properties.

Participatory modeling a purposeful learning process for action that engages the implicit and explicit knowledge of stakeholders to create formalized and shared representation(s) of reality.

Pathways standardized tools to track identified social or medical issues to a measurable outcome.

Pathways Community HUB Model a delivery system for care coordination services provided in a community setting with primary features: (1) core pathways, (2) the HUB itself, (3) a network of care coordination agencies, and (4) payments linked to outcomes.

Pathways client record (PCR) record of client's progress through pathway interventions that have been initiated for him/her.

Pathway structure refers to the manner in which states (steps) and transitions (from one state to another) of a pathway model relate to each other.

Pathway temporal behavior the succession in time of the pathway states (steps) including the time it takes to execute each step.

Perspective a focus on particular preselected aspects of a system.

Pragmatics refers to the manner in which data is to be used, including context and intent.

Public health surveillance the continuous and systematic collection, analysis, and interpretation of health-related data needed for the planning, implementation, and evaluation of public health practice.

Risk factors outcomes intervention that are achievable will depend to some degree on each patient's *initial conditions*, also called *risk factors*, in medical, social, and behavioral dimensions.

Scale-driven modeling   a modeling technique to exhibit a hierarchical organization from differences in temporal and spatial scales between phenomena of interest.

Semantics   refers to the meaning of the data independently of its use (both semantics and pragmatics are needed in a complete ontology).

Social determinants   the complex, integrated, and overlapping social structures and economic systems that are responsible for most health inequities, including the social environment, physical environment, health services, and structural and societal factors.

System entity structure (SES)   a declarative knowledge representation scheme that characterizes the structure of a family of models in terms of decompositions, component taxonomies, and coupling specifications and constraints.

System of systems (SoS)   a system composed of multiple complex systems.

Value-based healthcare system   a system that orchestrates its services to achieve the iron triangle objectives.

Zoonosis   contagious diseases that are transmitted from animals to humans.

Prologue
The "iron triangle"

This book is intended to fill a gap in the literature on healthcare that is becoming more evident as reform efforts proliferate: a holistic modeling and simulation (M&S) approach to value-based healthcare. The large majority of publications on M&S take specific perspectives that intend to improve the management efficiency of particular healthcare component systems. In contrast, this book develops a holistic approach to healthcare system of systems (SoS) by developing a multi-perspective modeling framework that supports a systems theory-based simulation methodology. Furthermore, this book is the first to apply such a framework to M&S of value-based healthcare.

The objectives of value-based healthcare can be broadly stated by the following equation:

$$Objectives = low\ cost + high\ quality + wide\ accessibility \qquad (\text{P.1})$$

The exact meaning of the attributes, cost, quality, and accessibility, can vary, as can their priority, or even applicability, in different contexts. Nevertheless, when we refer to measuring value, we mean some concrete formulation of increase in quality, while reducing cost, and increasing access. The importance of (P.1) becomes evident when we recognize that a healthcare service system is composed of a large number of distributed components that are interrelated by complex processes. Understanding the behavior of the overall system is becoming a major concern among healthcare managers and decision-makers intent on increasing value for their systems.

An Agency for Healthcare Research and Quality/National Science Foundation workshop [1] envisioned an ideal healthcare system that is unlike today's fragmented, loosely coupled, and uncoordinated assemblage of component systems. The workshop concluded, "An ideal (optimal) healthcare delivery system will require methods to model large-scale distributed complex systems." Improving the healthcare sector presents a challenge that has been identified under the rubric of SoS engineering [2], in that the optimization cannot be based on suboptimization of the component systems, but must be directed at the entire system itself. People with multiple health and social needs are high consumers of healthcare services and are thus drivers of high healthcare costs. The ability to provide the right information to the right people in real time requires a system-level model that identifies the various community partners involved and rigorously lays out how their interactions might be effectively coordinated to improve care for the neediest patients whose care costs the most.

The overarching objective of this book is to demonstrate an M&S methodology and framework to model the entire healthcare system as a loosely coupled distributed system [3]. The approach is to tackle a particular instance of such a framework through application to coordinated care. The initial goal was to develop a prototype of such a system model, as envisioned by the workshop, as an end in itself, and to demonstrate the applicability of the framework to myriad other health information technology (IT) problem areas. The criteria for such a model are that it should be

- flexible to meet variety of stakeholders' interests and variety of accountable care implementations;
- scalable to accommodate increases in scope, resolution, and detail;
- integrated as a SoS concept: system, components, and agent-based concepts extended to human behavioral limitations;
- enhanced with electronic medical records and health IT systems as needed to support coordinated care; and
- supported with services based on the model, e.g., patient tracking, medication reconciliation, and so forth.

Such a model will show how to systematically represent the behaviors of patients who require coordinated care interventions and the providers of such coordination services. This will render such behaviors amenable to healthcare system design and engineering.

Most of the work concerning healthcare system M&S in the literature is unit or facility specific. This is revealed in numerous research efforts published over many decades seeking to provide support for efficient healthcare system management. However, such perspective-specific efforts cannot provide the necessary big picture for a fine-grained understanding of issues. In this book, we take a holistic approach to healthcare systems M&S. We present a framework that encompasses common perspectives taken in the research literature but also goes beyond them toward their integration with additional perspectives that are becoming critical in today's environment. We propose a stratification of the levels of abstraction into multiple perspectives. In each of these perspectives, models of different components of healthcare systems can be developed and coupled together. Concerns from other perspectives are abstracted as parameters in such models. An important element of this approach is that we attempt to reflect the parameter values of other perspectives through explicit assumptions and simplifications. Consequently, the resulting top model within each perspective can be coupled with its experimental frame to run simulations and derive predictions of value outcomes for various alternatives tried. Components of the various perspectives are integrated to provide a holistic view of the healthcare problem and system under study. The resulting global model can be coupled with a holistic experimental frame to derive results that cannot be accurately addressed in any of the perspectives if taken alone.

Given its groundbreaking nature, the book is aimed primarily at the emerging healthcare systems research community. The book is a "must read" for such researchers as university professors and their graduate students with background in M&S in healthcare related disciplines and multidisciplinary fields such as industrial engineering, computer engineering, systems engineering, management science, and computer science. Another growing audience consists of researchers and developers of health policy alternatives in public and private health departments and research institutes. The book will also be of great interest to professionals in IT, policy and administration, public health, and law, medical technology developers whose devices must interoperate with others (an increasing important requirement as healthcare integration takes hold). Although primarily aimed at researchers, the holistic overview that the book takes to healthcare systems also makes it an attractive supplementary background resource for courses focused on more particular aspects of healthcare systems policy, design, or management. Furthermore, courses on methodological aspects of M&S as such can use the book as a source of concepts, methods, and tools that apply more generally and of examples of such concepts in specific application to the healthcare domain.

In the following paragraphs, a brief summary of the book is provided.

Part I of the book is organized so as to gradually provide the elements that lead to the design and implementation of our framework. In Chapter 1, an ontology is developed based on an extensive literature review, to provide healthcare systems M&S knowledge in a reusable and interoperable manner. An ontology is a formal specification of a conceptualization that can be used to rigorously define a domain of discourse in terms of classes/concepts, properties/relationships, and instances/individuals. The objective of the healthcare M&S ontology is to capture, from current practices in the domain, the core modeling components and their relationships, so as to assist the design of new models, while facilitating communication among systems engineers and ensuring the sharing of commonly understood semantics. We use the system entity structure ontological framework to formally express it. In Chapter 2, this ontology serves as a basis to define a multi-perspective approach to M&S of healthcare systems, which provides multiple levels of explanation for the same system. As using any perspective-specific model alone to analyze and design healthcare systems can be misleading, a methodology for a "loose" integration of all identified perspectives is defined to enable independent simulation processes built under different perspectives to exchange live updates of their mutual influences under a unifying umbrella. Such approach takes the results obtained closer to the reality of the interactions between health phenomena and helps stakeholders gain a holistic understanding of the whole healthcare system, while deriving more realistic decisions. This has the potential to significantly help toward more efficient management of the so-called iron triangle formulated by (P.1). Whereas Chapter 2 presents the multi-perspective modeling and holistic simulation framework in an informal way, Chapter 3 formalizes it using the discrete event system specification (DEVS) formalism. Key components of this paradigm include model and experimental frame (basic entities in M&S to be built and coupled together for M&S studies), as well as a hierarchy of systems specification to guide the design process of these components. DEVS rooting into systems theory and its uniqueness of providing a common model of computation to all discrete-event formalisms gives a sound theoretical background to the formal definition of the multi-perspective modeling and holistic simulation framework. A case study is presented to illustrate the application of the framework. The outbreak of Ebola in Nigeria in 2014 is the contextual challenge of this study, where models are built in isolation according to each perspective of the framework and then integrated together to form a complete whole. Chapter 4 focuses on methodological elements of engineering models under our framework. Visual modeling is considered in order to ease the conceptual stage, while multi-formalism modeling and participatory modeling are considered to accommodate various expertises and experiences, as well as to allow the use (and reuse) of models captured in different formalisms.

The focus in Part II of the book is to show how M&S can help design service infrastructure. To do this, we discuss the application of the DEVS formalism within SoS engineering. This allows the development of coordination models for transactions involving multiple disparate activities of component systems that need to be selectively sequenced to implement patient-centered coordinated care interventions. We show how such coordination concepts provide a layer to support a proposed IT as a learning collaborative SoS for the continuous improvement of healthcare. Chapter 5 delves into the Pathways Community HUB model in some depth as a basis for a more extended framework for care coordination. It expands upon a SoS engineering formalization and simulation modeling methodology for a more in-depth application of DEVS coordination pathways to reengineer healthcare service systems. In particular, we discuss a concept of coordination models for transactions that involve multiple activities of component systems and coordination pathways implementable in the DEVS formalism. We show how SoS concepts and pathway coordination models enable a formal representation of complex healthcare systems requiring collaboration, to enable implementation of criteria for measurement of outcome and cost. This leads to a pathways-based approach to coordination of such systems and to the proposed mechanism for continuous improvement of healthcare systems as learning collaborative SoS. Chapter 6 goes into more detail on a *Coordination Model* that abstracts essential features of the Pathways Community HUB model so that the kind of coordination it offers can be understood and employed, in a general SoS context. It presents a review of the pathways model and provides a formalization of pathways that serves as a basis for quality improvements in coordination of care. We discuss metrics defined based on the formalization and how such metrics can support simulation and analysis to provide insights into factors that influence care coordination quality of service and, therefore, health outcomes. We also discuss how the pathways model and its formalization will enable exploiting the emerging IT infrastructure to afford significantly greater coordination of care and fee-for-performance based on end-to-end outcomes. Chapter 7 discusses how to guide design, development, and evaluation of architectures for pathways-based coordination of care. A simulation model is presented that exemplifies this approach and is intended to predict return on investment for implementing pathways-based coordination as well as its sustainability over the long run. In order to more fully elaborate the framework to support, we point to needed expansion of the model to include dimensions such as client risk characteristics, the referral source of clients to the coordination program, the effect of incentives on service workers' performance, and alignment of pathways with payments. Chapter 8 presents an approach to ensuring pathways client data validity and improving the design of associated health information infrastructure for community-based coordination of care. Beyond making explicit the aspects in which data quality can fall short, the approach informs the design of decision support systems for community health workers and other participants in the Pathways Community HUB. We exploit the existence of pathways structures to organize the data using formal systems concepts to develop an approach that is both well defined and applicable to support general standards for certification of such organizations. Pathways structures appear to be an important principle not only for organizing the activities of coordination of care but also for structuring the data stored in electronic health records in the conduct of such care. We also show how it encourages design of effective decision support systems for coordinated care and suggest how interested organizations can set about acquiring such systems.

Part III of the book uses the foundation developed in Parts I and II to discuss the design and application of the framework in various contexts. Chapter 9 shows how the framework can support modeling the process that describes the patients' courses of treatment and interactions within the healthcare system in France. Such modeling allows agents in the system to better situate roles and point out the need to rebalance the coordination between actors who treat the patient. The maldistribution of human resources is particularly pointed out as a key factor of performance. We propose an M&S architecture that overcomes the coordination issue between healthcare stakeholders, in the case of inequitable distribution of physicians. The architecture is populated with data from National Institute of Statistics and Economic (INSEE) studies, community health workers, and some perspectives are given to validate by using data gathered from the rural area of Dordogne. Chapter 10 illustrates how an integrated approach to health can be achieved by using participatory modeling and the multi-perspective modeling framework. The case exhibited relates to the surveillance of a zoonosis in Vietnam, the highly pathogenic avian influenza.

The last three chapters broaden the lens to view the proposed framework in more encompassing contexts. Chapter 11 acknowledges that, given the complexities to be tackled in trying to provide evidence-based and simulation-verified proposals for healthcare delivery system reengineering and new system design, several prerequisites must be in place to enhance the probability of success of such reform efforts. Some major requirements are stated and shown to be addressed by the book's framework. Chapter 12 discusses the need to develop a suite of simulations to study specific approaches to value-based delivery of services. The focus is on coordinated care with the goal of enabling design and development of architectures that treat patients as agents interacting with systems and services that are coordinated using health information networks and interoperable electronic medical records. Success in this direction will contribute to the major global healthcare goal of solving the "iron triangle" of reducing cost while improving quality and increasing access. Our specific goal is to allow organizations, such as public health departments, hospital based, and emerging accountable care organizations (ACO), to evaluate specific coordinated care strategies, payment schemes, and trade-offs such as between quality and cost of coordinated care.

Finally, the epilogue of the book places the technical discussion of the book into global national healthcare perspectives. The DEVS comprehensive methodology underlies the development of pathways-based self-improving and efficient management architecture models. It proceeds along parallel paths of simulation model development, coordination and learning sub-models, testing of the sub-models in the simulation model, and implementation of the sub-models within actual healthcare environments. Progress in the development can proceed independently along segments where the paths do not intersect. We present case studies to summarize how the formalization in terms of DEVS enables temporal analysis that would be difficult to undertake with conventional biostatistics. The situations in Europe, Africa, and the United States are considered and compared.

Appendices at the end of the book provide supporting material for more in-depth reading and include links to videos and slides relating to topics of interest.

References

[1]   Valdez R.S., Ramly E., Brennan P.F. *Industrial and Systems Engineering and Health Care: Critical Areas of Research*. Rockville: AHRQ (Agency for Healthcare Research and Quality) Publication No. 10-0079; 2010.
[2]   Jamshidi M. (ed.). *Systems of Systems – Innovations for the 21st Century*. 1st Edition. New York: Wiley; 2008.
[3]   Mittal S., Risco-Martín J.L. *Net-Centric System of Systems Engineering with DEVS Unified Process*. Boca Raton, FL: CRC Press – Taylor & Francis Series on System of Systems Engineering; 2012.

Part I

Modeling and simulation framework for value-based healthcare systems

Overview—Regardless of the coordination of its activities, a healthcare system is composed of a large number of distributed components that are interrelated by complex processes. Understanding the behavior of the overall system is becoming a major concern among healthcare managers and decision-makers. This part presents a modeling and simulation framework to support a holistic analysis of healthcare systems through a stratification of the levels of abstraction into multiple perspectives and their integration in a common simulation framework. In each of the perspectives, models of different components of healthcare system can be developed and coupled together. Concerns from other perspectives are abstracted as parameters, i.e., we reflect the parameter values of other perspectives through explicit assumptions and simplifications in such models. Consequently, the resulting top model within each perspective can be coupled with its experimental frame to run simulations and derive results. Components of the various perspectives are integrated to provide a holistic view of the healthcare problem and system under study. The resulting global model can be coupled with a holistic experimental frame to derive results that cannot be accurately addressed in any of the perspective taken alone.

Chapter 1

Healthcare systems modeling and simulation

The issue of healthcare efficiency, i.e., the need to produce more and better care with fewer resources, is becoming a widely acknowledged concern among policy-makers and healthcare managers [1,2]. Such pressure is a result of the ever-increasing demand for care services due to the growth and aging of the world's population, coupled with the inherent complexity of health system management. The well-known three competing health priorities, including access, quality, and cost, represent the "iron triangle" such that any attempt at optimizing anyone of them would significantly deteriorate the others.

As healthcare stakeholders endeavor to balance these priorities, they have to face the fragmented, loosely coupled, and tightly cohesive essence of healthcare systems. Indeed, healthcare systems are fragmented, owing to a large number of components that are diverse, concurrent, and distributed while interrelated with intricate processes. Healthcare systems are loosely coupled, i.e., organized in a modular way such that separate and semiautonomous work units, e.g., neurology, cardiology, and ophthalmology are weakly linked to one another rendering difficult the flow of information within health organizations [3]. The loose coupling can be seen both at horizontal and vertical levels. At horizontal level, it is manifested between peers who are highly trained and self-directed with significant autonomy and can control their own tasks in their domain of expertise without necessarily consulting others before taking any action even though they may collaborate. At vertical level, it is also seen between health workers who are heads of the departments in directing their daily activities on their own and the management of health units such as hospital administrators who have limited information on what is done at the operational level. As a result of this decentralization, the components are focused on their own activities, such that the coordination of health services and workers is difficult since the autonomous work units have limited awareness of others' activities. Additionally, healthcare systems are tightly cohesive, i.e., they depend on one another for the delivery of health services that meet patients' needs. For example, when services required for a patient are dependent to more than one work unit, the collaboration between work units becomes a matter of necessity rather than of discretion. This unbalanced and brokenness leads to failing structures that obviously subject healthcare systems to crises of cost increase, poor quality, and inequality. For example, healthcare systems that are designed to deal only with patient needs without consideration of the interacting factors that can improve the health of the whole population within communities will probably fail in the long-term to satisfy the "iron triangle."

Understanding how well healthcare systems perform is a quite challenging task, and especially knowing where to start and where to end. There is a wide variety of healthcare systems around the world, and every country's healthcare system reflects its own history, politics, economy, and national values, that all vary to some degree. Depending on the country, healthcare may be so fragmented or, on the contrary, so integrated that it is very difficult to disentangle the contributions of the system's components. Thus, when evaluating overall health outcomes, each single intervention can be linked to a multitude of outcomes. This makes it very challenging to provide efficient management of health resources and services, such as developing strategic visions, while performing daily operations and maintaining the system's assets and records. Consequently, healthcare has become an attractive domain for scientific exploration using M&S.

A considerable amount of M&S research works has been devoted to healthcare in recent decades, seeking to address problems related to its management, including clinical and extra-clinical aspects. This chapter provides an overview of major M&S-based research contributions to healthcare systems management.

We adopt a reading grid for this literature review, which help us later consider an ontology for healthcare systems M&S. Such an ontology provides a basis for further development toward our framework.

1.1   Review of healthcare systems simulation

Considerable efforts have been made in relation to simulation-based study of healthcare systems, as evidenced by the huge amount of work published in recent years. Roberts [4] presented a tutorial of simulation modeling methods with a taxonomy of the use of computer simulation in healthcare that distinguishes the following two main categories: patient flow optimization and analysis and healthcare asset allocation. Also, Günal and Pidd [5] did a literature review of discrete event simulation for performance modeling in healthcare and distinguishes approaches between scheduling and patient flow at one side, and sizing and planning of beds, rooms, and staff at the other side. However, scheduling and capacity planning do not cover all the facets of healthcare M&S. Some other key facets are left behind, like the prediction of disease spread and the effect of healthcare interventions (e.g., restrictions on travels, information campaign, or quarantine). Additional surveys can be found in [6–9] among others.

In trying to browse the web of contributions in healthcare M&S in a structured way, we suggest a reading grid with the following criteria:

- Level of care (i.e., the kind of healthcare targeted by the M&S-based study, and the induced location and/or geographical coverage);
- Contextual challenge (i.e., the health crisis factor that motivates the M&S-based study, whether it has already happened or it is anticipated);
- Care objective (i.e., the overall aim of the M&S-based study);

- Described abstraction (i.e., what the simulation model specifies);
- Computed outcome (i.e., what the simulation model allows to determine);
- Modeling approach (i.e., how the abstractions are specified as a model); and
- Simulation technique (i.e., the time-management paradigm used to execute the model).

1.1.1 Levels of care

Care levels include primary care, secondary care, tertiary care, and home (and community) care. Additional categories can be found in the literature, which we can more or less categorized as subclasses of those mentioned here.

Primary care is a first point of consultation for patients, where professionals are general practitioners, family physicians, and nurses, who operate in multiple settings like primary care centers, provider offices, clinics, schools, colleges, prisons, and worksites. However, patients may be referred for secondary or tertiary care depending on the nature of the health condition. At this care level, Concierge Medicine appears as a new development in the healthcare system (especially in the United States) that provides comprehensive care in a timely manner [10]. Concierge medicine is an alternative to traditional medical practices that provides better care services to registered patients. In the traditional medical practice, thousands of patients are registered per doctor, while Concierge Medicine practice facilitates a limited number to hundreds of patients to register per doctor with some guaranteed revenue streams.

Secondary care more often is referred to as hospital units, where health specialists provide acute care, i.e., necessary treatments for a brief but serious illness, injury, or other health condition. This level of care also provides services such as skilled attendance during childbirth, intensive care, and medical imaging services. Health professionals working there are medical specialists including cardiologists, urologists, and dermatologists. As highlighted in [5], unit-specific M&S studies have been predominant in the literature, including but not limited to emergency department (ED) [11–19], outdoor patient department (OPD) [20–23], and inpatient facilities [5,24–26].

Tertiary care addresses specialized healthcare in advanced medical investigation and treatment, like intensive care [27,28], and complex medical and surgical interventions [29–33]. This care level is an advanced referential one that provides specialized care services mainly for inpatients referred from health professionals of primary or secondary health centers. Those services include cancer management, neurosurgery, cardiac surgery, plastic surgery, treatment for severe burns, advanced neonatology services, palliative, and other complex medical and surgical interventions.

Home care (often associated to *community care*) is concerned with public health interest, such as food safety surveillance, distribution of condoms, or needle-exchange campaign, usually outside of health facilities [34,35]. It also includes support to self-care, long-term care, assisted living, treatment for substance use disorders, and other types of health and social care services [36].

Some M&S studies have addressed healthcare issues regardless of the level of care [4,37,38].

1.1.2   Contextual challenges

The following two broad classes can be considered to categorize challenges usually mentioned as faced by healthcare systems under M&S studies:

1.   Volume and variability of demand and
2.   Burden and scarcity of supply.

As a market brings together demand of goods from consumers and its supply by suppliers, in this actual context, healthcare is mainly demanded by a population to improve health, while its supply deals with how resources, costs, and services are related to each other within a productive process. For example, in allocating health resources as intervention measures to control the spread of disease in a given population, the labor time of healthcare practitioners and the resources being allocated represent the supply of care, while the disease being propagated prompts demands of healthcare from individuals of that population. A campaign of vaccination can be seen as the productive process that combines resources to produce the expected service that in return influences the outbreak by reducing the illness attack rate. Likewise, the illness attack rate affects the allocation of the resources by consuming or reducing those resources.

1.1.2.1   Health demand and supply

The supply of healthcare can be regarded as a process by which resources such as personnel, equipment and buildings, and land and raw materials are transformed into services. Healthcare supply includes many different things such as labor time of various trained professionals: general practitioners, specialists, nurses, consultants, managers, medical technicians, pharmacists, and many others. Healthcare supply takes into account procedures and testing, like magnetic resonance imaging scans and laboratory analyses, operating theatres, pharmaceutical products, ability to manage waiting times, and budget like surplus, debt, available funds for investment, and other sources of income. One of the areas that health system expenditures increase remarkably these last decades is wage costs. While health system expenditures consist of investing in people, building, and equipment, wage costs represent approximately 65%–80% of renewable health expenditures in most countries. As such, healthcare systems are composed of three ways of paying healthcare practitioners: fee for service, capitation, and salary. General practitioners (GPs) are paid based on services provided in case of fee-for-service arrangements, while in capitation payment systems, GPs are paid for each patient registered in their list based on factors such as age and gender. In the case of salary mode of payment, governments employ and pay GPs on salary basis.

Meanwhile, demand for healthcare is driven by individuals in the community. Such demand is referred to as the amount of care a population needs and can be evaluated in terms of inpatient admissions with a high number of admissions indicating a high demand for a service, for example, a high number of admission seen in an aging population. Examples of healthcare seekers are military communities, civil societies, ageing population, or pregnant women. Another factor of demand is hospital catchment population that reflects the number of people who fall within the catchment area of a healthcare provider such as a clinic, healthcare practitioner office, or hospital. Consequently, demands expressed by patients who seek such care will be evaluated based on factors such as distance from the patient's household to the service provider, ease means of access, quality of care, and cost of care. Furthermore, average length of stay (LOS) is also considered in how healthcare demand is estimated, for example, in the case of good discharge planning, the corresponding demand maybe more, while it is less in the case of poor discharge planning when some scare resources are still in use. The demand of healthcare is also known to be relatively inelastic, that is, a person who is sick does not have other choice than to trade-off spending on other things just to purchase the medical care he needs. Such person can even be bankrupted as reported often in some countries.

1.1.2.2   Tensions on health systems

While healthcare demands increase, its supply however remains limited due to the scarcity of healthcare resources. Nowhere in the world is there a healthcare system that devotes enough resources to meet up with all demands of healthcare of its people.

The pressure from health demand may result from high volume of demands for health service [12,18,25] due to growing population or the outbreak of a health crisis. It may also result from

- variety of patients' priorities/categories [17], or
- change of health needs/wealth over time [13,39], or
- the inequal distribution of health demanders [29].

Similarly, the pressure from health supply may result from high cost burden of health services/products [12,29]. It may also result from resource scarcity, whether

1. financial, like limited budget [12] or donation shortage [29], or
2. material, like bed scarcity [17,24,27,30], or even
3. human, like staff unavailability [20,30,34].

1.1.3   Care objectives

As stated earlier, the "iron triangle" can be formulated as seeking to deliver quality services to the highest number of patients, while maximizing resources utilization and controlling health costs. We also know that the extent to which this is achieved has no simple answer, since the cursor must be placed somewhere in the space delineated by the three vertices of the "iron triangle," i.e.,

- reduction of cost [11] like reduction of capital (respectively operating) expenses such as acquisition of facilities/equipment (respectively salaries/training), or misuse of health resources [21,28,31];
- increase of accessibility, like shortening of waiting time to treatment/bed/etc., or widening of care timeliness to remote/unprivileged patients [27]; and
- improvement of quality, like lowering prevalence of obesity [40], enhancing survival chances [11,29], or decreasing LOS in healthcare units [24].

It is noteworthy that cost, quality, and access priorities are tightly correlated, both positively and negatively. Positive pairwise correlations exist between cost and accessibility—like more affordability of health services widens the accessibility to them [28,41], as well as between quality and cost—like shorter LOS of patients in health units save significant costs [24]. Similarly, more quality may lead to more access, like short stays of patients in health units make resources available for more admissions. At the same time, it is a known fact that negative pairwise correlations bind the same health priorities. Indeed, high quality care often leads to financial pressure, while higher accessibility results either in higher cost or less quality, and low cost often generates a loss of quality as illustrated by [27] who reported that intensive care unit (ICU) patients might experience a prolonged stay due to unavailable beds in the downstream stage.

1.1.4   Observed abstractions

Generally, healthcare M&S studies focus on supply and demand related abstractions. As noticed by [37] in a generic discussion on aspects of healthcare simulation, demand aspects relate to clinical, epidemiologic, and economic concerns. One can add that supply aspects also relate to the same concerns.

More generally, abstractions that healthcare M&S studies emphasize are of the following three kinds:

- Entities, ranging from macro to micro levels depending on concerns that are predominant in the study—clinical concerns emphasize patients, providers, and payers [42], while epidemiologic concerns focus on cells or/and population [19], and economic concerns mostly target society and population [43];
- Resources, including physical resources (e.g., rooms, beds, and equipment for clinical concerns, or drugs and vaccines for epidemiologic concerns), human resources (e.g., physicians, nurses, and staff for both clinical and epidemiologic concerns), financial resources (e.g., donations, taxes and out-of-pocket payments for economic concerns), and information resources (e.g., medical records for clinical concerns, or training and advertisement materials for epidemiologic concerns); and
- Processes, some related to clinical concerns (such as patient flows and pathways, or health services delivery), and others related to epidemiologic concerns (such as disease spreading), or economic concerns (such as health-financing schemes).

Entities are often captured by their stochastic/deterministic properties, like geographical distribution, disabilities, survival/recovery/aging rate [11,44], etc., and/or by their behavior/decision [33]. For example, an individual health seeker may have his/her behavior characterized by factors such as diet, smoking habit, alcohol use, sexual activity, education and income levels, and personal health data such as social security number, age, weight, height, gender, location, insurance coverage, etc.

Resources also are captured by their stochastic/deterministic properties, like staff distribution [12] or availability rate, bed occupancy [27], treatment room utilization [21], etc., and/or by their constraints of use [13,45]. Reliability of resources is also crucial to health planners and decision-makers in formulating their policies and regulations. Such policies include personal healthcare policy, pharmaceutical policy, vaccination policy, tobacco control policy, and breastfeeding promotion policy. Regulations concern prices and the use of public health facilities for private purpose. For example, information resources are used to evaluate the performance of healthcare in terms of availability, accessibility, quantity and use of health services, responsiveness of the system to users' needs, financial risk protection, and health outcomes. These outcomes can be seen in multidimensional perspective, and they are grouped in three major categories: (1) clinical outcomes, which refer to mortality, morbidity, intermediate clinical outcomes, symptom, and clinical events; (2) patient reported outcomes, which are health status such as pain, vitality, perceived well-being, health risk status, as perceived by the individual; and (3) economic outcomes, which consist of health service utilization, and cost per episode of care. While outpatient records are the bases for utilization data, hospital records are the bases for statistics on performance related to inpatient activities. These records include the number of beds, admissions, discharges, deaths, and the duration of stay. Progress toward national health systems performance depends on health resources defined as one of the core building blocks of health systems. A special focus is on health financing, which accounts for one of the largest areas of spending for both governments and individuals in the world. Healthcare systems are composed mainly of five sources of funding including general taxation, social health insurance, private health insurance, out-of-pocket payments (income-based), and donations to charities. Health insurance is provided through social insurance programs or private insurance companies to pay medical expenses for coverages like disability, long-term nursing, and custodial care needs for insured people. In such case, the person having health insurance will have to pay a price that is much lower than the normal price of healthcare. Health insurance may also be obtained based on employer and employee mandate in which case an employer must procure health insurance for its employees and their dependents. Under the individual mandate, workers are obliged to purchase health insurance either from private insurance or religious groups, and the poor are subsidized in their purchases through government taxation.

Processes are expressed by sequences of activities, like arrivals/treatment workflows [20], or spreading mechanisms [46]. Core processes in healthcare are clinical processes, educational processes, research processes, and environmental processes. Clinical processes are carried out through activities of healthcare professionals on subjects of care (patients) by which policies and plans are translated into the interventions of the public health system. Educational processes consist of mobilizing and educating communities. Through research processes, practitioners seek to identify causes for health problems in the population. Environmental processes refer to contamination outbreak surrounding a given population. A major concern is the expression of patient movements across the healthcare facility referred to as care pathway (CP). A CP is a sequence of treatments or services described as processes through which patients undergo during their medical journey. These processes are seen as key elements to evaluate the performance of healthcare providers and services since they reflect the amount of resources that are used, their order, and the time they are being allocated to patients. The movements of patients are monitored by a number of M&S studies that can be found in [14,18,23,26,30].

1.1.5   Measured/computed outcomes

Healthcare simulation models are used to carry out analyses for two major purposes: the first relates to scheduling and planning, while the second addresses prediction and projection. Moreover, both apply to each of the three categories of abstraction usually expressed, i.e., entities, resources, and processes. In other words, common measured/computed outcomes include the following:

- *Entity scheduling/planning*, e.g., capacity planning [5,25], patient-based discharge planning [24], or emergency residents scheduling [15];
- *Resources scheduling/planning*, e.g., bed/room allocation [30,45], requested service departure times [34], or nurse/physicians scheduling [16,17,47];
- *Processes scheduling/planning*, e.g., patient flow optimization [4] or surgeries coordination [31];
- *Entity prediction/projection*, e.g., population projection [43] or patient waiting time and LOS [18,20,28,29];
- *Resources prediction/projection*, e.g., staff utilization [12,20,23]; and
- *Processes prediction/projection*, e.g., patient throughput [18], or spreading of infections and communicable diseases [35,48].

1.1.6   Modeling approaches

Healthcare systems have been investigated using different modeling paradigms, including Petri Net [49], Cellular Automata [50], State Charts [32], and DEVS [51]. However, the following approaches have been largely predominant:

- Agent-based models [22,35,37,40,42,52]
- System dynamics models [41,44]
- Combination of both [33,39,46]

1.1.7   Simulation techniques

Frequently used methods include discrete event simulation [10,12,16,26,28–31,39], continuous simulation [34], and hybrid simulation [46]. Attempts have also been made to combine simulation with optimization techniques [12,15,17,30,31] or data science [16].

Let us make a point here on the difference we make between modeling approaches and simulation techniques, as the literature has coined various terms to qualify many of the terms used in this section and the former one. Table 1.1 gives an understanding from an M&S perspective, where modeling approaches are related to the *concepts* and *specifications* levels, and simulation techniques are related to the *operations* level.

Table 1.1   Modeling approaches versus simulation techniques

Concepts (*formalisms*)	DEVS, Petri Net, agents, etc.	Differential equations, system dynamics, etc.	Operations research methods, artificial intelligence methods, etc.
Specifications (*models*)	Discrete simulation models (*DisM*)	Continuous simulation models (*ContM*)	Algorithms (*Alg*)
Operations (*engines*)	Simulators	Integrators	Solvers

The concepts level, where the universe of discourse is set (such as the notions of state, event, concurrency, etc.), calls for formalisms and (more generally) methods to capture the required concepts in a symbolically manipulable way. While the M&S community traditionally distinguishes between discrete and continuous phenomena as regard to time management-related concepts, qualitative and quantitative computational approaches, such as operation research or artificial intelligence methods, rather focus on problem-solving steps and mechanisms. Some hybridization can happen at this level with the objective-driven need to deal with temporal considerations for the system under study while trying to find a solution to the problem under study. Such a situation happens for example when optimization techniques make use of simulation as a black box-type of evaluation function (exogenous hybridization, as regard to M&S), or when the requirement for a fine-grained understanding of the system entails both continuous and discrete phenomena be considered (endogenous hybridization, as regard to M&S).

At the specification level, the real-world system and problem under study is expressed as a model, using the universe of concepts adopted, i.e., discrete or continuous simulation model, or problem-solving algorithm. The literature has coined various terms to qualify the various possible hybridizations, such as DisM + ContM (often referred to as hybrid simulation), or DisM/ContM + Alg (often referred to as combined simulation), where "+" denotes a composition/mixing operation that can vary from loose to tight integration.

At the operations level, engines are built to execute the model defined at the immediate upper level. Such engines are often referred to in the M&S world as simulators and integrators (for respectively discrete and continuous operations), while solvers implement the algorithms defined in non-M&S-centered computational approaches.

1.1.8   Browsing the web of contributions

Keeping in mind the grid that has just been presented, let us review some significant contributions that can be found in the literature. As an exercise, discuss how each of these contributions relate to the items of the reading grid proposed.

Mes and Bruens [14] developed a generic modeling framework describing three major components including entities, resources, and processes to study an integrated emergency post (IEP) within a hospital at Almelo (the Netherlands).

Entity components are referred to the moving parts of the IEP such as patients, while resources comprise operating rooms (ORs), hospital beds, medical equipment, and medical staff. Processes are referred to sequence of services required by patients such as various treatments and are denoted by a CP for each patient going through IEP services. Such services include regular tasks, parallel tasks, and delay tasks. Sequence of activities that patients undergo starts from arrival processes and ends with treatment processes in the emergency room (ER).

Zeng *et al.* [18] presented a model of ED aiming at improving the quality of care in a community hospital that faces challenges of increase of patient visits, nursing workforce shortage, and long delays. The model was used to carry out analyses on patient throughput, waiting times, LOS, and staff and equipment utilizations. Sensitivity analysis as regards to the number of nurses on duty was carried out, and results showed that using such a model, the ED may require additional number of nurses to ensure a minimum waiting time and LOS for patients, while the authors of the ED model claimed that the model can still be used to analyze the effects of potential improvement policies in the ED.

Gavirneni *et al.* [10] developed a discrete-event simulation model to address the challenge of decision-making in Concierge Medicine. The developed model aims at improving healthcare efficiency at the primary care level while taking into account settings, such as physician settings, patient settings, and society settings. Results of the study have shown that concierge medicine is attractive to both patients and physicians and could lead to better health outcomes for the entire society.

Verma and Gupta [23] developed a model to improve the performance of OPDs, one of the most congested department in a general hospital in the state of Gujarat, India. While public health delivery in India is said to be in a very bad shape, the proposed model addresses critical issues within the OPD such as doctors' activities times including arrival time, time spent to examine admitted patients, break time, and time taken by doctors to treat patients depending on illness, as well as patient's arrival time, and number of personnel at registration counters. Surprisingly, the study showed that doctor's utilization time was rather below 100%, pointing out the lack of good management skills in the OPD. Thus, the problem of maintaining discipline and scheduling of staff was highlighted against the general opinion of staff shortage problem in the hospital. The study concluded that human resource utilization on a day-to-day basis affects significantly the performance of healthcare units.

A better planning is referred to as a remedy to deal with delayed discharges at tertiary level health unit leading to degrading health service factors such as disruption in patient flow, blocked beds, frustrated patients, and distressed unit staff. To address such obstacles, Khurma *et al.* [24] developed a patient-based discharge planning that reduces patients' LOS at a tertiary acute care hospital working with 412 physicians, 305 beds while providing care to an average of 120,000 patients per year. The focus of the study was more on the top-ranked medical units sending most patients (69.1%) to long-term care. As a result, the study showed a significant resemblance and promising improvement of 4.5 days deduction in the LOS at the acute care hospital. The authors reported that more people could stay in the hospital for lesser time, and this will result into considerable savings (in dollar values).

Morrice *et al.* [26] developed a discrete-event patient-centered surgical home simulation model for coordination of outpatient surgery process at the acute care facility for University Health System, Texas. The study was concerned with a system-level process analysis for the Anesthesia Preoperative Clinic (APC) which is the key clinic for system-wide coordination in outpatient surgery. The authors considered the problem of staffing and scheduling requirements for resources and patient flow process from patient arrivals, provider assessment, and provider wrap-up times within the APC. They reported that preparing outpatients for surgery requires inten-sive information, while medical records in the current system are often inefficient and fragmented among different providers that more often lead to surgery delays and cancellations. Thus, patients with more complicated medical conditions are referred to the APC for a preoperative assessment prior to their day of surgery. They concluded that the simulation results were given the go-ahead to be implemented, while a registered nurse has been added to the clinic staff to pilot the screening.

Fletcher and Worthington [13] presented a generic model of accident and emergency (A&E) department that captures key obstacles to its functioning. Such elements include process time (i.e., waiting time for a beds, decision to admit patients, diagnostics), resource constraints, and variability in demand. The devel-oped model also considers bottleneck factors faced by A&E department that are common to other hospital departments such as diagnostic processes of X-ray and blood test, inpatient beds in assessment units and ICUs that exert significant influences on A&E department. The authors concluded that the natural level of performance of the A&E department under study may be around 89% based on performance factors such as process time, resource constraints, variability and time of day and day of week, and demand.

Bountourelis *et al.* [27] addressed the challenge of bed scarcity at a large-scale ICUs in the Veterans Affairs Pittsburgh Healthcare System. The hospital under study was having 146 operating beds distributed among different departments such as medicine department, surgery department, neurology department, cardiology department, and critical care department. The intent of the study was to alleviate the burden of blocking patients on ICUs having limited capacity but are resource-intensive units that are designed for patients that require highest level of monitored cares. Such patients are medically able to leave the ICU but might experience a prolonged stay due to unavailable beds downstream. The authors concluded that based on the performed analysis, the results of their study can faithfully represent the level of blocking and bed occupancies in the hospital under study.

Cote [21] conducted a performance-based study of outpatient clinic depart-ment that addresses key factors such as patient flow time, examining room capa-city, i.e., examining room utilization and examining room queue length, and physician's activities. The outpatient clinic has four general outpatient services, three examining rooms, and 14 physicians working according to a shift with a nurse aide that has been allocated to each of them. The study was conducted based on the assumption that each patient can select only one primary care physician, while a choice of variables such as examining room capacity and arrival rate of the patients were independently considered. The author concluded that the assumption made based on a reduction in the number of examining rooms did not result in individual patient delays.

Sobolev *et al.* [32] used Statecharts to investigate on the perioperative processes of cardiac surgical care at the British Columbia hospital, Canada. Surgical care delivery is referred to as a reactive system due to its features to interrelate processes that produce events in concurrent activities. As such, the study considered three categories of patients such as elective patient, inpatient, and emergency patient that their CPs are intertwined with patients at the cardiac surgical care department within the hospital. The challenge of surgeon's allocation was also addressed based on their activities like diagnostic, preoperative, operative, and postoperative stages. Performance factors such as the availability of surgeons for consultations, scheduled operations, and on-call duties according to the rotation and vacation schedules in the service were considered. Statecharts specification paradigm was chosen because it extends the formalism of finite-state machines through notions of hierarchy, parallelism, and event broadcasting, for representing reactive systems, and therefore enables the representation of surgical care features in a rigorous manner.

Persson and Persson [31] developed a discrete-event simulation model to address the challenge of surgery department management with focus on both medical and economic constraints. The department of general surgery under study is overflowed with patients in waiting list divided into three medical priority groups according to surgical diseases. The study was motivated by a newly passed law in Sweden that states that patients who decide to receive surgery should not wait more than 90 days to be attended to. The study considers main factors such as patient's arrival, OR scheduling, and resource allocation related to surgery over time. The scheduling of surgeries was done based on medical priority, time spent in the queue by patients and available resources like ORs, surgeons, and postoperative beds, while two types of costs were considered including patient-related costs (outsourcing costs, rescheduling and cancellation costs, and surgery costs), extra bed costs, and overtime costs.

Weng *et al.* [16] developed a mixed method combining discrete-event simulation and data envelopment analysis (DEA) to assess the ED at hospital research center in Taiwan. The goal of the research was to find an adequate formula that provides a right number of health resources including physicians, nurses, and beds in the healthcare facilities such as resuscitation rooms, triage station, observation unit to maximize the efficiency of the ED. The ED department is equipped with beds distributed into the different subunits. Patients are modeled from their arrival at the ED until when they are either released from the ED or admitted into the hospital inpatient department for further treatment. A benchmarking approach was adopted using ED simulation model to generate different ED operation alternatives considering the available budget while DEA as a mathematical programming model is used to evaluate the relative efficiency of decision-making units. The authors concluded that their approach consolidates the benchmarking sets and provides better references when ED efficiency is considered.

Price *et al.* [28] proposed a discrete-event simulation model to examine facility layouts and scheduling plans that minimize idle equipment time while maximizing total patient throughput at the Maryland proton treatment center. The study aims at improving patient access to proton therapy, a highly expensive treatment. The treatment center is designed with each imaging and gantry rom servicing only one patient at a time with one cyclotron that provides protons for all different gantry rooms based on first come first served. The authors claimed that the study provides satisfactory results that can reduce the average total waiting time by over 55% at the center. The simulation model was run in discrete time considering patient flow processes' time including the amount of time waiting for entry into the imaging room and the amount of time waiting in the gantry room. The model was built to accommodate more complex interactions between patient, facility, and personnel. A simulation was run for 250,000 ticks with patients arriving whenever an imaging room became available, and the results showed that adding more gantry room leads to little throughput increase with longer patient wait times.

Choi *et al.* [20] conducted a research study that addresses performance factors of outpatient departments at King Abdulaziz University (KAU) Hospital. Such factors include patients waiting time, workload and pressure on clinic staff, and exceptions handling. The obstetric and gynecologic outpatient department at the KAU hospital is known to be one of the most congested departments with longer waiting times despite its 21 clinics (each operated by a consultant), a team of interns, residents, and nurses. The core components of the developed framework comprised three operation phases: preparation phase, service phase, and wrap-up phase, while each phase was handled by a workflow management system. Based on a test case performed on the outpatient department, the results led the authors to a conclusion that their study can significantly help assessing key performance factors related to patients waiting time, workload and pressure on clinic staff, and exceptions handling.

Topaloglu [15] developed a goal programming model to deal with emergency medicine residents (EMRs) scheduling problem that considers both hard and soft constraints in assigning day and night shifts to residents over a monthly planning horizon. It is widely acknowledged that ERs are stressful workplaces and shift work is more demanding than regular daytime work. As such, scheduling EMRs is one of the most difficult tasks among other groups of healthcare personnel. The authors used analytical hierarchy process into the proposed goal programming model to assign weights to the deviations in the proposed objective function and claimed that the research study is able to generate successful and high-quality monthly schedules in reasonable time considering all the constraints in the scheduling environment.

Ma and Demeulemeester [25] presented a multilevel integrative approach and capacity planning to match patient's demands and supplied resources. The study was directed both to patient volumes that can be taken care of at a hospital and the resource management. The research study consists of three planning phases including case mix planning phase, master surgery scheduling phase, and operational perfor-mance evaluation phase. The three stages interact in an iterative way to make sound decisions both on the patient case mix and the resource allocation. The authors argued that one of the reasons of the increase in total health expenditures is the incapability of hospitals to handle decision-making at strategic, tactical, and operational levels of hospital operations. Indeed, they noted that decisions regarding patient flows are based on the annual number of patients that can be treated per pathology group, while decisions regarding resources consist of the capacity requirement of each specialty within the hospital. They concluded a hospital is assumed to possess a fixed number of different departments while it is considered as a production system, in which the scarce resources are dynamically used to support the flow of patients.

Ahmed and Alkhamis [12] developed a discrete-event simulation model combined with optimization techniques to investigate the operation of an ED at a governmental hospital in Kuwait with focus on resource utilization. The authors reported that most often, health system managers and decision-makers face challenges of maximizing the utilization of their available resources while being at the same time constrained by high demands for service, high costs, and limited budget. The ED under study is a 24/7 working department, receives an average of 145 patients daily, and shares resources with other hospital services. Patients arriving at the ED are classified into three categories according to their conditions. The optimization problem aiming at maximizing patient throughput has been represented as a discrete stochastic optimization problem with two deterministic constraints and one stochastic constraint. The designed decision support tool is hoped to help decision-makers at the hospital to either evaluate different situations of staffing distribution or optimize the system for optimal staffing distribution at the ED unit.

While fight between life and death is always a hair's breath away, ED is a complex unit that requires a high skill of coordination between human and material resources. Yeh and Lin [17] developed a simulation model combined with a genetic algorithm (GA) to address nurses' schedule problems at the ED of Show-Chwan Memorial Hospital in Central Taiwan. The concerned ED is faced with management challenges of high patient acuity, hospital bed shortage, and radiology and lab delays that lead to overcrowding and staff unavailability. The simulation model was developed using eM-Plant, while the GA was developed using SimTalk. The authors considered ED processes such as triage, insurance procedures, recovery rooms, and diagnostic and respiratory therapy in their approach. They concluded that the quality care in the ED can be improved by making adjustments to the nursing schedules without increasing their number in the system.

Harper [45] developed a generic framework that integrates patient classification techniques for modeling of hospital resources. The author argued that the provision of healthcare services is perhaps one of the largest and most complex industries worldwide. As a result, one needs a sophisticated capacity models that take into account the complexity, uncertainty, variability, and limited resources to plan and manage the daily activities of a hospital system. The proposed framework was developed in response to the participating hospitals, including Reading, Portsmouth, and Southampton hospitals. Performance factors related to hospital resources such as hospital admission and discharge dates, time of arrival, LOS, emergency or elective status, and operation time of patients were considered. The model examined what-if scenarios for hospital beds, operating theatres, and use of human resources like nurses, doctors, and anesthetists. The author highlighted that the research work has proved to be very helpful in the planning and management of hospital beds, operating theatres, and workforce needs.

Ozcan *et al.* [30] combined discrete-event simulation model with optimization methods to study clinical pathway (path followed by an ill person through health-care facility) across surgery department at a public hospital. The authors identified critical activities and scarce resources that represent process bottlenecks both from patients and facility point of view. The simulation model considers the flow of patients going across the system as they compete against the same common resources of the specialty such as personnel, beds, and operating theatres, while the optimization model generates optimal OR allocation plans. A minimax optimization model was developed to generate optimal OR allocation plans.

Einzinger *et al.* [22] developed an agent-based model to study physicians' reimbursement schemes, a factor that influences physicians' treatment decisions in the Austrian healthcare sector. The authors defined two types of agents including patients and medical providers with main attributes such as epidemiology, service need, and provider utilization. In conclusion, they assert that the research study facilitates comparisons of different reimbursement systems in outpatient care while being useful for testing assumptions. However, they also reported that the model is limited to a number of chronic diseases that will fit into it.

Davis *et al.* [29] developed a discrete-event simulation model for alternative allocation strategies of kidney transplantation system, for improving policy allocation that gives more survival chance and quality of life to patients, and reducing geographic disparities in the United States. Kidney-transplantation challenges include high cost of services, donation shortage, and geographic disparities that more often result in considerable amount of waiting time and cause thousands of transplant patients to die each year. The kidney transplant system is divided into 11 regions of neighboring states with patients being registered in different waiting lists. The simulation study was driven by three main events including patient arrivals, patient deaths, and organ arrivals. Three system performances were considered including average waiting time, probability of death, and probability of transplant. The authors reported that there was not a significant difference between simulated and actual average waiting times from all scenarios performed while concluding that the study provides valid estimates of kidney-transplantation system outputs.

Lee *et al.* [34] developed a continuous-time dynamic model for controlling home care crew scheduling problem that minimizes delay and transportation costs between requested service time and starting service time. Homecare is referred to as a supportive care provided to individuals in their respective homes. Demand for long-term care in the United States has witnessed rapid increase causing the home care industry to face shortage of home caregivers. The authors used a complete graph to describe home care crew problems in which case vertex set corresponds to locations of residents and home care agency locations having caregivers while vertices set corresponds to travel time between two consecutive residents. A feedback control algorithm was used to obtain sequences of requested service departure times. The authors argued that the suggested study has great potential to solve a large-scale scheduling problem in a short time compared to other studies based on operation research approaches.

Charfeddine *et al.* [37] presented a conceptual framework for healthcare delivery systems with focus on two major components: population generating the demand for healthcare services and healthcare delivery network representing the organization of the healthcare system in order to satisfy the population demand. Population demand for healthcare services is expressed as the probability distribution through the stochastic modeling of the health state evolution of each person (represented as an agent), while the model of healthcare delivery network was based on a strategic mapping framework and agent-oriented modeling methodology. The authors argued that simulation studies focusing on population and demand aspects comprise economic, epidemiologic, and clinical modeling, while simulation studies focusing on healthcare delivery networks are directed toward modeling care processes, patient flows, and available resources within healthcare supply chains and facilities such as hospitals, clinics, and care units.

Healthcare managers are more often under financial pressure when trying to ensure the delivery of high quality care. However, the performance and quality of health systems ultimately depend on the quality and the motivation of health human resources. To tackle this challenge, Vanhoucke and Maenhout [47] developed a model based on four classes of performance indicators to characterize nurse scheduling problem instances. Such indicators include problem size, preference distribution measures, coverage requirements of the schedule, and incorporated time-related constraints.

Viana *et al.* [33] combined agent-based modeling (ABM) with system dynamics to address the problem of age-related macular degeneration (AMD) management that lead patients to interact with the eye clinic via appointment-scheduling processes. Individuals were modeled as agents in the population developing AMD that lead them to interact with the eye clinic via appointment-scheduling processes. System dynamics was used to model progressive sight loss from AMD, which affects agent eyes. The authors explored different scenarios from which new individuals are added weekly to initial individuals within a period of 1 year. The results of the study showed that improving the eye unit's capability by increasing the number of equipment will help more patients successfully complete their appointments. The authors reported that the integration ABM and system dynamics in a healthcare context is rare in the sense that the main conceptual challenges lie in designing those subcomponents and achieving their interactions.

Ramirez-Nafarrate and Gutierrez-Garcia [40] presented an agent-based simulation (ABS) framework that can help policy-makers to design meal menus and physical activity programs for school-age children that reduce the prevalence of obesity during childhood. It has been argued that child obesity is a public health problem for several countries as referred to the risk of diabetes, hypertension, sleep apnea, liver disease, stroke, and some types of cancer, leading to 25% higher health expenditure for an obese person than a person with a healthy weight and causing between 5% and 10% of the overall health expenditures in the United States. Excess caloric intake is mainly known to be the cause of child obesity because obese children consume too many calories without doing enough physical activity, whereas most of both the caloric intake and the caloric expenditure take place in school and at home. The authors argued on the proposed modeling framework that

ABS models are used because they allow analyzing a complex system with autonomous agents. Children were represented with eight attributes including age, gender, weight, height, body mass index, weight status category, daily caloric intake, and energy expenditure. The results of the study showed that the fraction of children with healthy weight increases significantly as they increase the intensity of their physical activity.

Bigus *et al.* [42] proposed a general multi-agent modeling framework for studying the impact of incentives on healthcare that allows to control the cost of health services and improve health by healthcare government and employers. The simulation model considered four components that are disease model, patient, medical intervention, and provider components. While patients drive demand for healthcare services, they seek to optimize the quality of life and maintain certain level of health. Based on "state abstraction" from the perspective of a Markov disease model, relevant disease states and the estimation of transition probabilities between disease states were automatically extracted. Statistical estimation of certain patterns of intervention was used for characterizing provider's behavioral model.

Ageing population is known as one of the major factors that influences both supply and demand for health and social care. Brailsford *et al.* [39] presented an integrated model of supply and demand of both health and social care of the United Kingdom (UK) health and social care system. UK society was modeled with statistical models using theories from social models of disability. An agent-based model of the demographics of aging and social care was constructed to investigate the effects of individual-level behaviors. A high-level system dynamics simulation model was developed to study health and social care at the institutional level. The authors reported that these three approaches are linked to build a suite of models which represents UK health and social care at multiple levels: population, individual, and institutional. They also concluded that demand for health is a function of need influenced by factors such as disability and disease, new technologies, and changes in levels of income and wealth, while supply for health is influenced by factors like demographic trends, economy, and policy environment.

Paleshi *et al.* [35] developed an ABS model of a pandemic within a generic US metropolitan area in order to study how the disease spreads and to prepare for handling the consequences by implementing intervention strategies. The proposed model consists of three main subroutines including the structure of the population, disease characteristics, and transmission of the disease between people. The population was structured into four age groups: less than or equal to 4 years old, 5–18 years old, 19–64 years old, and 65 years old or older. Two intervention strategies were examined including home confinement and school closure for the mitigation of infected individuals during the pandemic outbreak. The authors conducted 50 replications for each scenario and concluded that all intervention strategies have positive effects on the attack rate representing the percentage of infected people during the pandemic.

Zhang *et al.* [48] proposed a contact network-based study that incorporates different intervention strategies to assist policy-makers to make decisions for containing the spread of infectious diseases. The authors examined major health interventions including public health interventions that comprise pharmaceutical interventions like antiviral treatment and vaccination and non-pharmaceutical interventions such as social distancing, hand wash, and face mask. In an illustrative case study, they studied intervention strategies based on social distancing such as school closure and workforce shift for the mitigation of influenza spread in Singapore.

Kasaie *et al.* [52] developed an ABS model to study tuberculosis (TB) transmission dynamics and the role of various contact networks. People in the population were represented as agents in the model. TB natural history was modeled at the individual level using five main TB health states including susceptible, early latent TB, late latent TB, active TB, and recovered states. The population was structured into different groups including households, neighborhoods, and communities. Consequently, the authors defined three layers contact network comprising close contact, casual contact, and random contact capturing social relationships of each individual with the rest of the population. Close contacts represent contacts among household members, casual contacts are social relationships among friends in places such as bar, store, and school, and random contacts represent encounters of people at places such as bus stops and museums. The authors concluded that the study of timing and distribution of TB transmission allows understanding the population heterogeneity with regard to personal characteristics and different contact networks.

Population projection is related to public health issues, political decision-making, or urban planning. Models concerning population projections include microlevel models (focusing on a sample population) and macro-level models (projecting a total population by age, sex, and other characteristics). Bohk *et al.* [43] developed a probabilistic population projection model (PPPM) allowing detailed projections of a population. The proposed PPPM was based on macro-level projection model and integrated two variants: open type and limited type. The authors presented an illustration of open and limited PPPM types using data from the Federal Statistical Office of Germany and running 1,000 trials for each of the PPPM types.

Ng *et al.* [41] worked on system dynamics modeling to study healthcare affordability problem in Singapore by investigating different scenarios that evaluate the effectiveness and sustainability of policies over time. Major components of healthcare were considered including demand component, hospital resources component, and costing component and their respective relationships. Policies such as assigning a higher percentage of budget, changing migrant flow, differentiating subsidies according to income group, and shortening the length of hospital stay were tested. The authors concluded that the affordability problem will decrease significantly in the next three decades.

Ferranti and Freitas Filho [44] developed a system dynamics model to study risk factors for age-related cardiac diseases. The model considered key parameters such as growth rate, reserve rate, and aging rate. The authors reported that the results of the study have shown that maintaining good blood pressure and participating in physical activities have an impact on a person's lifespan and delaying mortality in the population.

Djanatliev *et al.* [46] developed a new approach called prospective health technology assessment that loosely integrates system dynamics and ABM within a hybrid simulation environment, to investigate the effects of implementing new technologies in healthcare systems. Major modules such as population dynamics, disease dynamics, healthcare, and healthcare financing were considered for the study. A use-case scenario with an innovative stroke technology as innovative health technologies was presented as having the power to improve the life quality of populations and to make healthcare more effective. The authors argued that their research effort has achieved an overall credibility from all domain experts including doctors, health economics, medical informatics, and knowledge management experts.

Brailsford and Shmidt [53] developed a discrete-event simulation model integrating both psychological and human behavior model to evaluate attendance for screening of diabetic retinopathy. Human behavior factors such as physical states, emotions, cognitions, and social status of the persons involved were considered. Each patient was modeled as individual entity with his or her own characteristics assigned with numerical attributes between 0 and 1 representing low, medium, or high value. Attributes like anxiety, disease knowledge, and educational level have been taken into account. The authors concluded that the model would have great potential value as a policy analysis tool to design efficient screening plan that will attract more nonattenders and improve the overall health of the population.

Roberts [4] presented a taxonomy of healthcare simulation that considers the following aspects: bed allocation and planning, admission control, room sizing and planning, patient flow, physician and healthcare staff scheduling, materials handling, and logistics. The underlying aspects are concerned with healthcare-management challenges including outpatient scheduling, inpatient scheduling and admissions, ED and specialist clinics, hospital departments like laboratory, radiology, surgery and recovery, pharmacy, and supply and support.

In a general classification, Barjis [38] presented healthcare simulation along four axes including clinical simulation (used for studying and analyzing the behavior of certain diseases), operational simulation (used for capturing and studying healthcare activities such as service delivery, healthcare operation, scheduling, and patient flow), managerial simulation (used as decision support tool for managerial purposes, strategic planning, and policy implementation), and educational simulation (used for training and educational purposes).

Onggo [36] presented a review on simulation modeling for the provision of social care services. Main components of social care services that were considered include demand, supply, delivery methods, and finance. The authors argued that health demand is generated by care users, and its planning is linked to population projection that is partly influenced by healthcare system. They reported that one of the key challenges in demand projections is its dependence on factors such as health, culture, and sociodemography.

Günal and Pidd [5] presented a review of performance modeling in healthcare simulation that considers models according to the objectives of the studies. The models address simulation aspects such as scheduling and patient flow, sizing and planning of beds, rooms, and staff. The authors highlighted that healthcare simulation based studies are unit specific, that is, their focus is on specific problems in individual units of healthcare systems. Additionally, those studies are facility-specific, that is, the models were built for specific hospitals and are hardly reused.

Brailsford [19] presented a review of applications of simulation in healthcare into three levels. Level 1 models refer to models at the cellular, organ, or system level of the human body, or disease models and are used to study clinical effectiveness of healthcare interventions. They are also used to study human health behaviors, and some disease spreading. Level 2 models are used to study activities of health unit (hospital department, clinic, or ER at operational or tactical level). Level 3 models also called strategic models are used for studying long-term problems.

Aboueljinane *et al.* [11] presented an extensive literature review on papers that deal both with timeliness and economic objectives to achieve analysis and improvement of EMS. The authors addressed commonly faced problems by EMS providers such as response time, recovery chances, and patients' disability. They presented key characteristics of EMS operations as follows: operations (processes describing central and external operations), decisions regarding EMS operations (long-term decisions, mid-term decisions, and short-term decisions), and performance measures associated with EMS operations such as timeliness, survival rate, and costs. Reducing significant expenses that concern capital such as acquisition of facilities, emergency vehicles, equipment and communication devices, and operating costs like salaries, training, and maintenance was widely discussed. The authors argue that a successful modeling of demand in EMS systems considers three key characteristics: the arrival distribution, the geographical distribution, and the priority of calls.

Many more significant works can be found in the literature, since one of the areas that M&S has gained a tremendous popularity in these last decades is the domain of healthcare. All of these efforts cannot be reported here, although we have endeavored to provide a sufficiently representative overview of the directions taken in these studies.

1.2   Lessons learnt in a nutshell

The predominant characteristic that healthcare M&S must face is its systems-of-systems (SoS) nature with the central organizational shortcoming (i.e., coordination that rises no higher than pairwise interactions).

As summarized in Table 1.2, healthcare systems are loosely coupled in structure, but tightly cohesive in required functionality. The system is organized around pairwise interactions between physicians and patients with discrete maladies. Interactions between physicians are discretional rather than institutionalized, and clinical care is very weakly influenced by community determinants of health.

Therefore, dealing with the SoS-nature of healthcare where organizational structure does not meet functional objectives constitutes a key requirement for M&S of healthcare.

Moreover, literature review (as summarized by Table 1.3) leaves no doubt on the fact that

1. healthcare M&S has multiple facets and
2. there is a lack of ontology that structures all the knowledge about existing models in a hierarchy allowing one to easily derive new models.

Table 1.2   Characteristics of healthcare SoS

Property	Definition	Comment
Fragmented	Large number of components that are diverse, concurrent, and distributed while interrelated with intricate processes	
Loosely coupled	Weakly linked to one another rendering difficult the flow of information within health organizations	*Horizontal level:* highly trained autonomous peers control own tasks in their domain of expertise *Vertical level:* decentralization of authority with limited awareness of others' activities
Tightly cohesive	Care providers depend on one another for the delivery of health services in delivery chains that are not well defined	System is designed to deal only with patient needs without consideration of the interacting factors that affect both the patient's and that of the community

Table 1.3   Aspects dealt within healthcare M&S studies

Aspect	Definition	Detail
Level of care	The kind of healthcare targeted by the M&S-based study and the induced geographical coverage	• Primary • Secondary • Tertiary • Home and community
Contextual challenge	The health crisis factor that motivates the M&S-based study, whether it has already happened or it is anticipated	• Demand volume and variability • Supply scarcity and cost
Care objective	The overall aim of the M&S-based study	• Reduction of cost • Increase of accessibility • Improvement of quality
Described abstraction	What the simulation model specifies	• Entities, at macro/micro levels • Resources • Processes
Computed outcome	What the simulation model allows to determine	• Scheduling/planning • Prediction/projection
Modeling approach	Used to specify the model	• Agent-based models • Systems dynamics • Others
Simulation technique	Used to execute the model	• Discrete event simulation • Continuous simulation • Hybrid simulation

1.3   Ontology for healthcare systems simulation

The review we just presented suggests one can benefit from a large set of research results and tools when building new healthcare simulation models to conduct new M&S-based studies. However, in order to "not reinvent the wheel" and reuse existing works in a disciplined way, an ontological support would be of great interest.

Why an ontology?

Durak *et al.* [54] defined an ontology as a vocabulary of terms and specification of their meaning including definitions and indications of how concepts are interrelated while collectively imposing a structure on a domain by constraining the possible interpretations of terms. Hofmann *et al.* [55] define it as an unambiguous and machine understandable description used to categorize concepts and the relationships among them within a particular knowledge domain. Zeshan and Mohamad [56] define an ontology as a hierarchy of important concepts in a domain and description of the properties of each concept.

As such, ontologies are used not only to structurally define knowledge within a given domain but also to assist in communication between humans and achieve interoperability between software systems. Many fields have developed ontologies. In M&S too, considerable efforts have been made using ontologies to address modeling challenges like model reuse, composability, and interoperability.

For practical applications, Tolk and Blais [57] reported that, "If a formal specification concisely and unambiguously defines concepts such that anyone interested in the specified domain can consistently understand the concept's meaning and its suitable use, then that specification is an ontology."

Within the context of M&S composability and interoperability, the objective of ontologies is to document the conceptualization, in a way that machines and computers can not only read the result but also make sense out of it in the context of their applications. Hence, using ontologies in large-scale and complex M&S application provides notable benefits such as promoting the "do not reinvent the wheel" principle while reducing development cost and time, leading to increased quality and risk reduction.

1.3.1   Ontology in M&S

Ontology-based study has been widely proposed to address modeling challenges in large-scale and complex systems simulation. Notable examples are discrete event system modeling ontology—DESO [58], and web-based ontology for discrete event modeling ontology—DeMO [59].

Seeking to address the lack of ontologically well founded conceptual modeling language for Discrete Event Simulation Engineering, Guizzardi and Wagner [58] presented a foundational ontology for discrete event system modeling called DESO that addresses an agreed-upon precise definition of common concepts in discrete event simulation such as entity, object, event, and state. Furthermore, DESO provides some basic properties for the evaluation of general-purpose discrete event simulation languages which are soundness, completeness, lucidity, and laconicity.

Similarly, Miller *et al.* [59] developed a web-based ontology for discrete event modeling called DeMO capturing knowledge about discrete event domain such as event-scheduling, activity-scanning, and process-interaction, known as three main worldviews for discrete-event simulation modeling. Consequently, DeMO was built based on four types of discrete event models which are state, event, activity, and process oriented. It is reported that the usage of DeMO provides several possible benefits such as browsing, querying, service discovery, components, hypothesis testing, research support, mark-up language, and facilitates collaboration.

Tolk and Turnitsa [60] developed conceptual modeling based on ontological spectrum that captures the data representation in a computable way for supporting composability and interoperability of information exchanged between distributed systems. Conceptual models (CMs) are organized in a trichotomy relationship based on ontological entities representation rooted in system-entity definition to support multi-resolution modeling in a service-oriented context.

Most often, two common fundamental types of modeling errors occur when modeling complex systems. These errors are system description errors and model translation errors. To address these errors, McGinnis *et al.* [61] used a formal system modeling language based on OMG SysML (Object Management Group's Systems Modeling Language) for creating a usable ontology for a formal representation of knowledge and formal model transformation technologies for model automation. According to the authors, SysML has a significant benefit to support both ontology definition and specific models development.

According to Ezzell *et al.* [62], the purpose of ontology is to structurally define knowledge about a topic. This purpose led to the development of a methodology based on domain ontologies using a human-interface layer to construct dynamic models with their corresponding three-dimensional (3D) visualizations. The proposed ontologies are being manifested in the user interface to address the needs in education and medical training by integrating simulation with traditional teaching methods. An example of cardiovascular physiology model construction was presented to show how a 3D visualization is created after augmenting the visualized ontology with new attributes.

In order to address the challenge of model reuse in large and complex systems, Durak *et al.* [54] adopted an ontology-based approach to trajectory simulation called TSONT (trajectory simulation ontology) that facilitates conceptual interoperability in trajectory simulation. According to the authors, a successful interoperability at the system implementation level is achievable through meaningful composability at the conceptual level. A conceptual platform-independent model was constructed using model-driven engineering concepts based on the domain ontology combined with the high level architecture approach to achieve the underlying interoperability. The work was built upon MATSIX—MATLAB® 6 DOF trajectory simulation framework using a model transformation tool for transforming an OWL ontology (that captures the domain knowledge of a trajectory simulation, specifically the static structure of the simulation, behavior model, and the definition of interfaces) into a UML (Unified Modeling Language) class diagram. The PUMA federate simulation object model resulting from the model transformation was then used to compute the flight path and flight parameters of munitions.

While seeking to overcome the limitations of the current practices for composition of models and simulation systems, Wang *et al.* [63] proposed a framework that describes the levels of conceptual interoperability model (LCIM) as well as a descriptive and prescriptive model. The authors argued that the proposed LCIM was derived from many research efforts in various publications that dealt with different aspects on LCIM. Achieving conceptual interoperability and composability was reported to be a difficult problem. Therefore, seven levels of interoperability were differentiated, starting from level 0 (representing no interoperation) to level 6 (indicating conceptual interoperability), while intermediate levels such as level 1 represents technical interoperability, level 2 represents syntactic interoperability, level 3 represents semantic interoperability, level 4 represents pragmatic interoperability, and finally level 5 represents dynamic interoperability.

Benjamin *et al.* [64] developed a knowledge-driven framework for semantic simulation application integration (KSAI) and interoperability that captures ontological information in a context involving multiple domains, modeling paradigms, and software tools. KSAI is based on two categories for simulation application integration including design time integration and run time integration and is supported by five interrelated activities:

1. Defining a simulation project scope,
2. Assessing and filling knowledge gaps,
3. Performing integration assessment,
4. Generating executable information exchanges, and
5. Running integrated simulation.

Balci *et al.* [65] discussed how reusability and composability can be achieved using a simulation CM to facilitate their design. Conceptual constructs, being independent of any design strategy and execution requirements, can be reused and composed at all possible levels of abstraction and promote the "do not reinvent the wheel" principle by providing significant benefits such as reduced development cost and time, effective use of subject matter expertise, increased quality, and reduced risk. A CM was defined as a repository of high-level conceptual constructs and knowledge specified in a variety of communicative forms such as animation, audio, chart, diagram, drawing, equation, graph, image, text, and video intended to assist in the design of any type of large-scale complex M&S application. Hence, through reusability, an artifact is expected to be reused multiple times whether it has been developed in isolation or not. Using composability, an artifact is constituted by combining things, parts, or elements.

While investigating on epistemic and normative aspects of ontologies for M&S, Hofmann *et al.* [55] categorized ontologies for M&S into two classes: methodological ontologies and referential ontologies. The former defines modeling methods and simulation techniques while the later represents real world systems to be simulated.

Arguably, ontologies are to be considered as formal specifications as well as means of knowledge representation. An advantage of being formal is that it provides an easy processing mechanism by computers in order to logically deduce higher order relationships between concepts. Referential ontologies are more powerful than taxonomies or glossaries as concepts are not only being categorized but also being interrelated to each other. As such, the strength of ontologies is based on their precision such as precision for a common terminology, precision for a common logical structure of conceptual relations, and precision for the denotation of concepts as far as possible by definitions. Precision, in this case, is essential for knowledge exchange and knowledge reuse such as compatibility with other ontologies. Hence, a notable benefit of using ontologies is to increase the interoperability of models while ontology-driven modeling significantly contributes in reducing time spent for finding the most appropriate formalism.

1.3.2 Ontology in healthcare M&S

In the field of healthcare, Okhmatovskaia *et al.* [66] introduced ontology for simulation modeling of population health (SimPHO), an explicit machine-readable specification of a domain of knowledge integrating both aspects of taxonomy and vocabulary in a form of logical axioms. SimPHO was reported to have a supporting set of software tools intended to facilitate simulation model development, validation, comprehension, and reuse. The authors argued that ontologies for modeling and simulation from findings are mainly domain-independent, while SimPHO describes the content of the model in the specific context of the simulation modeling in population health. SimPHO combines existing knowledge from medical ontologies and vocabularies and is implemented in OWL (Web Ontology Language) using the Protégé-OWL editor. Its domain coverage and scope include discrete event simulations, micro simulation models of population health (individual levels), and a wide range of diseases, their risk factors and outcomes, demographic characteristics of the population, healthcare associated costs and measures of disease burden. To characterize health simulation models using SimPHO, one should consider three major parts: the general high-level definition, the characterization of the content of the model in terms of the relevant domain of knowledge, and the technical specification of modeling details. The high-level description provides a top-level class called simulation and a number of subclasses to represent different model categories based on logical axioms and class properties, while the content of the model in health domain terms is focused on healthcare systems and public health policies. The technical details of the modeling that have been formalized by SimPHO include concepts such as data types and measurement scales of state variable, procedures for initializing state variables, model parameters, methods and data sources used for their derivation, and causal relationships between states and events. The authors presented a first prototype software application of SimPHO called Ophiuchus, implemented as a web-based application. It has been reported that after performing a preliminary informal assessment with a mixed group of 20 users, the feedback that was received from the subjects was positive. The authors claimed that SimPHO provides a conceptual framework for clear, unambiguous, multifaceted representation of simulation models of population health and serves as a foundation for a set of software tools intended to facilitate model comprehension, validation, and reuse.

Silver *et al.* [67] developed an ontology-driven simulation model that promotes relationship between domain ontology and simulation ontology through an existing domain ontology called problem-oriented medical records ontology in healthcare domain to derive simulation model as ontology instances. The resulting simulation models are then translated into executable simulation models that can be used by simulation tools. The authors based their arguments of mapping domain ontology into simulation ontology on four different worldviews of the modelers including state oriented, event oriented, activity oriented, and process oriented as suggested by the DeMO as well as the influence of the application domain. The selection of a worldview modeling determines the choice of a particular modeling formalism.

Zeshan and Mohamad [56] presented domain ontology for information technology (IT)-based healthcare systems that support knowledge sharing between devices and actors during the diagnostic process of patients in EDs. Through a methodology called "methontology," the constructing of the proposed ontology was achieved following step-by-step guidelines that include entity extraction, taxonomy formation, relationships, and the axioms to add logical expressions using Protégé software for consistency checking at the evaluation phase.

Puri *et al.* [68] proposed ontology mapping and alignment to integrate ontologies from heterogeneous sources together and to support data integration and analysis. As an example, patient information and domains of healthcare information were derived from different sources like electronic health records, personal health records, Google Health, and Microsoft HealthVault to provide a common vocabulary that enables interoperability and resolves ambiguity.

In this book, we propose the ontology for healthcare systems simulation (O4HCS). O4HCS is not meant to establish a unified vocabulary, such as formalizing and reorganizing healthcare terminologies and taxonomies. Instead, it is meant *to capture and share a common understanding of the abstractions necessary/ used for the simulation of the entire healthcare domain (beyond-unit specific and facility-specific modeling)*. When developing O4HCS, it is essential that we provide, at some general level, a formal way to capture all the knowledge that might be in the range of healthcare M&S that the ontology is likely to be used for [69]. For this reason, we use the system entity structure (SES) framework [70].

1.3.3   System entity structure ontological framework

SES enables fundamental representation of hierarchical modular model providing a design space via the elements of a system and their relationships in hierarchical and axiomatic manner. It is a declarative knowledge representation scheme that characterizes the structure of a family of models in terms of decompositions, component taxonomies, and coupling specifications and constraints. SES supports development, pruning, and generation of a family of hierarchical simulation models. It is a formal ontology framework, axiomatically defined, to represent the elements of a system (or world) and their relationships in hierarchical manner.

Figure 1.1 provides a quick overview of the nodes and relationships involved in a SES. Entities represent things that have existence in a certain domain. They can have variables, which can be assigned a value within given range. An aspect

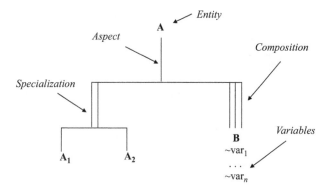

Figure 1.1   Basic SES construction

expresses a way of decomposing an object into more detailed parts and is a labeled decomposition relation between the parent and the children. Multi-aspects are aspects for which the components are all of the one kind. A specialization represents a category or family of specific forms that a thing can assume. It is a labeled relation that expresses alternative choices that a system entity can take on.

SES is targeted to support the plan-generate-evaluate process in simulation-based systems design:

- The plan phase recaptures all the intended objectives of the modeler.
- The generate phase reproduces a candidate design model that will meet the initial objectives.
- The evaluate phase assesses the performance of the generated model through simulation.

As such, SES organizes a family of alternative models from which a candidate model can be generated, selected, and evaluated through system design repeatedly until the model meets an acceptable objective.

While complex systems are composed of large components and their structural knowledge can be broken down and systematically represented in SES, their behaviors can be specified in either atomic or coupled models and saved in a model base (MB, an organized library) for later use. Once the models are saved, they can be retrieved and reused to design complex systems.

1.3.4   Ontology for healthcare systems simulation

Figure 1.2 displays the O4HCS, a SES-specified hierarchy of modeling concepts elaborated to support the design of holistic healthcare simulation models. It formalizes a healthcare system as a whole made up of one or various facets, each of which being a production system, a consumption system, or a coordinating system between production and consumption.

There is a reason why we use the terminology of "production" instead of "supply" and "consumption" instead of "demand." As previously mentioned, demand and supply, in healthcare-related literature, usually refer respectively to patients seeking for care services and providers of these services. With the concept

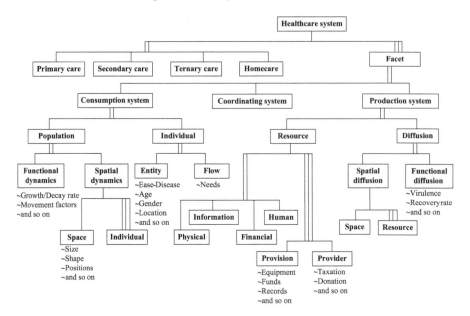

Figure 1.2   Ontology for healthcare systems simulation (O4HCS)

of "Production," we also include health phenomena that generate or amplify health concerns, such as disease spreading, as well as mechanisms used to supply healthcare, such as vaccination or information diffusion. Similarly, with the concept of "consumption," we extend health demand beyond the individual patients, and we include the population as a dynamic entity.

As noticed in [39], literature contains numerous models for the demand and supply of healthcare services, although these models have mainly focused on specific conditions (and in some cases on specific locations). More recent works have focused on the coordination dimension [71,72].

1.3.4.1   Health consumption

Healthcare consumption is made by individuals and populations seeking care in times of need. Hence, M&S models of consumption systems focus on these health consumers and the dynamics of their demands, and abstract all other processes by parameters (which are represented by variables in the SES diagram). Two facets appear, population models and individual models, corresponding respectively to macro and micro approach to healthcare-consumption modeling.

A population model is related to births, deaths, and demographic flows such as immigration and emigration. It is expressed

- either as a functional dynamics model, which is formulated in the shape of equations, such as Leslie matrix [73] or predator–prey model [74], with parameters influencing the model's dynamics (like growth or decay rate of the targeted population, movement-related parameters, etc.) or

- as an emerging phenomenon (i.e., spatial dynamics model) composed of individuals geographically located in a space model (e.g., a cellular automata), which has variables such as the size (e.g., the number of cells) and the shape (e.g., rectangular/circular/hexagonal cells).

An individual can be modeled in two ways:

- An autonomous entity with specific attributes (e.g., abilities and disabilities, age, gender, location) and a goal-driven behavior, including social dimensions. Typical examples are agent-based models [29,36,37,40,75].
- A flow that captures scenarios the individual can undergo (e.g., patient flow, care pathways), parameterized by his needs (such as required health services/ resources).

1.3.4.2   Health production

Similar to consumption models, healthcare production models focus on health producers and provision, while abstracting by parameters all other processes. Healthcare production has two facets: resource models and diffusion models.

The first facet deals with how resources are transformed into services [21,23,24,27,30,32,33,76–78]. Resources (and therefore services) can be physical (beds, rooms, etc.), human (physicians, nurses, etc.), financial, or information. Models of such transformation explicitly describe the dynamics of providers, or provision or both:

- Provider models make use of parameterized variables such as taxation rates and amount of donations. Examples include the economic model of health funding, which can be tax/out-of-pocket/insurance-based, or the information system that provides health data.
- Provision models use variables such as equipment (physical provision), funds (financial provision), records (information provision), and workforces (human provision).

The second facet of healthcare production deals with the (generation and) diffusion of health phenomena, whether positive or negative [52,79]. Positive phenomena (like vaccination campaign) produce ease, while negative ones (like disease spreading) produce disease. Diffusion processes are described as either spatial or functional phenomena:

- Spatial models explicitly describe space (e.g., cellular automata-based models), and the resources involved (e.g., the cellular automata's cells can be models of their own);
- Functional models formulate the dynamics of the diffusion process in the form of mathematical equations, such as compartmental models [80], with parameters related to assumptions done on the diffusion factors (like the virulence of a disease, or the recovery rate of infected individuals).

1.3.4.3   Health coordination

Care coordination can be seen as cross-organization coordination managing the entities and resources of existing ones. It is needed to the extent that existing organization is lacking. Pathways, as detailed in the second part of this book, are means to do that coordination [81].

Let us have here a look on clinical pathways placed within the context of M&S.

Generally speaking, clinical CPs consist of algorithms (usually informally presented as a flow chart with conditional branching) that delineate the overall structure of decision-making for treating a specific medical condition. CPs are referenced in the literature under a variety of names such as clinical pathways, critical pathways, and clinical process models, and there is no single, widely accepted definition, although key characteristics have been extracted from over 200 articles [82]. Since the introduction in the 1990s, CPs have become widespread in hospital healthcare management [83].

A comprehensive analysis of 27 published studies compared outcomes and costs for hospitals that used clinical pathways with those that do not. It found CP use was correlated with benefits in reduction of in-hospital complications, decreased LOS, and reduction in hospital costs [84]. A recent large-scale study for knee surgery found reduction in LOS and avoidance of treatment complications for hospitals using CPs versus nonusers [85]. Based on the concept analysis of de Bleser *et al.* [82], the European Pathway Association derived an all-inclusive definition of CP [86]: "A care pathway is a complex intervention for the mutual decision-making and organization of care processes for a well-defined group of patients during a well-defined period."

Defining characteristics of CPs include

- An explicit statement of the goals and key elements of care based on evidence, best practice, and patients' expectations and their characteristics;
- The facilitation of the communication among the team members and with patients and families;
- The coordination of the care process by coordinating the roles and sequencing the activities of the multidisciplinary care team, patients, and their relatives;
- The documentation, monitoring, and evaluation of variances and outcomes; and
- The identification of the appropriate resources.

"The aim of a care pathway is to enhance the quality of care across the continuum by improving risk-adjusted patient outcomes, promoting patient safety, increasing patient satisfaction, and optimizing the use of resources" [87]. Not all studies indicate successful application of CPs. Rotter *et al.* [84] noted that although use of pathways tended to improve documentation, nevertheless, poor reporting prevented the identification of characteristics common to application of successful pathways. Shi *et al.* [88] enumerate eight types of factors that can influence results. They consider factors in CP design, execution, and evaluation (CP design: inclusion of all participating disciplines, applicability to intended medical condition, flexibility incorporated in pathway specification; CP execution: training of participants, continuous improvement, psychological influences, computer-support; and CP evaluation: consideration of multiple factors).

CPs, originally described in paper form, are being implemented in computerized form in which they can support a variety of functions [89]. Recent interest in computerization of clinical pathways has stimulated considerable work in treatment of pathways from computer science and software engineering perspectives.

To help understand the nature of the variegated contributions, we consider a clinical pathway as a mathematical system model that is being designed and manipulated to support reengineering an existing real-world clinical process. This allows taking a systems engineering approach in which the system is modeled and simulated before being implemented in reality (i.e., operationalized in a hospital environment).

Generally, this approach takes the following phases:

1. Determine objectives
   Clarify requirements (specify the decisions that model should support), values (how to measure the model outputs), and weights (how to weight the measures).
2. Gather relevant data
   Find the right data and validate it to make sure it is representative of the system.
3. Construct model
   Choose a model formalism to express it, infer its structure and/or calibrate it with data gathered in the previous step, and validate the model against unused data or newly gathered relevant data.
4. Simulate model
   Formulate alternative decisions and run simulation experiments to get the model's evaluation of these alternatives.
5. Implement model
   Select highly ranked alternative and reengineer current pathway implementation to operationalize the model.

Table 1.4 organizes some of the aspects subject to M&S in terms of these phases. Some of the aspects of pathway formalization are exploited in a generic architecture for execution of CPs capable of adapting to individual patient variations [90]. Its holistic IT solution comprises an inference engine (operating on a CP-based rule set) assisted by a semantic infrastructure (based on existing disease and business ontologies) supporting adaptation and reconfiguration during the execution.

1.4 Model base for healthcare systems M&S

SES introduces two mechanisms to allow interactive or automatic generation of an executable simulation model: the MB and the pruning process. The MB is a repository where basic models of the SES tree with a predefined input/output interface are organized. Pruning is the process of extracting from the SES tree a specific system configuration (called PES for pruned entity structure), resolving the choices in aspect, multi-aspect, and specialization relations (i.e., selecting particular subsets of aspects, cardinalities of multi-aspects, and instances of specializations), and assigning values to the variables.

Table 1.4   Phases of modeling and simulation of clinical care pathways

Phase	Sub-aspects and references in literature
Determine objectives	Pathway personalization to provide care plans [91] Pathway customization [92] Support consensus formation [93] Care standardization [94] Identify process bottlenecks [30]
Gather relevant data	Process and time dependency mining [95–97] Variation monitoring [98] Use clustering and multidimensional scaling [99] Use similarity-based patient traces [94] Observe individual patient treatment and waiting times [30]
Construct model:	Learn Patterns for Markov Model [100,101] Model based on Ontology [102–104], extended to incorporate patient state, intervention and time [105] Semantic-based workflow model [106] Normative Semiotics Model [91] Witness Software [30]
Simulate model	Verify and validate behavior [93] Observe critical activities and scarce resources [30]
Implement model	Operationalize Pathway [107] Manage workflow [106–108] Manage CP variance [107,109,110] Intelligently reconfigure [98] Evaluate patient satisfaction [111]

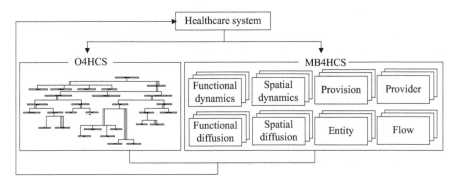

Figure 1.3   SES/MB to healthcare systems M&S

Consequently, the MB for healthcare systems M&S as derived from O4HCS (which we call MB4HCS for Model Base for Healthcare Systems) shall include a large spectrum of models, such as theoretical models provided in the literature ranging from functional dynamics models to spatial dynamics models (from bottom left to bottom right of the O4HCS hierarchy). As depicted in Figure 1.3, the idea is to follow the plan-generate-evaluate process employed in simulation-based systems design.

The plan phase enables to define the objectives of the study, the generate phase provides a synthesis of a candidate design model selected in the MB to meet the design objectives, and the evaluate phase evaluates the performance of the generated model within its experimental frame to derive useful results. The overall cycle is repeated until a satisfactory design is obtained. More on the MB4HCS models will be discussed in the next chapter.

1.5 Summary

The need to produce more healthcare services despite the scarcity of resources is becoming a widely acknowledged concern among policy-makers and healthcare managers worldwide. Yet, decision-making concerning questions related to healthcare systems performance such as the extent to which the system achieves its mission, has no clear or simple answers. The diversity of healthcare components and the complex relationships that exist between them impose the limitations of understanding the overall system by separating components and studying them in isolation, especially in a case like healthcare industry where "everything affects everything else" [112].

Consequently, one of the areas where M&S has gained a tremendous popularity in these last decades is the domain of healthcare. Healthcare systems M&S has a broad application around different disciplines such as clinical simulation, operational simulation, managerial simulation, and educational simulation. However, the domain of healthcare is characterized by a high degree of complexity and a diversity of facets, and modelers are often confronted with the challenge of formulating a simulation model that captures this complexity in a systematic and manageable manner. This chapter did a review of frequently used M&S-related concepts to address the problem. We learnt that clinical and epidemiologic simulations are predominant in the literature.

As it turns out, conducting new healthcare simulation studies while benefiting from the tremendous knowledge produced in this domain by many research efforts over the years calls for a structured knowledge base to be used as a repository of models, as well as to guide the engineering process. We then turn to SES, a hierarchical knowledge-representation framework for high-level ontology construction [65], which has an underlying plan-generate-evaluate process for simulation-based systems design. The resulting SES-specified O4HCS is a hierarchy of modeling concepts elaborated to support the design of holistic healthcare simulation models. It introduces the notion of "health production," which encompasses the traditional concept of health supply, and similarly the notion of "health consumption" is also introduced, which encompasses the traditional concept of health demand.

As reported in [39], demand for health is a function of need influenced by factors such as disability and disease, new technologies, changes in levels of income and wealth, while supply for health is influenced by factors like demographic trends, economy, and policy environment. Therefore, studying any of these facets, without looking in depth into influencing factors, might not be sufficient for healthcare decision-makers to achieve an efficient M&S-based management of their systems.

From what we discussed, we argue that there is a need to provide a simulation-based study framework that could reliably specify the various facets of healthcare systems as an integrated whole while reflecting at the same time how these components interact among themselves. Such a framework will definitely serve as a foundation for measuring healthcare systems performances in terms of objective as formulated by the "iron triangle" (i.e., wider access to quality care with less expense). This concern is the overall goal of this book, and we will address the design of its base architecture in the next chapter.

References

[1]   Shin S.Y., Balasubramanian H., Brun Y., Henneman P.L., and Osterweil L.J. Resource Scheduling through Resource-Aware Simulation of Emergency Departments. *Proceedings of the 5th International Workshop on Software Engineering in Health Care*, San Francisco, CA, USA, 20–21 May 2013. pp. 64–70.

[2]   Eklund F.J. *Resource Constraints in Health Care – Case Studies on Technical, Allocative and Economic Efficiency*. Doctoral Dissertation Series 2008/3, Department of Industrial Engineering and Management, Helsinki University of Technology, 2008.

[3]   Pinelle D., and Gutwin C. Loose Coupling and Healthcare Organizations: Deployment Strategies for Groupware. *Computer Supported Cooperative Work (CSCW)*. 2006; 15(5–6):537–572.

[4]   Roberts S.D. Tutorial on the Simulation of Healthcare Systems. In Jain S., Creasey R.R., Himmelspach J., White K.P., and Fu M. (eds.). *Proceedings of the Winter Simulation Conference*, Phoenix, AZ, USA, 11–14 Dec 2011. pp. 1408–1419.

[5]   Günal M.M., and Pidd M. Discrete Event Simulation for Performance Modelling in Health Care: A Review of the Literature. *Journal of Simulation*. 2010; 4(1):42–51.

[6]   Thorwarth M., and Arisha A. *Application of Discrete-Event Simulation in Healthcare: A Review*. Dublin Institute of Technology Reports 3, 2009.

[7]   Katsaliaki K., and Mustafee N. Applications of Simulation within the Healthcare Context. *Journal of the Operational Research Society*. 2011; 62(8): 1431–1451.

[8]   Almagooshi S. Simulation Modelling in Healthcare: Challenges and Trends. *Procedia Manufacturing*. 2015; 3:301–307.

[9]   Powell J.H., and Mustafee N. Widening Requirements Capture with Soft Methods: An Investigation of Hybrid M&S Studies in Healthcare. *Journal of the Operational Research Society*. 2016; 68(10):1211–1222.

[10]   Gavirneni S., Kulkarni V., Manikas A., and Karageorge A. Concierge Medicine: Adoption, Design, and Management. *Proceedings of the Winter Simulation Conference – Simulation: Making Decisions in a Complex World*, Washington, DC, USA, 8–11 Dec 2013. pp. 2340–2349.

[11] Aboueljinane L., Sahin E., and Jemai Z. A Review on Simulation Models Applied to Emergency Medical Service Operations. *Computers and Industrial Engineering*. 2013; 66(4):734–750.

[12] Ahmed M.A., and Alkhamis T.M. Simulation Optimization for an Emergency Department Healthcare Unit in Kuwait. *European Journal of Operational Research*. 2009; 198(3):936–942.

[13] Fletcher A., and Worthington D. What is a "Generic" Hospital Model? A Comparison of "Generic" and "Specific" Hospital Models of Emergency Patient Flows. *Health Care Management Science*. 2009; 12(4):374–391.

[14] Mes M., and Bruens M. A Generalized Simulation Model of an Integrated Emergency Post. In Laroque C., Himmelspach J., Pasupathy R., Rose O., and Uhrmacher A.M. (eds.). *Proceedings of the Winter Simulation Conference*, Berlin, Germany, 9–12 Dec 2012. pp. 1–12.

[15] Topaloglu S. A Multi-Objective Programming Model for Scheduling Emergency Medicine Residents. *Computers & Industrial Engineering*. 2006; 51(3):375–388.

[16] Weng S.J., Tsai B.S., Wang L.M., Chang C.Y., and Gotcher D. Using Simulation and Data Envelopment Analysis in Optimal Healthcare Efficiency Allocations. In Jain S., Creasey R.R., Himmelspach J., White K.P., and Fu M. (eds.). *Proceedings of the Winter Simulation Conference*, Phoenix, AZ, USA, 11–14 Dec 2011. pp. 1295–1305.

[17] Yeh J.-Y., and Lin W.-S. Using Simulation Technique and Genetic Algorithm to Improve the Quality Care of a Hospital Emergency Department. *Expert Systems with Applications*. 2007; 32(4):1073–1083.

[18] Zeng Z., Ma X., Hu Y., Li J., and Bryant D. A Simulation Study to Improve Quality of Care in the Emergency Department of a Community Hospital. *Journal of Emergency Nursing*. 2012; 38(4):322–328.

[19] Brailsford S.C. Advances and Challenges in Healthcare Simulation Modeling: Tutorial. *Proceedings of the 39th Winter Simulation Conference: 40 years! The Best is Yet to Come*, Washington, DC, USA, 9–12 Dec 2007. pp. 1436–1448.

[20] Choi B.K., Kang D., Kong J., *et al.* Simulation-Based Operation Management of Outpatient Departments in University Hospitals. *Proceedings of the Winter Simulation Conference – Simulation: Making Decisions in a Complex World*, Washington, DC, USA, 8–11 Dec 2013. pp. 2287–2298.

[21] Cote M.J. Patient Flow and Resource Utilization in an Outpatient Clinic. *Socio-Economic Planning Sciences*. 1999; 33(3):231–245.

[22] Einzinger P., Popper N., Breitenecker F., Pfeffer N., Jung R., and Endel G. The GAPDRG Model: Simulation of Outpatient Care for Comparison of Different Reimbursement Schemes. *Proceedings of the Winter Simulation Conference – Simulation: Making Decisions in a Complex World*, Washington, DC, USA, 8–11 Dec 2013. pp. 2299–2308.

[23] Verma S., and Gupta A. Improving Services in Outdoor Patient Departments by Focusing on Process Parameters: A Simulation Approach. *Proceedings of the Winter Simulation Conference – Simulation: Making Decisions in a Complex World*, Washington, DC, USA, 8–11 Dec 2013. pp. 2250–2261.

[24] Khurma N., Salamati F., and Pasek Z.J. Simulation of Patient Discharge Process and its Improvement. *Proceedings of the Winter Simulation Conference – Simulation: Making Decisions in a Complex World*, Washington, DC, USA, 8–11 Dec 2013. pp. 2452–2462.

[25] Ma G., and Demeulemeester E. A Multilevel Integrative Approach to Hospital Case Mix and Capacity Planning. *Computers & Operations Research*. 2013; 40(9):2198–2207.

[26] Morrice D., Wang D.E., Bard J., Leykum L., Noorily S., and Veerapaneni P. A Simulation Analysis of a Patient-Centered Surgical Home to Improve Outpatient Surgical Processes of Care and Outcomes. *Proceedings of the Winter Simulation Conference – Simulation: Making Decisions in a Complex World*, Washington, DC, USA, 8–11 Dec 2013. pp. 2274–2286.

[27] Bountourelis T., Ulukus M.Y., Kharoufeh J.P., and Nabors S.G. The Modeling Analysis and Management of Intensive Care Units. In *Handbook of Healthcare Operations Management*. New York: Springer; 2011. pp. 153–182.

[28] Price S., Golden B., Wasil E., and Zhang H.H. Optimizing Throughput of a Multi-Room Proton Therapy Treatment Center via Simulation. *Proceedings of the Winter Simulation Conference – Simulation: Making Decisions in a Complex World*, Washington, DC, USA, 8–11 Dec 2013. pp. 2422–2431.

[29] Davis A., Mehrotra S., Friedewald J., and Ladner D. 2013. Characteristics of a Simulation Model of the National Kidney Transplantation System. *Proceedings of the Winter Simulation Conference – Simulation: Making Decisions in a Complex World*, Washington, DC, USA, 8–11 Dec 2013. pp. 2320–2329.

[30] Ozcan Y.A., Tànfani E., and Testi A. A Simulation-Based Modeling Framework to Deal with Clinical Pathways. In Jain S., Creasey R.R., Himmelspach J., White K.P., Fu M. (eds.). *Proceedings of the Winter Simulation Conference*, Phoenix, AZ, USA, 11–14 Dec 2011. pp. 1190–1201.

[31] Persson M., and Persson J.A. Health Economic Modeling to Support Surgery Management at a Swedish Hospital. *Omega*. 2009; 37(4):853–863.

[32] Sobolev B., Harel D., Vasilakis C., and Levy A. Using the Statecharts Paradigm for Simulation of Patient Flow in Surgical Care. *Health Care Management Science*. 2008; 11(1):79–86.

[33] Viana J., Rossiter S., Channon A.R., Brailsford S.C., and Lotery A.J. A Multi-Paradigm, Whole System View of Health and Social Care for Age-Related Macular Degeneration. In Laroque C., Himmelspach J., Pasupathy R., Rose O., and Uhrmacher A.M. (eds.). *Proceedings of the Winter Simulation Conference*, Berlin, Germany, 9–12 Dec 2012. pp. 1070–1081.

[34] Lee S., Kang Y., and Prabhu V.V. 2013. Continuous Variable Control Approach for Home Care Crew Scheduling. *Proceedings of the Winter Simulation Conference – Simulation: Making Decisions in a Complex World*, Washington, DC, USA, 8–11 Dec 2013. pp. 2262–2273.

[35] Paleshi A., Evans G.W., Heragu S.S., and Moghaddam K.S. Simulation of Mitigation Strategies for a Pandemic Influenza. In Jain S., Creasey R.R., Himmelspach J., White K.P., and Fu M. (eds.). *Proceedings of the Winter Simulation Conference*, Phoenix, AZ, USA, 11–14 Dec 2011. pp. 1340–1348.

[36] Onggo B.S. Simulation Modeling in the Social Care Sector: A Literature Review. In Laroque C., Himmelspach J., Pasupathy R., Rose O., and Uhrmacher A.M. (eds.). *Proceedings of the Winter Simulation Conference*, Berlin, Germany, 9–12 Dec 2012. pp. 739–750.

[37] Charfeddine M., and Montreuil B. Integrated Agent-Oriented Modeling and Simulation of Population and Healthcare Delivery Network: Application to COPD Chronic Disease in a Canadian Region. *Proceedings of the Winter Simulation Conference*, Baltimore, MD, USA, 5–8 Dec 2010. pp. 2327–2339.

[38] Barjis J. Healthcare Simulation and its Potential Areas and Future Trends. *SCS M&S Magazine*. 2011; 2(5):1–6.

[39] Brailsford S., Silverman E., Rossiter S., *et al.* Complex Systems Modeling for Supply and Demand in Health and Social Care. In Jain S., Creasey R.R., Himmelspach J., White K.P., and Fu M. (eds.). *Proceedings of the Winter Simulation Conference*, Phoenix, AZ, USA, 11–14 Dec 2011. pp. 1125–1136.

[40] Ramirez-Nafarrate A., and Gutierrez-Garcia J.O. An Agent-Based Simulation Framework to analyze the Prevalence of Child Obesity. *Proceedings of the Winter Simulation Conference – Simulation: Making Decisions in a Complex World*, Washington, DC, USA, 8–11 Dec 2013. pp. 2330–2339.

[41] Ng A.T.S., Sy C., and Li J. A System Dynamics Model of Singapore Healthcare Affordability. In Jain S., Creasey R.R., Himmelspach J., White K.P., and Fu M. (eds.). *Proceedings of the Winter Simulation Conference*, Phoenix, AZ, USA, 11–14 Dec 2011. pp. 1–13.

[42] Bigus J.P., Chen-Ritzo C-H., and Sorrentino R. A Framework for Evidence-Based Health Care Incentives Simulation. In Jain S., Creasey R.R., Himmelspach J., White K.P., and Fu M. (eds.). *Proceedings of the Winter Simulation Conference*, Phoenix, AZ, USA, 11–14 Dec 2011. pp. 1103–1116.

[43] Bohk C., Ewald R., and Uhrmacher A. Probabilistic Population Projection with James II. *Proceedings of the Winter Simulation Conference*, Austin, TX, USA, 13–16 Dec 2009. pp. 2008–2019.

[44] Ferranti J., and Freitas Filho P. Dynamic Mortality Simulation Model Incorporating Risk Indicators for Cardiovascular Diseases. In Jain S., Creasey R.R., Himmelspach J., White K.P., and Fu M. (eds.). *Proceedings of the Winter Simulation Conference*, Phoenix, AZ, USA, 11–14 Dec 2011. pp. 1263–1274.

[45] Harper P.R. A Framework for Operational Modelling of Hospital Resources. *Health Care Management Science*. 2002; 5(3):165–173.

[46] Djanatliev A., German R., Kolominsky-Rabas P., and Hofmann B.M. Hybrid Simulation with Loosely Coupled System Dynamics and Agent-Based Models for Prospective Health Technology Assessments. In Laroque C., Himmelspach J., Pasupathy R., Rose O., and Uhrmacher A.M. (eds.). *Proceedings of the Winter Simulation Conference*, Berlin, Germany, 9–12 Dec 2012. pp. 770–781.

[47] Vanhoucke M., and Maenhout B. On the Characterization and Generation of Nurse Scheduling Problem Instances. *European Journal of Operational Research*. 2009; 196(2):457–467.

[48]   Zhang T., Lees M., Kwoh C.K., Fu X., Lee G.K.K., and Goh R.S.M. A Contact-Network-Based Simulation Model for Evaluating Interventions Under What-If Scenarios in Epidemic. In Laroque C., Himmelspach J., Pasupathy R., Rose O., and Uhrmacher A.M. (eds.). *Proceedings of the Winter Simulation Conference*, Berlin, Germany, 9–12 Dec 2012. pp. 1–416:12.

[49]   Salimifard K.K., Hosseinee S.Y., and Moradi M.S. Improving Emergency Department Processes using Computer Simulation. *Journal of Health Administration*. 2014; 17(55):62–72.

[50]   White S.H., Martin del Rey A., and Rodriguez Sanchez G. Using Cellular Automata to Simulate Epidemic Diseases. *Applied Mathematical Sciences*. 2009; 3(20):959–968.

[51]   Pérez E., Ntaimo L., Bailey C., and McCormack P. Modeling and Simulation of Nuclear Medicine Patient Service Management in DEVS. *Simulation*. 2010; 86(8–9):481–501.

[52]   Kasaie P., Dowdy D.W., and Kelton W.D. An Agent-Based Simulation of a Tuberculosis Epidemic: Understanding the Timing of Transmission. *Proceedings of the Winter Simulation Conference – Simulation: Making Decisions in a Complex World*, Washington, DC, USA, 8–11 Dec 2013. pp. 2227–2238.

[53]   Brailsford S., and Schmidt B. Towards Incorporating Human Behaviour in Models of Health Care Systems: An Approach Using Discrete Event Simulation. *European Journal of Operational Research*. 3002; 150(1):19–31.

[54]   Durak U., Oğuztüzün H., Köksal Algin C., and Özdikiş Ö. Towards Interoperable and Composable Trajectory Simulations: an Ontology-Based Approach. *Journal of Simulation*. 2011; 5(3):217–229.

[55]   Hofmann M., Palii J., and Mihelcic G. Epistemic and Normative Aspects of Ontologies in Modelling and Simulation. *Journal of Simulation*. 2011; 5(3):135–146.

[56]   Zeshan F., and Mohamad R. Medical Ontology in the Dynamic Healthcare Environment. *Procedia Computer Science*. 2012; 10:340–348.

[57]   Tolk A., and Blais C. Taxonomies, Ontologies, Battle Management Languages – Recommendations for the Coalition BML Study Group. *Proceedings of the Spring Simulation Interoperability Workshop*, Simulation Interoperability Standards Organization, San Diego, CA, 2005. Paper 05S-SIW-007.

[58]   Guizzardi G., and Wagner G. Towards an Ontological Foundation of Discrete Event Simulation. *Proceedings of the Winter Simulation Conference*, Baltimore, MD, USA, 5–8 Dec 2010. pp. 652–664.

[59]   Miller J.A, Baramidze G.T., Sheth A.P., and Fishwick P.A. Investigating Ontologies for Simulation Modeling. *Proceedings of the 37th Annual Simulation Symposium*, Arlington, VA, USA, 18–22 Apr 2004. pp. 55–71.

[60]   Tolk A., and Turnitsa C.D. Conceptual Modeling of Information Exchange Requirements Based on Ontological Means. *Proceedings of the 39th Winter Simulation Conference: 40 years! The Best is Yet to Come*, Washington, DC, USA, 9–12 Dec 2007. pp. 1100–1107.

[61] McGinnis L., Huang E., Kwon K.S., and Ustun V. Ontologies and Simulation: A Practical Approach. *Journal of Simulation.* 2011; 5(3):190–201.

[62] Ezzell Z., Fishwick P.A., Lok B., Pitkin A., and Lampotang S. An Ontology-Enabled User Interface for Simulation Model Construction and Visualization. *Journal of Simulation.* 2011; 5(3):147–156.

[63] Wang W., Tolk A., and Wang W. The Levels of Conceptual Interoperability Model: Applying Systems Engineering Principles to M&S. *Proceedings of the Spring Simulation Multiconference*, Orlando, FL, USA, 14–19 Sep 2003. Article #168.

[64] Benjamin P., Akella K., Verma A., Gopal B., and Mayer R. A Knowledge-Driven Framework for Simulation Application Integration. *Journal of Simulation.* 2011; 5(3):166–189.

[65] Balci O., Arthur J.D., and William F.O. Achieving Reusability and Composability with a Simulation Conceptual Model. *Journal of Simulation.* 2011; 5(3):157–165.

[66] Okhmatovskaia A., Buckeridge D., Shaban-Nejad A., *et al.* SIMPHO: An Ontology for Simulation Modeling of Population Health. In Laroque C., Himmelspach J., Pasupathy R., Rose O., and Uhrmacher A.M. (eds.). *Proceedings of the Winter Simulation Conference*, Berlin, Germany, 9–12 Dec 2012. pp. 883–894.

[67] Silver G.A., Hassan O.A.H., and Miller J.A. From Domain Ontologies to Modeling Ontologies to Executable Simulation Models. *Proceedings of the 39th Winter Simulation Conference: 40 years! The Best Is Yet to Come*, Washington, DC, USA, 9–12 Dec 2007. pp. 1108–1117.

[68] Puri C.A., Gomadam K., Jain P., Yeh P.Z., and Verma K. Multiple Ontologies in Healthcare Information Technology: Motivations and Recommendation for Ontology Mapping and Alignment. *International Conference in Biomedical Ontologies (ICBO)*, Buffalo, NY, USA, 26 Jul 2011. pp. 367–369.

[69] Partridge C., Mitchell A., and De Cesare S. Guidelines for Developing Ontological Architectures in Modelling and Simulation. In *Ontology, Epistemology, and Teleology for Modeling and Simulation*. Berlin: Springer; 2013. pp. 27–57.

[70] Zeigler B.P. *Multifacetted Modelling and Discrete Event Simulation.* London: Academic Press Inc.; 1984.

[71] Redding S., Conrey E., Porter K., Paulson J., Hughes K., and Redding M. Pathways Community Care Coordination in Low Birth Weight Prevention. *Maternal and Child Health Journal.* 2014; 18(6):1–8.

[72] Zeigler B.P., Carter E.L. Redding S.A., and Leath B.A. Pathways Community HUB: A Model for Coordination of Community Health Care. *Population Health Management.* 2014; 17(4):199–201.

[73] Leslie P.H. On the Use of Matrices in Certain Population Mathematics. *Biometrika.* 1945; 33(3):183–212.

[74] Voltera V. Variations and Fluctuations of the Number of Individuals in Animal Species Living Together. In Chapman R.N. (ed.). *Animal Ecology.* New York, NY: McGraw-Hill; 1931. pp. 31–113.

[75]   Fishbein M., and Ajzen I. *Belief, Attitude, Intention, and Behavior: An Introduction to Theory and Research*. Reading: Addison-Wesley; 1975.

[76]   Kuhl M.E. A Simulation Study of Patient Flow for Day of Surgery Admission. In Laroque C., Himmelspach J., Pasupathy R., Rose O., and Uhrmacher A.M. (eds.). *Proceedings of the Winter Simulation Conference*, Berlin, Germany, 9–12 Dec 2012. pp. 1–7.

[77]   Marmor Y.N., Rohleder T.R., Huschka T., Cook D., Thompson J., and Clinic M. A Simulation Tool to Support Recovery Bed Planning for Surgical Patients. In Jain S., Creasey R.R., Himmelspach J., White K.P., and Fu M. (eds.). *Proceedings of the Winter Simulation Conference*, Phoenix, AZ, USA, 11–14 Dec 2011. pp. 1333–1339.

[78]   Findlay M., and Grant H. An Application of Discrete-Event Simulation to an Outpatient Healthcare Clinic with Batch Arrivals. In Jain S., Creasey R.R., Himmelspach J., White K.P., and Fu M. (eds.). *Proceedings of the Winter Simulation Conference*, Phoenix, AZ, USA, 11–14 Dec 2011. pp. 1166–1177.

[79]   Dibble C. Effective Real-Time Allocation of Pandemic Interventions. *Proceedings of the Winter Simulation Conference*, Baltimore, MD, USA, 5–8 Dec 2010. pp. 2211–2220.

[80]   Hethcote H. The Mathematics of Infectious Diseases. *SIAM Review*. 2000; 42(4):599–653.

[81]   Zeigler B.P. How Can Modeling and Simulation Help Engineering of System of Systems? In Traoré M.K. (ed.). *Computational Frameworks: Systems, Models and Applications*. London: ISTE Press – Elsevier; 2017. pp. 1–46.

[82]   de Bleser L., Depreitere R., de Waele K., Vanhaecht K., Vlayen J., and Sermeus W. Defining Pathways. *Journal of Nursing Management*. 2006; 14(7): 553–563.

[83]   Pearson S.D., Goulart-Fisher D., and Lee T.H. Critical Pathways as a Strategy for Improving Care: Problems and Potential. *Annals of Internal Medicine*. 1995; 123(12):941–948.

[84]   Rotter T., Kinsman L., James E.L., *et al.* Clinical Pathways: Effects on Professional Practice, Patient Outcomes, Length of Stay and Hospital Costs. *Cochrane Database of Systematic Reviews.* 2010; 3:CD006632.

[85]   Husni M.E., Losina E., Fossel A.H., Solomon D.H., Mahomed N.N., and Katz J.N. Decreasing Medical Complications for Total Knee Arthroplasty: Effect of Critical Pathways on Outcomes. *BMC Musculoskelet Disord*. 2010; 11:160. doi: 10.1186/1471-2474-11-160.

[86]   European Pathway Association. http://www.e-p-a.org/index2.html [accessed Jan 16, 2014].

[87]   Vanhaecht K., De Witte K., and Sermeus W. *The Impact of Clinical Pathways on the Organization of Care Processes*. PhD Dissertation, Katholieke Universiteit Leuven, 2007.

[88]   Shi J., Su Q., and Zhao Z. Critical Factors for the Effectiveness of Clinical Pathway in Improving Care Outcomes. *International Conference on Service Systems and Service Management*, IEEE, 2008. pp. 1–6.

[89] Date M., Tanioka T., Yasuhara Y., *et al*. The Present Conditions, Problems and Future Direction of the Server-Controlled Clinical Pathway System Development in Psychiatric Hospitals. *International Conference on Natural Language Processing and Knowledge Engineering*, 2010. pp. 1–8.

[90] Alexandrou D.A., Skitsas I.E., and Mentzas G.N. A Holistic Environment for the Design and Execution of Self-Adaptive Clinical Pathways. *IEEE Transactions on Information Technology in Biomedicine*. 2010; 15(1): 108–118.

[91] Li W., Liu K., Li S., and Yang H. Normative Modeling for Personalized Clinical Pathway Using Organizational Semiotics Methods. *International Symposium on Computer Science and Computational Technology*, Volume 2, Shanghai, China, 20–22 Dec 2008. pp. 3–7.

[92] Li W., Liu K., Li S., and Yang H. An Agent Based Approach for Customized Clinical Pathway. *International Forum on Information Technology and Applications*, Volume 2, Chengdu, China, 15–17 May 2009. pp. 468–472.

[93] Zhang Y., Liu K., and Cui G. Consensus Forming in Clinical Pathway Development: Norm Based Modeling and Simulation. *International Conference on Computational Intelligence for Modeling Control & Automation*, Vienna, Austria, 10–12 Dec 2008. pp. 931–936.

[94] Huang Z., Dong W., Duan H., and Li H. Similarity Measure between Patient Traces for Clinical Pathway Analysis: Problem, Method, and Applications. *IEEE Journal of Biomedical and Health Informatics*. 2013; 18(1):4–14.

[95] Lin F.-R., Chou S.-C., Pan S.-M., and Chen Y.-M. Mining Time Dependency Patterns in Clinical Pathways System Sciences. *Proceedings of the 33rd Annual Hawaii International Conference*, Maui, HI, USA, 7 Jan 2000. pp. 8.

[96] Fernández-Llatas C., Meneu T., Benedí J.M., and Traver V. Activity-Based Process Mining for Clinical Pathways. *32nd Annual International Conference of the IEEE Engineering in Medicine and Biology Society*, Buenos Aires, Argentina, 31 Aug–4 Sep 2010. pp. 6178–6181.

[97] Iwata H., Tsumoto S., and Hirano S. Data-Oriented Construction and Maintenance of Clinical Pathway Using Similarity-Based Data Mining Methods. *IEEE 12th International Conference on Data Mining Workshops*, Brussels, Belgium, 10 Dec 2012. pp. 293–300.

[98] Du G., Jiang Z., Diao X., Ye Y., and Yao Y. Modelling, Variation Monitoring, Analyzing, Reasoning for Intelligently Reconfigurable Clinical Pathway. *IEEE/INFORMS International Conference on Service Operations, Logistics and Informatics*, Chicago, IL, USA, 22–24 Jul 2009. pp. 85–90.

[99] Tsumoto S., Hirano S., and Iwata H. Data-Oriented Maintenance of Clinical Pathway using Clustering and Multidimensional Scaling. *IEEE International Conference on Systems, Man, and Cybernetics*, Seoul, South Korea, 14–17 Oct 2012. pp. 2596–2600.

[100] Lin F.-R., Hsieh L.-S., and Pan S.-M. Learning Clinical Pathway Patterns by Hidden Markov Model System Sciences. *Proceedings of the 38th Annual Hawaii International Conference*, Big Island, HI, USA, 6 Jan 2005. pp. 142a.

[101] Elghazel H., Deslandres V., Kallel K., and Dussauchoy A. Clinical Pathway Analysis using Graph-Based Approach and Markov Models. *2nd International Conference on Digital Information Management*, Volume 1, Lyon, France, 28–31 Oct 2007. pp. 27–284.

[102] Hurley K.F., and Abidi S.S.R. Ontology Engineering to Model Clinical Pathways: Towards the Computerization and Execution of Clinical Pathways. *20th IEEE International Symposium on Computer-Based Medical Systems*, Maribor, Slovenia, 20–22 Jun 2007. pp. 536–541.

[103] Zhen H., Li J.-S., Yu H.-Y., Zhang X.-G., Suzuki M., and Araki K. Modeling of Clinical Pathways based on Ontology. *2nd IEEE International Symposium on IT in Medicine & Education*, Volume 1, Jinan, China, 14–16 Aug 2009, pp. 1170–1174.

[104] Abidi S.R., and Abidi S.S.R. An Ontological Modeling Approach to Align Institution-Specific Clinical Pathways: Towards Inter-Institution Care Standardization. *25th International Symposium on Computer-Based Medical Systems*, Rome, Italy, 20–22 Jun 2012. pp. 1–4.

[105] Fan W., Lu X., Huang Z., Yu W., and Duan H. Constructing Clinical Pathway Ontology to Incorporate Patient State, Intervention and Time. *4th International Conference on Biomedical Engineering and Informatics*, Volume 3, Shanghai, China, 15–17 Oct 2011. pp. 1697–1701.

[106] Ye Y., Jiang Z., Yang D., and Du G. A Semantics-Based Clinical Pathway Workflow and Variance Management Framework. *IEEE International Conference on Service Operations, Logistics and Informatics*, Volume 1, Beijing, China, 12–15 Oct 2008. pp. 758–763.

[107] Ye Y., Diao X., Jiang Z., and Du G. A Knowledge-Based Variance Management System for Supporting the Implementation of Clinical Pathways. *International Conference on Management and Service Science*, Wuhan, China, 20–22 Oct 2009. pp. 1–4.

[108] Du G., Jiang Z., and Diao X. The Integrated Modeling and Framework of Clinical Pathway Adaptive Workflow Management System Based on Extended Workflow Nets (EWF-nets). *IEEE International Conference on Service Operations, Logistics and Informatics*, Volume 1, Beijing, China, 12–15 Oct 2008. pp. 914–919.

[109] Ye Y., Jiang Z., Diao X., and Du G. Knowledge-Based Hybrid Variance Handling for Patient Care Workflows Based on Clinical Pathways. *IEEE/ INFORMS International Conference on Service Operations, Logistics and Informatics*, Chicago, IL, USA, 22–24 Jul 2009. pp. 13–18.

[110] Li Y., Hua S., Li S., and Ma Q. The Variance Analysis of Clinical Pathway in Children with Bronchopneumonia. *International Conference on Human Health and Biomedical Engineering*, Jilin, China, 19–22 Aug 2011. pp. 1260–1263.

[111]   Shi J., Su Q., Zhang G., Liu D., Zhu Y., and Zue L. Hidden Impact of Clinical Pathway on Patient Satisfaction – A Controlled Trial in Elective Cesarean Section. *7th International Conference on Service Systems and Service Management*, Tokyo, Japan, 28–30 Jun 2010. pp. 1–6.

[112]   Brailsford S.C., Desai S.M., and Viana J. Towards the Holy Grail: Combining System Dynamics and Discrete-Event Simulation in Healthcare. *Proceedings of the Winter Simulation Conference*, Baltimore, MD, USA, 5–8 Dec 2010. pp. 2293–2303.

Chapter 2

Multi-perspective architecture for holistic healthcare M&S

In the previous chapter, we introduced O4HCS (ontology for healthcare systems simulation), an ontology that reflects a hierarchy of modeling concepts and a description of the relationships between them. In this chapter, we will derive a disciplined stratification of abstractions for healthcare systems modeling. That way, we lay the basis for the design and engineering approach of this book with a modeling framework of four perspectives that can serve to develop healthcare simulation models at each level of abstraction and couple them together. As a consequence, the simulation from one perspective abstracts all realities concerning the rest of the perspectives, while connecting different perspectives simultaneously takes into consideration all realities. We show how this feature is capable of providing multiple levels of explanation (and is computationally more efficient as regards to complexity) than simulating the system at one level at a time. The novelty of our approach is that notable components of the healthcare system are modeled as autonomous systems that can influence and be influenced by their environments.

But first of all, where are these four layers coming from?

We have seen in the previous chapter that healthcare simulation models often emphasize three types of abstraction: entities, resources, and processes. We have also seen that the kind of analysis carried on is of two sorts: scheduling/planning and projection/prediction. Both sorts of analysis can be performed for each of the abstractions emphasized. However, a classification of healthcare models along the six possibilities of combining these criteria (i.e., two analysis approaches applied to three kinds of abstraction) would not be of great help. Indeed, most often, a same model may compute outcomes related to one or more analysis approach(es) applied to one or more kind of abstraction. As an example, a bed allocation model may compute bed occupancy (i.e., resource-related) as well as patients waiting time (i.e., entity related) and health service throughput (i.e., process related).

We rather turn to O4HCS for a better understanding of the whole picture.

2.1 Multi-perspective modeling of healthcare systems

While examining O4HCS carefully, one can notice that it has epistemological top levels and ontological bottom levels. If its levels are numbered starting from 0 with the top concept (i.e., healthcare system), then the frontier is between level 3 and level 4. Indeed, due to the cascade of specialization that links the Facet model to all other models downward, one can reasonably argue that ontological (i.e., real-world) models are the final specialized models of O4HCS (i.e., functional dynamics, spatial dynamics, entity, flow, provision, provider, spatial diffusion, and functional diffusion models), while the other entities are epistemological (i.e., mental) models. Consequently, we can consider population, individual, resource, and diffusion as defining the four layers of the most basic epistemological stratification of abstraction.

Another way of seeing this is by scrutinizing the true meaning of the SES-based ontology. Recall that the SES pruning process allows us to extract from the O4HCS a specific system configuration by selecting particular subsets of aspects, cardinalities of multi-aspects, and instances of specializations and assigning values to the variables. As such, a healthcare system can have various facets, some may focus on population dynamics (PD) (whether functional or spatial) or individual behavior (IB) (whether entity-based or flow-based), while some others may focus on resource allocation (RA) (whether provider-based or provision-based or both), or health phenomenon diffusion (whether spatial or functional). Obviously, a study may limit itself to only one facet, or extend to involve more facets. We know from the literature review done in Chapter 1 that most often, facets are studied in isolation.

Consequently, we distinguish four fundamental perspectives that simulation models develop, either, one at a time, or by combining two or more of them. The layers of our framework, as presented in Figure 2.1, cover the full set of healthcare concerns, which, though interrelated, are often treated separately and the impact of other concerns on any one of them being approximated by parameters.

We place this stratification of abstractions in the context of the hierarchy of systems specification introduced by [1]. That way, each perspective can be seen as encompassing a family of questions that can be formulated through dedicated experimental frames [2]. Consequently, models can be developed within each perspective and coupled together. The resulting top model in each perspective can be coupled with its experimental frame to derive results specific to this perspective. More on the methodological aspects will be provided in Chapter 4.

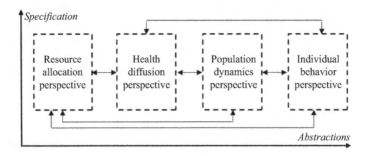

Figure 2.1   Multi-perspective approach to healthcare systems M&S

While this feature provides multiple levels of explanation for the same system, there is also a need to encompass the influences of perspectives on one another. While the dashed boxes in Figure 2.1 depict independent simulations in the different perspectives, the double arrows represent the live exchanges of information between them. *The idea is to allow for the transmissions of the outputs of the simulations in one perspective to provide live feedbacks to the simulation parameters in other perspectives where required.* We have defined an integration mechanism to enable such exchange, which we detail later in this chapter.

Before we delve into such details, let us first have a look on the perspective-specific models highlighted by the O4HCS hierarchy. Obviously, we will not cover all possible models in these perspectives, but only a few predominant theoretical models. In addition, we completely exclude from the scope of our work all models in which the information on temporal events are lost (such as time-free optimization models), since simulation is core to the framework presented.

2.1.1 Health diffusion models

The health diffusion (HD) perspective is described by Figure 2.2 (an excerpt of O4HCS).

The HD perspective covers simulation studies of ease/disease spreading. It falls under health production, as defined in O4HCS. HD-specific experimental frames are coupled to HD models, in order to derive answers for questions such as the forecasted proportion of individuals in a population according to their health status or the patterns of contamination areas from given initial conditions [3–6]. Two kinds of HD models exist: functional (also referred to in the literature as equation-based models) and spatial.

2.1.1.1 Functional diffusion models

Predominant theoretical functional diffusion models in literature, include SIR [7] and its derived SIS, SIRS, SEIR, SIRQ, MSEIR, etc. models [8]. These models

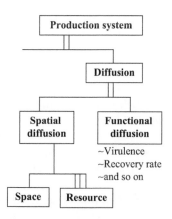

Figure 2.2   Health diffusion models

consider a targeted population as a whole (hence their characterization as macro models) and describe the evolution of compartments of individuals (hence their characterization as compartmental models) over time. Compartments are stratifications defined along a given health status, where each compartment denotes the number or individuals having the same status:

- *S* is the number of susceptible individuals in the population,
- *E* is the number of exposed individuals (susceptible individuals become exposed before being infected),
- *I* is the number of infectious individuals,
- *R* is the number of recovered (or removed) individuals (such individuals cannot be infected anymore, due to immunity or death),
- *Q* is the number of individuals in quarantine,
- *M* is the number of individuals having a passive immunity,
- and so on.

As such, compartmental models are flexible, in that new compartments can be defined as new assumptions arise. Also, various topologies of the network of compartments are possible (i.e., SIR, SIS, SIRS, SEIR, SIRQ, MSEIR, etc.), based on the nature of the ease/disease diffusion under study (a knowledge provided by health experts) and assumptions made on the characteristics of the target population (such as mobility, distribution by age/wealth, behavior-related risks, etc.).

The SIR model considers a linear transition flow *S→I→R,* where the arrows are labeled with specific transition rates. The model introduced by Kermack and McKendrick [7], which has several variants in the literature, is expressed by the following equations:

$$\frac{dS}{dt} = -\frac{\beta SI}{N} \tag{2.1}$$

$$\frac{dI}{dt} = \frac{\beta SI}{N} - \gamma I \tag{2.2}$$

$$\frac{dR}{dt} = \gamma I \tag{2.3}$$

where

- *N* is a fixed population, i.e., $N = S(t) + I(t) + R(t)$.
- β is the infection rate, which is assumed to be equal for all susceptible individuals; the infection process between susceptible and infectious individuals is referred to as the Law of Mass Action, a widely accepted idea that the rate of contact between two groups in a population is proportional to the size of each of the groups concerned [9]; another assumption is that the fraction of contacts by an infected is *S/N.*
- γ is the "recovery or death" rate (therefore, $1/\gamma$ is the mean infective period); assumptions on equations (2.2) and (2.3) are that the population leaving the susceptible class is equal to the number entering the infected class, and the

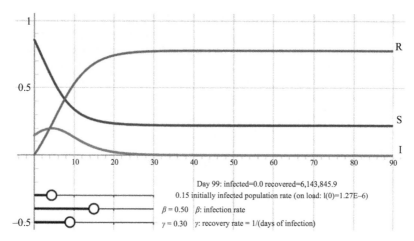

Figure 2.3   SIR trajectory for β = 0.5, γ = 0.3, and initial infected population rate of 15%

population leaving the infected class is equal to the number entering the recovery/removed class.

It can be noted that if *S*, *I*, and *R* are taken as population proportions, then *N* can be dropped from the equations. Figure 2.3 shows the trajectory obtained for specific values of parameters, using JSXGraph, a JavaScript library for function plotting in a web browser [10].

The original SIR model also assumes that the infection and recovery rates are much faster than the timescale of births and deaths, and therefore, these factors are ignored in this model. To include them, the previous equations can be changed to the following equations:

$$\frac{dS}{dt} = -\frac{\beta SI}{N} + \lambda N - \mu S \tag{2.4}$$

$$\frac{dI}{dt} = \frac{\beta SI}{N} - \gamma I - \mu I \tag{2.5}$$

$$\frac{dR}{dt} = \gamma I - \mu R \tag{2.6}$$

where

- The population is not fixed anymore, due to births and deaths, i.e., $N(t) = S(t) + I(t) + R(t)$,
- λ is the birth rate; therefore, λN represents the arrival of new susceptible individuals into the population, and
- μ is the death rate (therefore, *R* is restricted to recovered individuals, and γ to recovery rate).

The SIS model derives from the SIR model by assuming that individuals are immediately susceptible once they have recovered (i.e., there is no immunity to the ease/disease phenomenon considered). Consequently, the linear transition flow is $S \rightarrow I \rightarrow S$, with corresponding transition rates as depicted by (2.7), (2.8) and (2.9), (2.10) given hereafter, respectively, for the case death and birth rates are ignored or not:

$$\frac{dS}{dt} = -\frac{\beta SI}{N} + \gamma I \tag{2.7}$$

$$\frac{dI}{dt} = \frac{\beta SI}{N} - \gamma I \tag{2.8}$$

$$\frac{dS}{dt} = -\frac{\beta SI}{N} + \lambda N - \mu S + \gamma I \tag{2.9}$$

$$\frac{dI}{dt} = \frac{\beta SI}{N} - \gamma I - \mu I \tag{2.10}$$

The SIRS model is an extension of the SIR model that allows members of the recovered compartment to return to the susceptible compartment. The SEIR model takes into consideration a period of latency, giving an additional compartment (E) of individuals who, after being contaminated by infected individuals, have a probability to be infectious or not. The SEIS model is a modification of the SEIR model such that individuals that have recovered return to being susceptible again, closing the loop of the transition flow. The MSIR model takes into consideration the fact that some individuals are born with a passive (i.e., inherited) immunity, giving an additional compartment (M, for maternally derived immunity) of individuals who are not born into the susceptible compartment but are immune for some period of their life due to genetic/natural protection. The MSEIR model extends the MSIR model with the consideration of a latency period. An MSEIRS model is similar to the MSEIR, but the immunity in the R class is temporary, so that R individuals return to the S compartment when the temporary immunity ends.

There are many other derivations of the SIR model [11]. Parameters are extremely important in all these models, since they allow to capture assumptions on the characteristics of both the population and the ease/disease phenomenon (like age distribution and its impact on contacts between individuals, or the virulence of a disease as regards to the fact that its mode of transmission is by air, by body contact, or by inoculation). They also allow to inject into the model the effect of intervention strategies or the environment. Examples include the seasonal nature of the infection rate for some diseases (such variability can be expressed with a linear/exponential/quadratic time-varying parameter instead of a constant one) or decisionmaking such as mass vaccination or restriction of social contacts.

A key indicator in these models is the so-called basic reproduction number, denoted by R_0, a threshold quantity which determines the force, i.e., long-term transmission potential of the diffusion phenomenon, provided there is no heterogeneity in the contact rate. It is also interpreted as the average number of secondary cases caused by a primary infectious individual. As stated by Trottier and Philippe [11], if $R_0 < 1$, the diffusion dies out, while if $R_0 > 1$, there is an epidemic of the phenomenon in the population, and in cases where $R_0 = 1$, the phenomenon becomes endemic, meaning it remains in the population at a consistent rate. However, neglecting the variability of the contact pattern in the population is a very strong assumption. In more realistic scenarios, R_0 is not a constant value.

Functional diffusion models are to be considered more useful in explanatory modeling than to mimic real-world system as closely as possible. As such, they capture the main features of the dynamic of HD phenomena, given some assumptions expressed through parameters. Due to the inherent nonlinearity of these models, the values of these parameters can make the difference between regular cyclic variations of incidence and chaos [12].

For small isolated populations, heterogeneities among individuals (such as variation in risks of exposure, infection, and other factors) are often better known. In such cases, stochastic models are preferred to compartmental models, the latter being referred to as deterministic models [13,14]. Stochastic models rely on linear statistical specifications that are based on Newtonian physics and which consider the diffusion phenomena to have attained a stable state, where interactions are either ignored or considered second-order processes. One of the most known of them is the Reed-Frost model [15], a chain-binomial model based on the SIR distinction of individuals along their health status, which is often described according to a discrete time scenario. In such scenario, short infectious periods (also called generations) are preceded/followed by longer latent periods. The occurrence of event in a given generation depends only on the state of the epidemic in the previous generation, thus giving a Markov model with conditional event probabilities, as described by the following equation:

$$
\begin{aligned}
P(Y_{j+1} = y_{j+1} | X_0 &= x_0, Y_0 = y_0, \dots, X_j = x_j, Y_j = y_j) \\
&= P(Y_{j+1} = y_{j+1} | X_j = x_j, Y_j = y_j) \\
&= \binom{x_j}{y_{j+1}} (1 - q^{y_j})^{y_{j+1}} (q^{y_j})^{x_j - y_{j+1}}
\end{aligned}
\tag{2.11}
$$

where

- X_j and Y_j denote the number of susceptible and infected individuals,
- q is the probability for a susceptible individual to get infected; different susceptible individuals in a given generation become infected independently of one another, and
- $X_{j+1} = X_j - Y_{j+1}$, meaning that a given susceptible individual of generation j remains susceptible in generation $(j+1)$ if this individual escapes infection from all infected individuals of generation j.

Stochastic models can be analyzed statistically. A survey of such models is given in [16], and a thorough analysis can be found in [17]. Andersson and Britton [16] give also several links to literature on deterministic and stochastic models, the two major categories of functional HD models.

2.1.1.2 Spatial diffusion models

As the spatial population structure is key to a more accurate estimation of the rate of contact between individuals, spatial diffusion models emphasize the representation of space and the mapping of diffusion pathogens (i.e., human/material/information resources that trigger the production of ease/disease) onto it.

Cellular automata (CA) models [18] are widely used to explicitly express spatially located interactions. A CA model is a finite set of interconnected cells, which represent discrete spatial domains. For health spatial diffusion representation, resources (human, material, etc.) characterized by health-related properties are located into cells. In fact, two conceptual approaches to map resources onto space are possible. In one way, space and resources can be integrated such that resources appear as simply defining the properties of the cell containing them. In the other way, resources are autonomous agents [19,20], with proper decision-making rules. Different geometries can be adopted for cells (circle, triangle, rectangle, octagon, etc.), which often define the topology of the CA. The concept of neighboring (or connectivity) reflects the local interactions between resources within cells, and different types of neighboring are possible, such as the so-called von Neumann neighborhood, Moore neighborhood, etc.

Regarding simulation, all cells of a CA model evolve synchronously by the application of rules, at each time step, where the next state of a resource within a cell depends on its current state and the other resources within its neighborhood. In the case of an agent-based resource representation, such rules are specific to each agent, while in the case of a cell-based resource representation, these rules are defined at the global level and apply equally to all cells. Rules of special interest are the ones related to health status evolution (like the set of conditions under which a susceptible individual gets infected, depending on the health status distribution in the neighborhood) and the motion-related ones as well (such as travels, migrations, distribution of health resources, etc.).

White *et al.* [21] presents a CA-based disease spreading model, where each cell of the CA is considered to represent a square area of the land in which the epidemic is propagating in a population and the state $s^t_{a,b} \in [0, 1]$, of any cell (a,b) at any time t, represents the ratio of infected population to the total population of the cell. The local state transition rule is defined for each cell by the following equation:

$$s^{(t+1)}_{a,b} = g\left((1 - P(t))s^t_{a,b} + \left(1 - s^t_{a,b}\right)\left[\varepsilon s^t_{a,b} + \sum_{\alpha,\beta \in V^*} \mu^{(a,b)}_{\alpha\beta} s^t_{a+\alpha, b+\beta}\right]\right) \qquad (2.12)$$

where

- $V^* = \{(\alpha,\beta)| \ a - 1 \le a \le a + 1, \ b - 1 \le \beta \le b + 1\} - \{(a,b)\}$ is the neighborhood,
- $P(t) = 0.2t + 0.2$ is a measure of the infected population that has recovered from the disease within the last time step,
- g is a discretization function that returns a value in [0,1], and
- $\mu^{(a,b)}_{\alpha\beta} = c^{(a,b)}_{\alpha\beta} \times m^{(a,b)}_{\alpha\beta} \times v$, where $c^{(a,b)}_{\alpha\beta}$ and $m^{(a,b)}_{\alpha\beta}$ are abstractions of connections/links and movement of infected people, respectively, between cell (a,b) and its neighboring cells in V^*, and v is the virulence of the epidemic.

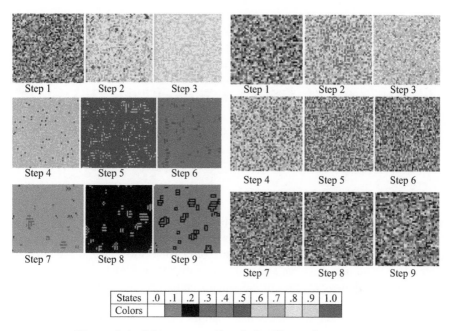

States	.0	.1	.2	.3	.4	.5	.6	.7	.8	.9	1.0
Colors											

Figure 2.4   Disease spread with fixed/varied parameters

Figure 2.4 displays two scenarios of the CA model's trajectory. Each of the scenarios shows the levels of spread in nine successive time steps. At the left-hand side, the model is executed with the following constant assumptions of the parameters: $\varepsilon = 0.4$, $c_{\alpha\beta}^{(a,b)} = 1$, $m_{\alpha\beta}^{(a,b)} = 0.4$ and $v = 0.4$. At the right-hand side, numbers are randomly generated between 0 and 1 to be dynamic values of $m_{\alpha\beta}^{(a,b)}$ (therefore capturing a dynamic change in the pattern of movement between cells). For simplicity, state values are rounded to one place of decimal to have a finite state set {0.0, 0.1, 0.2, 0.3, 0.4, 0.5, 0.6, 0.7, 0.8, 0.9, 1.0}.

The color codes chosen for cells states in successive time steps are given at the bottom of Figure 2.4. We see that with fixed parameters, the disease spread appears to subside in successive time steps from step 3 with most cells having infection rates of below 30% at the ninth time step, while with varied movement pattern, the rate of disappearance of the disease is slower and the infection rates within most cells are still above 50%.

Space representation in spatial diffusion models can also be implicit, rather than explicit. This is useful when the spatial connectivity evolves over time, instead of being fixed once for all. The so-called network models [22–25] provide a surrogate for spatial structure through a graph representing the population's contact structure. Each of the nodes of the graph represents an individual. An edge is placed between any pair of individuals who come into contact with one another, labeled with an encounter rate, hence giving an explicit description of the variability of contacts. That way, each node relates to a network of encountered agents, which

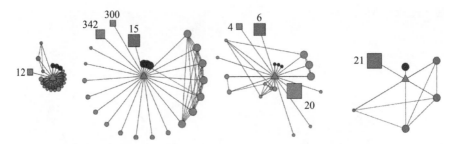

Figure 2.5   Individual contact heterogeneity as a network model [27]

varies from one simulation step to the next, according to the probabilistic distribution of encounters. If nodes are agents (rather than passive resources), they can make decisions impacting their health status, or be subject to a natural evolution of their health condition, between any two consecutive steps [26].

Figure 2.5 shows examples of individual-centric networks reported by Danon *et al.* [27], that result from a survey they conducted, and from which they characterize the distribution of contacts that captures these forms. They include, from left to right, a school girl aged 12 years, a female flight attendant aged 22 years, a male fire fighter aged 44 years, and a retired male aged 62 years. The individual is represented by the orange central triangle, while circles represent individual contacts, and squares represent groups of contacts (with size of group indicated). Red color indicates home contact, while blue indicates work/school contact, yellow indicates travel contact, and green indicates other contacts. The larger a symbol is, the longer is the contact duration, while the closer a symbol is, the more frequent the contact. The authors argue that such a work helps to determine the types of realistic network structure that should be assumed in studies of infection transmission, leading to better interpretations of epidemiological data.

In a comparative study between functional diffusion models and agent-based spatial diffusion models, Ozmen *et al.* [26] argue that the magnitude and velocity of a HD phenomenon (a flu epidemic in their study case) depends critically on the selection of modeling principles, assumptions of disease process, and the choice of time advance.

2.1.2   Resource allocation models

The RA perspective encompasses all scheduling and planning problems, mostly in the context of limited resource provisions, to meet the healthcare demand. It falls under health production, as defined in O4HCS (see Figure 2.6). RA models are used to answer questions formulated through RA-specific experimental frames. Examples of such questions are the occupancy rate of beds in a surgical unit, the average waiting time in an emergency department, or the optimal scheduling of healthcare activities [28–35]. Two kinds of RA models exist: provision and provider.

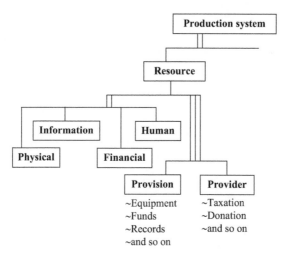

Figure 2.6 Health resource allocation models

2.1.2.1 Provision models

Provision models focus on how health services are delivered, including the organization, dependencies, and activities of resources. Parameters are used to capture the structure of the demand (such as arrival of patients or requirements for delivery). As such, healthcare systems are considered as production systems. A large variety of case-specific models in the literature falls under this category. However, most of them can be related more or less to queueing theory models [36] and differ mainly by the formalism used (state charts, Petri net, queuing network, DEVS, agents, etc.).

The essence of queueing models is that a number of services are offered at different nodes of a network, each node allowing customers to queue, waiting for the service provided in that node, as illustrated by Figure 2.7. The provision of service at a node is materialized by one or many concurrently operating servers. When a customer is serviced at one node, it can join another node and queue for service or leave the network. To capture salient characteristics of a queuing node, Kendall [37] introduced a notation as a three-part code *a/b/c,* where *a* specifies the interarrival time distribution of customers, *b* the service time distribution at the node, and *c* the number of servers. Specific letters are used to represent standard probabilistic distributions (e.g., M is used for exponential distribution, D for deterministic times, and G for general distribution). Examples of models defined according to this notation are M/M/1 (for exponential arrival of customers, exponential service law, and a single server) and M/D/3 (for exponential arrival of customers, deterministic service times, and three identical servers). The notation can be extended with extra letters to cover other queueing models. In the basic queueing model, customers arrive one by one, and they are always allowed to enter the system, there are no priority rules and customers are served in order of arrival (FIFO, for first in first out). In extended cases, one or more of these assumptions do

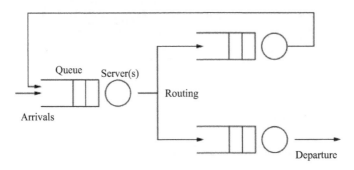

Figure 2.7   Provider models

not hold. Possibilities include a limited length of queue, and customers served according to a different discipline, such as LIFO (for last in first out) or along priorities defined depending on categories of customers. Moreover, customers' routing can be stochastically/deterministically defined. A stochastic routing reduces the queuing model to a Markov chain.

Queueing models are well suited to the application of scheduling and planning techniques [38–44]. Under some assumptions, analytical results can be derived. Relevant performance measures in such analysis include the distribution of the waiting time and the sojourn time of a customer at each node, the distribution of the number of customers in the system (including or excluding the one or those in service), the distribution of the amount of work in the system, and the distribution of the busy period of servers at each node. The first significant results in this area were established by Jackson [45], Little [46], Jackson [47], and Baskett *et al.* [48]. In more complex cases, discrete-event simulation is required. In a survey of applications of queuing theory in the field of healthcare, Fomundam and Herrmann [49] summarized a range of results established for systems at different scales, including heath and regional healthcare systems, in the following areas: waiting time and utilization analysis, system design, and appointment systems. They emphasized on salient aspects such as the conflict between minimizing the time that health consumers (like patients) have to wait and maximizing the utilization of the health resources (like doctors, nurses, beds, etc.), or the impact of the so-called reneging phenomenon (i.e., when a customer waiting in a queue decides to forgo the service because he does not wish to wait any longer) on the performances of the healthcare system. Variable arrival rates, priority queueing discipline, and blocking (i.e., when a queuing system places a limit on queue length) are other key influencing factors [50].

2.1.2.2   Provider models

Provider models focus on who delivers health services, including actors representing the government or involved in the health financing system and/or the health records system, or resource providers, etc. Entities in such models are often modeled as agents with decisionmaking capabilities that interact with other agents and

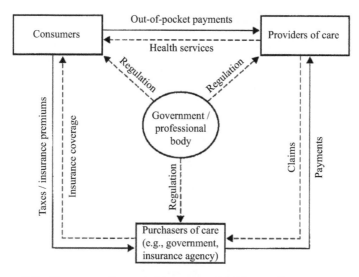

Figure 2.8   Structure of a social insurance healthcare finance system [51]

objects of their environment. In a compendium of quantitative techniques for healthcare financing, Cichon *et al.* [51] presented the generic structure reproduced in Figure 2.8 that can guide the design of such models in the case of a social insurance healthcare finance system.

Agents used to represent health providers who do not need to have a sophisticated level of intelligence (such as emotion, cognition, beliefs, etc.). As such, they define models known as agent-based models (ABM), in opposition to multi-agent systems (MASs) where the social dimension of agents is more elaborated (as we will see with entity-based IB models presented in the next section). In other words, although both categories of model rely on the principle of a collection of autonomous decision-making components, the difference lies in the extent to which autonomy is forged.

2.1.3   Individual behavior models

The IB perspective covers the studies of social behavior in relation to how its components (such as educational level, physical state, emotion, cognition, decision, etc.) affect the willingness/ability of an individual to effectively access available healthcare services. It falls under health consumption, as defined in O4HCS (see Figure 2.9). IB-specific experimental frames address questions such as the relationship between sociocultural decisions and health status of individuals, or the evaluation of life strategies in the context of competition/selection and scarcity of resources [52–54]. Two kinds of IB models exist, as recalled in Figure 2.9: entity and flow.

2.1.3.1   Entity models

MASs are among the most used entity-based individual models. A MAS [19,55] is a collection of interacting autonomous discrete entities, each acting (both on its

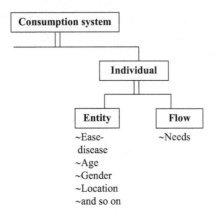

Figure 2.9   Individual behavior models

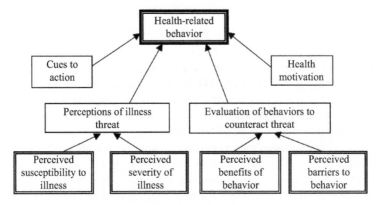

Figure 2.10   Health belief model [56]

local environment and other agents) in its proper way and based on its social knowledge upon itself, the local environment, and the other agents.

A reference model is HBM, the health belief model [56]. As shown in Figure 2.10, it has four basic constructs (perceived susceptibility, severity, benefits, and barriers), which though not formalized in a standard way, are intuitive and easy to understand. Variables of the model, as well as connections between them, have to be defined by the model engineer.

Another well-known entity model is PECS (physics, emotion, cognition, status), an architecture that incorporates state variables to represent physical state, emotion, cognition, and social status of an individual [57]. Two patterns of dynamics are defined, called "reactive" and "deliberative." Reactive behavior operates at the emotional level, whereas deliberative behavior operates at the cognitive level. Figure 2.11 depicts the basic PECS architecture. The components "Sensor" and "Perception" correspond with the input of the agent system. The internal state of the

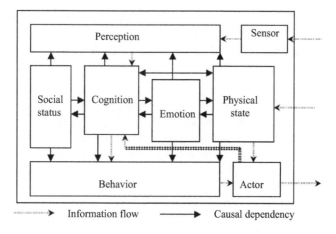

Figure 2.11   Basic structure of a PECS agent [58]

system is distributed over the components "cognition," "emotion," "physical state," and "social status." State transitions can be specified by discrete transition rules or algebraic/differential equations. The component "Actor" corresponds to the output of the system and is provided by the component "behavior" with specific task descriptions.

Conner and Norman [59] thoroughly examine most common social cognition models to health behaviors, including the well-known health locus of control model [60]; the theory of planned behavior [61]; and the beliefs, desires, and intentions architecture [62]. Agent models, including MAS and ABM, provide structural solutions, while the behavioral dimension (i.e., the temporal aspects) is left to the model builder. Consequently, agents can encapsulate models of other types, such as equation/graph/network-based representations. Therefore, *they provide an organizational architecture that can ease glueing heterogeneous abstractions* [63].

2.1.3.2   Flow models

Rather than focusing on the structural aspect of an individual, flow models concentrate on the behavioral dimension. A *Patient flow* models the flow of activities related to an individual through a health unit. It is a structured sequence of events, activity timings, staff and resource requirements, locations, problems, and decisions that a patient will experience during a pathway of care. As such, it encompasses all clinical and extra clinical processes of attending to a patient, from the admission into the medical facility to the discharge. Patient flow is not about the *what* of care decisions, but about the *how* (healthcare services are accessed), *where* and *when* (treatment and assessment is available), and *who* (it is provided by) of care provision [64].

The impact of patient flow optimization on healthcare quality and cost has been demonstrated in numerous research works, with focus made at different levels [65], including patient planning and control, patient group planning and control, resource planning and control, patient volume planning and control, and strategic (i.e., long-term policy) planning. Several formalisms have been used (various patient flow models are shown in Figure 2.12), which differ mainly by the fact that abstractions are either centered on activities, or information, or communications, or a mix:

- Activity-based flow models capture the flow of control, as in simple deterministic/stochastic flowcharts [64,66] or in PERT charts [67].
- Information-based flow models emphasize on the flow of information, like Business record [68], or one of the IDEF family languages [69].
- Communication-based flow models capture the collaborations between tiers, as DEMO [70] or role activity diagram does [71].
- Hybrid flow models emphasize on establishing a pairwise combination of the previous models or even an integration of all of them, as in Business Process Modeling Notation based efforts [72,73].

Patient flow optimization methods range from traditional optimization techniques such as the Job shop [74] scheduling approach (the basic formulation of which is that we are given n jobs $J_1, J_2, \ldots, J_n$ of varying processing times, which need to be scheduled on m machines with varying processing power, while trying to minimize the makespan, i.e., the total time that elapses from the beginning to the end), and its numerous variants (such as flow shop, hybrid flow shop, etc.), to techniques stemming from Management science, such as critical path method, Lean Six Sigma, business process redesign, and the theory of constraints. See [75] for a general review of these methods. As usual, when analytical methods fall short due to some assumptions not satisfied, discrete event simulation counterparts of these models are considered.

A patient flow model is highly parametrized by essence, since the flow routing is dependent on several factors external to the individual represented, such as care and decisions taken by physicians, or uncertainties inherent to the healthcare system [28,34]. In this context, queueing theoretic models can also be used to describe the movement of patients, but with a transactional approach. It means that the routing information is not carried by the nodes of the queueing model as done in provision models, but by customers of the queue, whereas the decision taken by health workers and the uncertainties upon health resources are captured in the nodes. Performance of these queueing networks may be analyzed either through analytical queueing formulations, Markov chain analysis or by computer simulation of these networks [66,76–78]. Based on an extensive review, Bhattacharjee and Ray [79] propose a generic framework for patient flow modeling and performance analysis of healthcare delivery processes in hospitals.

2.1.4   *Population dynamics models*

The PD perspective comprises all studies of the dynamics in the population of a community (immigration, emigration, birth, death, etc.). It falls under health consumption, as defined in O4HCS (see Figure 2.13). PD-specific experimental frames

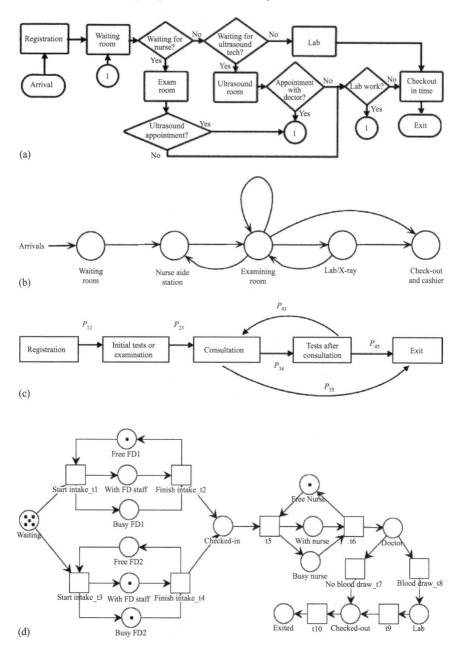

Figure 2.12   Various categories of patient flow models: (a) chart-like [80], (b) compartmental [81], (c) Markovian [79], and (d) transactional [82]

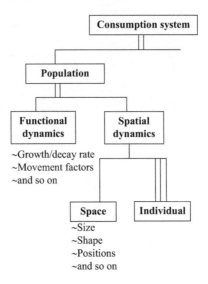

Figure 2.13   Population dynamics models

formulate summary mappings to answer questions like the forecasted distribution of a population by gender, social status or age range, or the impact of a species strategy on the encapsulating ecosystem [83–87]. Two kinds of PD models exist: functional dynamics and spatial dynamics.

As such, PD models show similarities with HD models, in that some predominant formalisms are commonly used for both models, including differential equations (for functional models), and agents and CA (for spatial models). However, there are significant differences between the semantics behind these models. Also, it is usual to find models that integrate both PD and HD aspects such that it is not easy to draw a clear line between processes from both perspectives. While such integrations are of great benefit for a better understanding of systems of interest, a major challenge with them is that they do not allow multiple levels of explanation, like the capacity to separately study each of the processes involved while taken all of them into account together.

2.1.4.1   Functional dynamics models

The first principle to PD is widely regarded in the population ecology as the Malthusian model [88], also known as the exponential growth model. Essentially, this model assumes members of a population reproduce at a steady rate, based on birth and death rates. The population either increases or decreases exponentially, depending on whether the proportionality constant for the birth rate is greater (respectively lesser) than that for the death rate. A general form of the model can be expressed as the following time-dependent equation:

$$P(t) = P(0)e^{Kt} \tag{2.13}$$

Figure 2.14 Areas under growth and decay trends in Malthusian model

where

- $P(0)$ is the initial population size (i.e., at time $t = 0$), and
- K is the net growth rate (sometimes called Malthusian parameter), i.e., birth rate – death rate; in the original version of the model, only the birth rate was considered; it is also important to notice that the net growth rate can incorporate the rates of immigrations in and out of the population.

Figure 2.14 shows areas under the growth (respectively decay) curve, in the case of a positive (respectively negative) rate ($+0.001$, respectively -0.001). This model gives an accurate description of the growth of certain ecological populations (e.g., bacteria and cell cultures). However, it does not take into account impacting factors such as competition for limited resources and space, specificities of the environment, and disasters. Several variants of the Malthusian model have been proposed to address such issues [89–92].

The Prey—Predator model [93] is another famous functional PD model, also known as the Lotka—Volterra model. In its simplest form, it is expressed by the following equations:

$$\frac{dx}{dt} = \alpha x - \beta xy \tag{2.14}$$

$$\frac{dy}{dt} = \delta xy - \gamma y \tag{2.15}$$

where

- x is the number of individuals of some prey population,
- y is the number of individuals of some predator population,
- α is the birth rate of the prey population,
- β is the rate of predation upon the prey, i.e., the rate at which the prey population decays from predation,
- δ is the rate of predation upon the predator, i.e., the rate at which the predator population grows from predation, and
- γ is the death rate of the predator population.

Figure 2.15 displays the well-known phase-space pattern of evolution of this model, for specific values of the parameters ($\alpha = 2/3$, $\beta = 4/3$, and $\gamma = \delta = 1$). The

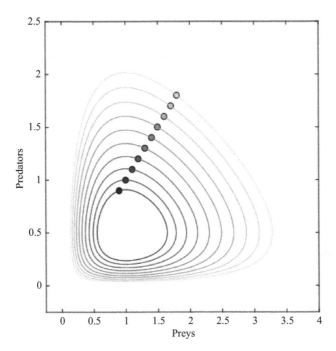

Figure 2.15   Phase-space plot of the Prey-Predator model ($\alpha = 2/3$, $\beta = 4/3$,
$\gamma = \delta = 1$)

phase-space representation is a parametric approach with one axis representing the
number of prey and the other axis representing the number of predators for all times
(therefore, time is eliminated from the equations, leading to one single differential
equation). Small circles on orbits represent prey and predator initial conditions
from $x = y = 0.9$ to 1.8, in steps of 0.1.

Several simplifying assumptions are made in this model, including the continuous approximation of variables that are discrete in nature (as for functional
diffusion models seen previously), the homogeneity of prey and predators subpopulation (i.e., all preys behave similarly, and so do all predators), and the isolated
nature of the system (i.e., no migration in or out), as well as its deterministic nature.
Consequently, numerous variants have been proposed for the Prey-Predator model
[94,95], including the well-known competitive Lotka–Volterra equations [96,97]
and the Arditi–Ginzburg equations [98,99].

The cohort-component model is also a well-known PD model formalized in
matrix algebra by Leslie [85], which allows to divide a population into subpopulations
of common characteristics (such as age, sex, location, etc.). That way, each subpopulation can be defined by specific dynamics rates (such as survival/death/fertility
rates). The model can be formulated by the following equation:

$$N(t+1) = L \cdot N(t) \tag{2.16}$$

where

- $N(t)$ is the population vector at time t, with $N_i(t)$ representing the count of individuals of subpopulation i, and m being the number of subpopulations considered and
- L is the Leslie matrix, with L_{ij} being the fraction of individuals that survives from subpopulation i at t to subpopulation j at $t + 1$; L_{ij} is used to encode evolution parameters, such as the fecundity rate, aging rate and death rate per subpopulation; the developed counterpart of (2.16) gives $N_i(t+1) = \Sigma_{k=1,m} L_{ik} N_k(t)$.

In most of its applications in the literature, the population is stratified into age groups [100,101], and the Leslie matrix has the birth/fecundity rates of each age group on the top row, while it has survival probabilities on a lower diagonal, as described by the following equation:

$$
\begin{pmatrix} n_0^{(t+1)} \\ n_1^{(t+1)} \\ n_2^{(t+1)} \\ \vdots \\ n_T^{(t+1)} \end{pmatrix} = \begin{pmatrix} f_0 & f_1 & f_2 & \cdots & f_T \\ p_0 & 0 & 0 & \cdots & 0 \\ 0 & p_1 & 0 & \cdots & 0 \\ \vdots & \vdots & \vdots & \ddots & \vdots \\ 0 & 0 & 0 & p_{T-1} & 0 \end{pmatrix} \begin{pmatrix} n_0^{(t)} \\ n_1^{(t)} \\ n_2^{(t)} \\ \vdots \\ n_T^{(t)} \end{pmatrix}
\tag{2.17}
$$

The Logistic model [102,103], another widespread functional dynamics model, relies on the idea that population growth rates are regulated by the density of a population such that whether the population density is high or low, PD returns the population density to a point of equilibrium, which is called homeostasis and is defined as K. The dynamics of this model is described by the following equation:

$$
\frac{dN}{dt} = rN\left(1 - \frac{N}{K}\right)
\tag{2.18}
$$

where

- N is the population count,
- K is the parameter expressing the equilibrium point; it is called the *carrying capacity,* and is usually interpreted as the amount of available resources expressed in the number of individuals that these resources can support, and
- r is the reproduction rate (therefore, r/K is interpreted as the competition rate).

As such, this model has three possible outcomes:

- Population increases from an initial count ($<K$) to a plateau K (this is called logistic curve).
- Population decreases from an initial count ($>K$) to a plateau K.
- Population does not change from an initial count ($= K$ or 0).

2.1.4.2 Spatial dynamics models

CA are used to model also spatial PD, with cells containing individuals [104–107]. In classic CA-based spatial dynamics models, individuals are just defined by their properties, while in agent-based spatial dynamics model, each individual is an agent located in a cell. Transition rules include local influences on growth, death and decision to move, which are influenced by local factors such as density rates, individual household factors, and neighborhood positive/negative attractors [108].

Alternative to cellular models are metapopulation models [86,109], which hierarchical structure can be used as a substitute for spatial structure. A meta-population model, also known as regional/multi-regional model [110], is a network model that represents several populations as the nodes of a graph, and the migration relationships between them as edges between corresponding nodes. The dynamic inside each node is described by equations, including but not limited to aspects related to housing market, job market, land market, and transportation, whereas migrations are captured as instantaneous streams between nodes that depend on factors such as the attractivity of a node, the distance between nodes, and the accessibility of a node. Usually, metapopulation models do not accommodate for dynamic change of links between components of the network, even if this is technically doable. Such a constraint sets a clear difference between metapopulation models and network models used for HD phenomena seen previously (where links are dynamically created/destroyed as individuals undergo various social encounters).

As for all the other theoretical models presented in this chapter, the purpose of spatial dynamics models is not to forecast actual future values of the system, but rather to learn about the relative impacts of alternative assumptions and interventions.

2.1.5 *Analysis of a hospital home care service delivery model*

An example of application of the multi-perspective modeling framework to an existing model may help the reader get an overall view of the approach. Figure 2.16 illustrates how the O4HCS characterizes a model developed in a recent paper [111] by showing the pruning that applies to generate the model components.

The model developed in [111] aims to evaluate the effectiveness of a home hospital service, optimize the current configuration given existing constraints and evaluate potential future scenarios. Home hospital services provide some hospital level services at the patient's residence. The model uses a combined discrete event simulation, ABM and geographical information system (GIS) to assess the system effects of different demand patterns, appointment scheduling algorithms, varying levels of resource on patient outcomes and impact on hospital visits.

The model considers both the production and consumption entities at the top level but does not explicitly deal with coordination among such entities. The focus within production is on RA involved in health provision where the hospital is modeled as a queueing network which is embedded in an agent with vehicles also modeled as agents. Nurses and doctors are mentioned but not modeled. Agents provide the organizational architecture that glues all abstractions in this model, as

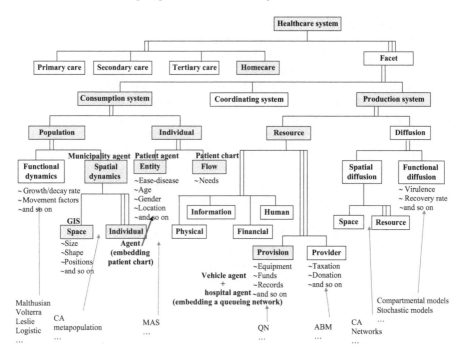

Figure 2.16 *Pruning of O4HCS SES for a hospital home care health service delivery model. Grey boxes are entities that have been instantiated through pruning. Bold texts outside boxes are components of the model. Bottom texts are models presented in Sections 2.1.1–2.1.4*

advocated by [63] and mentioned in Section 2.1.3.1, and also illustrated in Chapter 9 of this book. The focus within consumption is on PD through the generation of patients based on a GIS. The PD center on municipalities modeled as spatial models, with GIS for space representation, and individuals generated, each of which being a flow model (patient chart) embedded in an entity model (agent).

Although this model concerns home care, there is no indication of the coordination entity at the top level of the ontology. Coordination can be envisioned at

- the hospital level (either in coordinating the treatment capacities, or the vehicles, or both resources)
- the patient level (home/hospital visits)
- both levels (taking into account simultaneously patient requests, availability of resources and the need for optimization of cost/distance/length of stay)

2.2 Holistic healthcare systems M&S

The review done in the previous section clearly underlines the much parameterized nature of all these theoretical models developed through decades of research efforts that can apply to health-related perspectives identified in O4HCS. In practice, M&S processes in each of the identified perspectives are executed in isolation, i.e., without recourse to the processes from other perspectives. In reality, however, processes usually have mutual influences. For instance, when there is an epidemic in a community (HD perspective), it will naturally affect the provisions and allocations of the human and infrastructural healthcare resources in the health centers within the community (RA perspective) and the migrations of people into and out of the community (PD perspective). The interrelation of perspectives is echoed in various research efforts, such as [112] who argued that health demand planning is linked to population projection, which itself depends on factors such as healthcare system, culture, and sociodemography. However, building a monolithic highly detailed mega-model (that involves all these factors) is not considered viable [113].

To allow a holistic simulation, which encompasses isolated perspective-specific simulations and their mutual influences, without a drastic increase of complexity, we suggest an integration mechanism to enable live exchanges of information between models from different perspectives. However, while models within the same perspective are coupled the classic way (i.e., outputs to inputs) to form larger models within the same perspective, models from distinct perspectives relate in a different way. Indeed, the parameters of a focused model in a given perspective are fed by the outputs of models from other perspectives. In other words, these outputs provide a disaggregated understanding of the phenomena approximated by the parameters of the focused model. Technically, this is realized by creating a model the activity of which is to translate outputs received from the other models into new values for the parameters of the focused model.

This approach is very comparable to the one introduced in [114], where the model used to realize the integration is called a bridging model. However, there is a major difference in that we do not allow here the output of a model in a given perspective to feed the input of another model in a different perspective. The reason is that inputs and outputs of models are defined based on the perspective envisioned, which also set the family of objectives of the corresponding M&S study. Any process, which output can feed such inputs or which input can be fed by such outputs, is an abstraction within that perspective. Abstractions from other perspectives are solely captured by model parameters.

Figure 2.17 schematizes the technical difference between "coupling" and "integration" in the context of our framework. A more formal description will be provided in the next chapter. By coupling the output of a disease-spreading model to the input of an integrator, we create a coupled model in the HD perspective. The role of this integrator is to interpret the outputs received from the disease model and translate it into new values for the parameters of a PD model. The integrator will then call the method of the PD model to modify its parameters. Similarly, the PD model is coupled to an integrator that translates its output to values for the parameters of the disease-spreading model. A holistic model of the healthcare system is obtained by introducing appropriate integrators between perspective-specific models.

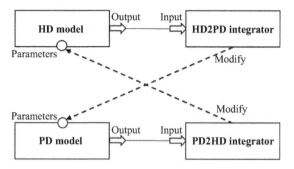

Figure 2.17   Holistic approach to healthcare systems M&S

Let us now discuss two well-known model concepts this integration relates to. They are

- aggregation and disaggregation
- dynamic change of structure

2.2.1   Aggregation and disaggregation

In M&S, the concept of resolution refers to the level of detail of a model. Although a relative concept, it promotes the idea that a model (or an abstraction in the model) can be represented at different levels of detail (also referred to as fidelity). A greater insight about the phenomenology of a given process requires an increase of resolution. As such, a given parameter in a model translates a decrease of the resolution of some external influencing process(es).

In multiresolution modeling, the dynamic change of resolution is called aggregation (from high-resolution entity/process to low-resolution entity/process) or disaggregation (the reverse operation), and the problem of linking simulations at different levels of resolution is known as the aggregation–disaggregation problem [63,115–117].

Against this background, the holistic integration approach presented here is an aggregation/disaggregation technique, where low-resolution processes (i.e., parameters) aggregates high-resolution processes (i.e., the feeding models). Since, live feedbacks are given in both directions between perspective-specific models, both models can be seen as disaggregation of each other's external processes. The aggregation/disaggregation operations are realized by the integrators components. As stated by Davis [118], a recurring question is whether it is legitimate (and desirable) to disaggregate and reaggregate processes during the course of a given simulation run. Translated within our framework, the question is how to define the integrators.

One must notice that the parameters of a model may express external processes operating at a different timescale from the one of the model. Therefore, integrators are models where issues, such as the how and validity of scale transfer, are addressed. The correlation (whether linear, quadratic, polynomial, or a more complex relationship) between outputs of some models and parameters of others is either an a priori knowledge or need to be established. The establishment of such a knowledge may require a different complete simulation study be conducted, and interpolations be derived from several simulation experiments, quantities of data collected, and statistical analysis be performed [119].

2.2.2   Dynamic change of structure

A parameterized model is nothing more than a family of models having some common behavioral patterns but different structures. Each assignment of specific values to the components of the parameters vector corresponds to a single model of the family. Let us clarify this, with the simple patient flow example previously seen in Figure 2.12(c). If the assignment of values to the parameters vector ($p_{12}, p_{23}, p_{34}, p_{35}, p_{43}, p_{45}$) is such that $p_{34} = 0$ and all other parameters are $\neq 0$, then the system obtained can undergo only four possible states instead of five and has three possible transitions instead of six. Other assignments can give other state sets with various transition sets. Consequently, by giving live feedback to perspective-specific models in our integration scheme (i.e., through parameters), we realize dynamic change of structure of the models that receive these feedbacks.

A general framework for dynamic change of structure is given in [120] from DEVS perspective. We will come back to the formal aspects of the holistic integration approach in the next chapter.

2.3   Summary

In this chapter, we have introduced a framework based on four perspectives, which allows multi-perspective modeling and holistic simulation of healthcare systems. Interactions between models of different perspectives are realized by dynamic update of models through output-to-parameter integration during concurrent simulations. Such an approach provides multiple levels of explanation for the same system, while offering, at the same time, an integrated view of the whole. The novelty of our approach is that notable components of the healthcare system are modeled as autonomous systems that can influence and be influenced by their environments. The resulting global model can be coupled with a holistic experimental frame to derive results that could not be accurately addressed in any of the perspective taken alone.

Literature review shows a huge number of research papers in the area of M&S applied to Healthcare management. Table 2.1 shows a representative sample of such contributions.

Many of these efforts concentrate on one of the four generic perspectives we have identified (i.e., RA, HD, IB, and PD). Some of them integrate two or three of these perspectives. To our knowledge, none of them integrates the four perspectives in one holistic approach.

A more formal approach to the multi-perspective modeling and holistic simulation framework is provided in the next chapter, upon which a methodology will be built and presented in the subsequent chapter.

Table 2.1   Benchmark of integrated healthcare M&S frameworks

Healthcare M&S frameworks	Resource allocation	Health diffusion	Population dynamics	Individual behavior
[29]	✓			✓
[30]	✓	✓		✓
[31]	✓			✓
[3]		✓	✓	
[4]		✓		✓
[28]	✓			✓
[5]		✓	✓	
[6]		✓		✓
[32]	✓			✓
[33]	✓			✓
[84]			✓	✓
[34]	✓			✓
[58]			✓	✓
[35]	✓			✓
[121]				
[122]	✓	✓	✓	

References

[1]   Zeigler B.P. *Theory of Modeling and Simulation.* New York: John Wiley & Sons; 1976.

[2]   Zeigler B.P. *Multifacetted Modelling and Discrete Event Simulation.* London: Academic Press Inc.; 1984.

[3]   Bisset K.R., Eubank S., and Marathe M.V. High Performance Informatics for Pandemic Preparedness. In Laroque C., Himmelspach J., Pasupathy R., Rose O., and Uhrmacher A.M. (eds.). *Proceedings of the Winter Simulation Conference,* Berlin, Germany, 9–12 Dec 2012. pp. 804–815.

[4]   Macal C.M., David M.Z., Dukic V.M., *et al.* 2012. Modeling the Spread of Community-Associated MRSA. In Laroque C., Himmelspach J., Pasupathy R., Rose O., and Uhrmacher A.M. (eds.). *Proceedings of the Winter Simulation Conference,* Berlin, Germany, 9–12 Dec 2012. pp. 73.1–73.12.

[5]   Okhmatovskaia A., Finès P., Kopec J., and Wolfson M. SIMPHO: An Ontology for Simulation Modeling of Population Health. In Laroque C., Himmelspach J., Pasupathy R., Rose O., and Uhrmacher A.M. (eds.). *Proceedings of the Winter Simulation Conference,* Berlin, Germany, 9–12 Dec 2012. pp. 883–894.

[6]   Ferranti J., and Freitas Filho P. Dynamic Mortality Simulation Model Incorporating Risk Indicators for Cardiovascular Diseases. In Jain S., Creasey R.R., Himmelspach J., White K.P., and Fu M. (eds.). *Proceedings of the Winter Simulation Conference,* Phoenix, AZ, USA, 11–14 Dec 2011. pp. 1263–1274.

[7]   Kermack W.O., and McKendrick A.G. A Contribution to the Mathematical Theory of Epidemics. *Proceedings of the Royal Society of London Series A.* 1927; 115: 700–721.

[8]   Hethcote H. The Mathematics of Infectious Diseases. *SIAM Review*. 2000; 42(4): 599–653.

[9]   Daley D.J., and Gani J. *Epidemic Modeling: An Introduction*. New York: Cambridge University Press; 2005.

[10] Gerhäuser M., Valentin B., and Wassermann A. JSXGraph – Dynamic Mathematics with JavaScript. *International Journal for Technology in Mathematics Education*. 2010; 17(4): 211–215.

[11] Trottier H., and Philippe P. Deterministic Modeling of Infectious Diseases: Theory and Methods. *The Internet Journal of Infectious Diseases*. 2001; 1(2), ISSN 1528–8366.

[12] Pool R. Is It Chaos, or Is It Just Noise? *Science*. 1989; 243: 25–28.

[13] Bailey N. *The Mathematical Theory of Infectious Diseases*. New York: Griffin & Co; 1975.

[14] Anderson R.M., and May R.M. *Infectious Diseases of Humans - Dynamic and Control*. Oxford: Oxford University Press; 1992.

[15] Abbey H. An Examination of the Reed Frost Theory of Epidemics. *Human Biology*. 1952; 24: 201–233.

[16] Andersson H., and Britton T. *Stochastic Epidemic Models and Their Statistical Analysis*. New York: Springer; 2000.

[17] Andersen P., Borgan O., Gill R., and Keiding N. *Statistical Models Based on Counting Processes*. New York: Springer-Verlag; 1993.

[18] von Neumann J. The General and Logical Theory of Automata. In Taub A.H. (ed.). *Collected Works*. Vol. 5. New York: Macmillan; 1963. pp. 288–328.

[19] Ferber J. *Multi-Agent Systems – An Introduction to Distributed Artificial Intelligence*. Boston: Addison-Wesley; 1999.

[20] Bousquet F., Bakam I., Proton H., and Le Page C. Cormas: Common-Pool Resources and Multi-Agent Systems. *Lecture Notes in Artificial Intelligence*. 1998; 1416: 826–838.

[21] White S.H., Martin del Rey A., and Rodriguez Sanchez G. Using Cellular Automata to Simulate Epidemic Diseases. *Applied Mathematical Sciences*. 2009; 3(20): 959–968.

[22] Molina C., and Stone L. Modelling the Spread of Diseases in Clustered Networks. *Journal of Theoretical Biology*. 2012; 315: 110–118.

[23] Lloyd A.L., and May R.M. How Viruses Spread among Computers and People. *Science*. 2001; 292: 1316–1317.

[24] Newman M.E.J. Ego-centered Networks and the Ripple Effect. *Social Networks*. 2003; 25: 83–95.

[25] Keeling M.J., and Eames K.T.D. Networks and Epidemic Models. *Journal of the Royal Society Interface*. 2005; 2(4): 295–307.

[26] Ozmen O., Nutaro J.J., Pullum L.L., and Ramanathan A. Analyzing the Impact of Modeling Choices and Assumptions in Compartmental Epidemiological Models. *Simulation: Transactions of the Society for Modeling and Simulation International*. 2016; 92(5): 459–471.

[27] Danon L., House T.A., Read J.M., and Keeling M.J. Social Encounter Networks: Collective Properties and Disease Transmission. *Journal of the Royal Society Interface*. 2012; 9: 2826–2833.

[28] Harper P.R. A Framework for Operational Modelling of Hospital Resources. *Health Care Management Science.* 2002; 5(3): 165–173.

[29] Augusto V., and Xie X. Modelling and Simulation Framework for Health Care Systems. *IEEE Transactions on Systems, Man and Cybernetics, Part A: Systems and Humans.* 2014; 44(1): 30–46.

[30] Viana J., Rossiter S., Channon A.R., Brailsford S.C., and Lotery A.J. A MultiParadigm, Whole System View of Health and Social Care for Age-Related Macular Degeneration. In Laroque C., Himmelspach J., Pasupathy R., Rose O., and Uhrmacher A.M. (eds.). *Proceedings of the Winter Simulation Conference*, Berlin, Germany, 9–12 Dec 2012. pp. 1070–1081.

[31] Zulkepli J., Eldabi T., and Mustafee N. 2012. Hybrid Simulation for Modelling Large Systems: An example of Integrated Care Model. In Laroque C., Himmelspach J., Pasupathy R., Rose O., and Uhrmacher A.M. (eds.). *Proceedings of the Winter Simulation Conference*, Berlin, Germany, 9–12 Dec 2012. pp. 758–769.

[32] Fletcher A., and Worthington D. What Is a "Generic" Hospital Model? A Comparison of "Generic" and "Specific" Hospital Models of Emergency Patient Flows. *Health Care Management Science.* 2009; 12(4): 374–391.

[33] Bountourelis T., Ulukus M.Y., Kharoufeh J.P., and Nabors S.G. The Modeling Analysis and Management of Intensive Care Units. In *Handbook of Healthcare Operations Management.* New York: Springer; 2011. pp. 153–182.

[34] Cote M.J. Patient Flow and Resource Utilization in an Outpatient Clinic. *SocioEconomic Planning Sciences.* 1999; 33(3): 231–245.

[35] Ahmed M.A., and Alkhamis T.M. Simulation Optimization for an Emergency Department Healthcare Unit in Kuwait. *European Journal of Operational Research.* 2009; 198(3): 936–942.

[36] Gelenbe E., and Pujolle G. *Introduction to Queueing Networks.* New York: Wiley; 1987.

[37] Kendall D.G. Stochastic Processes Occurring in the Theory of Queues and their Analysis by the Method of the Imbedded Markov Chain. *The Annals of Mathematical Statistics.* 1953; 24(3): 338–354.

[38] Worthington D.J. Queueing Models for Hospital Waiting Lists. *The Journal of the Operation Research Society.* 1987; 38: 413–422.

[39] Albin S.L., Barrett J., Ito D., and Mueller J.E. A Queueing Network Analysis of a Health Center. *Queueing Systems.* 1990; 7: 51–61.

[40] Khan M.R., and Callahan B.B. Planning Laboratory Staffing with a Queueing Model. *European Journal of Operational Research.* 1993; 67: 321–331.

[41] Siddhartan K., Jones W.J., and Johnson J.A. A Priority Queuing Model to Reduce Waiting Times in Emergency Care. *International Journal of Health Care Quality Assurance.* 1996; 9: 10–16.

[42] Tucker J.B., Barone J.E., Cecere J., Blabey R.G., and Rha C. Using Queueing Theory to Determine Operating Room Staffing Needs. *Journal of Trauma.* 1999; 46: 71–79.

[43] Gorunescu F., McClean S.I., and Millard P.H. Using a Queueing Model to Help Plan Bed Allocation in a Department of Geriatric Medicine. *Health Care Management Science.* 2002; 5: 307–312.

[44] Bruin A.M., Rossum A.C., Visser M.C., and Koole G.M. Modeling the Emergency Cardiac In-patient Flow: An Application of Queuing Theory. *Health Care Management Science.* 2007; 10: 125–137.

[45] Jackson J.R. Networks of Waiting Lines. *Operations Research.* 1957; 5(4): 518–521.

[46] Little D. A Proof of the Queueing Formula $L = \lambda W$. *Operations Research.* 1961; 9: 383–387.

[47] Jackson J.R. Jobshop-like Queueing Systems. *Management Science.* 1963; 10(1): 131–142.

[48] Baskett F., Chandy K.M., Muntz R.R., and Palacios F.G. Open, Closed and Mixed Networks of Queues with Different Classes of Customers. *Journal of the ACM.* 1975; 22(2): 248–260.

[49]   Fomundam S., and Herrmann J. *A Survey of Queuing Theory Applications in Healthcare.* ISR Technical Report 24, 2007.

[50] Alexopoulos C., Goldsman D., Fontanesi J., Kopald D., and Wilson J.R. Modeling Patient Arrivals in Community Clinics. *Omega.* 2008; 36(1): 33–43.

[51] Cichon M., Newbrander W., Yamabana H., *et al. Modelling in Health Care Finance – A Compendium of Quantitative Techniques for Healthcare Financing.* Geneva, Switzerland: International Labour Office and International Social Security Association; 1999.

[52] Charfeddine M., and Montreuil B. Integrated Agent-oriented Modeling and Simulation of Population and Healthcare Delivery Network: Application to COPD Chronic Disease in a Canadian Region. *Proceedings of the Winter Simulation Conference*, Baltimore, MD, USA, 5–8 Dec 2010. pp. 2327–2339.

[53] Ramirez-Nafarrate A., and Gutierrez-Garcia J.O. An Agent-Based Simulation Framework to Analyze the Prevalence of Child Obesity. *Proceedings of the Winter Simulation Conference - Simulation: Making Decisions in a Complex World*, Washington, DC, USA, 8–11 Dec 2013. pp. 2330–2339.

[54] Kasaie P., Dowdy D.W., and Kelton W.D. An Agent-Based Simulation of a Tuberculosis Epidemic: Understanding the Timing of Transmission. *Proceedings of the Winter Simulation Conference - Simulation: Making Decisions in a Complex World*, Washington, DC, USA, 8–11 Dec 2013. pp. 2227–2238.

[55] Fishbein M., and Ajzen I. *Belief, Attitude, Intention, and Behavior: An Introduction to Theory and Research.* Reading: Addison-Wesley; 1975.

[56] Becker M.H. The Health Belief Model and Sick Role Behavior. *Health Education Monographs.* 1974; 2: 409–419.

[57] Schmidt B. *The Modelling of Human Behaviour.* San Diego: Society of Computer Simulation Publications; 2000.

[58] Brailsford S., and Schmidt B. Towards Incorporating Human Behaviour in Models of Health Care Systems: An Approach Using Discrete Event Simulation. *European Journal of Operational Research.* 2003; 150: 19–31.

[59] Conner M., and Norman P. *Predicting Health Behaviour: Research and Practice with Social Cognition Models.* Buckingham: Open University Press; 2005.

[60] Wallston K.A. Hocus-Pocus, the Focus Isn't Strictly on Locus: Rotter's Social Learning Theory Modified for Health. *Cognitive Theory and Research.* 1992; 16:183–199.

[61] Ajzen A. The Theory of Planned Behaviour. *Organizational Behavior and Human Decision Processes.* 1991; 50: 179–211.

[62] Rao A.S., and Georgeff M.P. BDI Agents: From Theory to Practice. *Proceedings of the First International Conference on Multi-Agent Systems*, San Francisco, CA, USA, 12–14 Jun 1995. pp. 312–319.

[63] Yilmaz L., and Oren T.I. Dynamic Model Updating in Simulation with Multimodels: A Taxonomy and a Generic Agent-Based Architecture. In Bruzzone A.G., and Williams E. (eds.). *Proceedings of the Summer Computer Simulation Conference*, San Jose, CA, USA, 25–29 Jul 2004. pp. 3–8.

[64] The Health Foundation. *Improving Patient Flow.* Learning Report, London, UK, Apr 2013.

[65] Vissers J., and Beech R. *Health Operations Management: Patient Flow Statistics in Health Care.* New York: Routledge, Taylor & Francis Group Publishers; 2005.

[66] Hall R., Belson D., Murali P., and Dessouky M. Modeling Patient Flows through the Healthcare System. In Hall R.W. (ed.). *Patient Flow: Reducing Delay in Healthcare Delivery.* New York: Springer; 2006. pp. 1–44.

[67] Schrijvers G., van Hoorn A., and Huiskes N. The Care Pathway Concept: Concepts and Theories - An Introduction. *International Journal of Integrated Care.* 2012; 12(6):e192.

[68] Nigam A., and Caswell N.S. Business Artifacts: An Approach to Operational Specification. *IBM Systems Journal.* 2003; 42(3): 428–445.

[69] Mayer R.J., Painter M.K., and de Witte P.S. *IDEF Family of Methods for Concurrent Engineering and Business Re-engineering Applications.* Texas: College Station University Knowledge Based Systems Laboratory; 1992.

[70] Swenson K.D. Demo: Cognoscenti Open Source Software for Experimentation on Adaptive Case Management Approaches. *IEEE International Enterprise Distributed Object Computing Workshop*, Ulm, Germany, 1–2 Sep 2014. pp. 402–405.

[71] Ould M.A. *Business Process Management: A Rigorous Approach.* Swindon: The British Computer Society; 2005.

[72] Zacharewicz G., Frydman C.S., and Giambiasi N. G-DEVS/HLA Environment for Distributed Simulations of Workflows. *Simulation.* 2008; 84: 197–213.

[73] Bazoun H., Bouanan Y., Zacharewicz G., Ducq Y., and Boye H. Business Process Simulation: Transformation of BPMN 2.0 to DEVS Models. *Proceedings of the Symposium on Theory of Modeling & Simulation - DEVS Integrative M&S Symposium*, Society for Computer Simulation International, Tampa, Florida, 13–16 Apr 2014. Article 20.

[74] Garey M.R. The Complexity of Flowshop and Jobshop Scheduling. *Mathematics of Operations Research.* 1976; 1(2): 117–129.

[75] Hamel G. *The Future of Management.* Boston: Harvard Business School Press; 2007.

[76] Creemers S., and Lambrecht M. Modeling a Hospital Queueing Network. In Boucherie R.J., Van Dijk N.M. (eds.). *Queueing Networks: A Fundamental Approach*. International Series in Operations Research and Management Science, Vol. 154. USA, New York: Springer-Verlag; 2011. pp. 767–798.

[77] Marshall A., Vasilakis C., and El-Darzi E. Length of Stay-Based Patient Flow Models: Recent Developments and Future Directions. *Health Care Management Science*. 2005; 8(3): 213–220.

[78] Zhao L., and Lie B. Modeling and Simulation of Patient Flow in Hospitals for Resource Utilization. *Simulation Notes Europe*. 2010; 20(2): 41–50.

[79] Bhattacharjee P., and Ray K. Patient Flow Modelling and Performance Analysis of Healthcare Delivery Processes in Hospitals: A Review and Reflections. *Computers & Industrial Engineering*. 2014; 78: 299–312.

[80] Lenin R.B., Lowery C.L., Hitt W.C., Manning N.A., Lowery P., and Eswaran H. Optimizing Appointment Template and Number of Staff of an OB/GYN Clinic - Micro and Macro Simulation Analyses. *BMC Health Services Research*. 2015; 15: 387, doi: 10.1186/s12913–015–1007–9.

[81] Mardiah F.P., and Basri M.H. The Analysis of Appointment System to Reduce Outpatient Waiting Time at Indonesia's Public Hospital. *Human Resource Management Research*. 2013; 3(1): 27–33.

[82] Carter J. Petri Nets and Clinical Information Systems, Part III: Modeling Concepts and Tips. *EHR Science - Explorations in the Design and Implementation of Clinical Information Systems*, 5 Nov 2012, Available at https://www.ehrscience.com/ [Accessed on 14 May 2018].

[83] Allen P.M. Evolution, Population Dynamics, and Stability. *Proceedings of the National Academy of Sciences of the United States of America*. 1976; 73: 665–668.

[84] Bohk C., Ewald R., and Uhrmacher A. Probabilistic Population Projection with James II. *Proceedings of the Winter Simulation Conference*, Austin, TX, USA, 13–16 Dec 2009. pp. 2008–2019.

[85] Leslie P.H. On the Use of Matrices in Certain Population Mathematics. *Biometrika*. 1945; 33(3): 183–212.

[86] Sheppard E. Urban System Population Dynamics: Incorporating Non-linearities. *Geographical Analysis Columbus*. 1985; 17(1): 47–73.

[87] Sikdar P.K., and Karmeshu P.K. On Population Growth of Cities in a Region: A Stochastic Nonlinear Model. *Environment and Planning A*. 1982; 14: 585–590.

[88] Malthus T.R. *An Essay on the Principle of Population*. New Haven: Yale University Press (The 1803 Edition); 1798.

[89] Usher D. The Dynastic Cycle and the Stationary State. *The American Economic Review*. 1989; 79: 1031–1044.

[90] Chu C.Y.C., and Lee R.D. Famine, Revolt, and the Dynastic Cycle: Population Dynamics in Historic China. *Journal of Population Economics*. 1994; 7: 351–378.

[91] Turchin P., and Korotayev A. Population Density and Warfare: A Reconsideration. *Social Evolution & History*. 2006; 5(2): 121–158.

[92] Komlos J., and Artzrouni M. Mathematical Investigations of the Escape from the Malthusian Trap. *Mathematical Population Studies*. 1990; 2: 269–287.

[93]   Volterra V. Variations and Fluctuations of the Number of Individuals in Animal Species Living Together. In Chapman R.N. (ed.). *Animal Ecology.* New York: McGraw-Hill; 1931. pp. 31–113.

[94]   Metz J.A.J., Geritz S.A.H., Meszèna G., Jacobs F.J.A., and Van Heerwaarden J. S. Adaptive Dynamics, a Geometrical Study of the Consequences of Nearly Faithful Reproduction. In van Strien S.J., and Verduyn Lunel S.M. (eds.). *Stochastic and Spatial Structures of Dynamical Systems.* Amsterdam: North-Holland; 1996. pp. 183–231.

[95]   Hofbauer J., and Sigmund K. *Evolutionary Games and Population Dynamics.* Cambridge: Cambridge University Press; 1998.

[96]   Bomze I.M. Lotka–Volterra Equation and Replicator Dynamics: A Two-dimensional Classification. *Biological Cybernetics.* 1983; 48: 201–211.

[97]   Bomze I.M. Lotka–Volterra Equation and Replicator Dynamics: New Issues in Classification. *Biological Cybernetics.* 1995; 72: 447–453.

[98]   Arditi R., and Ginzburg L.R. Coupling in Predator-Prey Dynamics: Ratio Dependence. *Journal of Theoretical Biology.* 1989; 139: 311–326.

[99]   Arditi R., and Ginzburg L.R. *How Species Interact: Altering the Standard View on Trophic Ecology.* New York: Oxford University Press; 2012.

[100]   Charlesworth B. *Evolution in Age-structured Population.* Cambridge: Cambridge University Press; 1980.

[101]   Caceres M.O., and Caceres-Saez I. Random Leslie Matrices in Population Dynamics. *Journal of Mathematical Biology.* 2011; 63(3): 519–556.

[102]   Verhulst P.F. Notice sur la Loi que la Population Poursuit dans son Accrois-sement. *CorrespondanceMathematique etPhysique.* 1838; 10: 113–121.

[103]   McKendrick A.G., and Pai K. XLV. - The Rate of Multiplication of MicroOrganisms: A Mathematical Study. *Proceedings of the Royal Society of Edinburgh.* 191; 231: 649–653.

[104]   Batty M. *Cities and Complexity - Understanding Cities with Cellular Auto-mata, Agent-Based Models, and Fractals.* Cambridge: MIT Press; 2005.

[105]   Liu Y., and Phinn S.R. Modelling Urban Development with Cellular Automata Incorporating Fuzzy-Set Approaches. *Computers, Environment and Urban Systems.* 2003; 27: 637–658.

[106]   Torrens P.M. A Geographic Automata Model of Residential Mobility. *Environment and Planning B: Planning and Design.* 2007; 34: 200–222.

[107]   Vancheri A., Giordano P., Andrey D., and Albeverio S. Urban Growth Processes Joining Cellular Automata and Multiagent Systems. Part 1: Theory and Models. *Environment and Planning B: Planning and Design.* 2008; 35: 723–739.

[108]   Dabbaghiana V., Jackson P., Spicer V., and Wuschke K. A Cellular Auto-mata Model on Residential Migration in Response to Neighborhood Social Dynamics. *Mathematical and Computer Modelling.* 2010; 52: 1752–1762.

[109]   Hanski I.A., and Gilpin M.E. *Metapopulation Biology: Ecology, Genetics, and Evolution.* Waltham: Academic Press; 1997.

[110]   Benenson I. Modeling Population Dynamics in the City: From a Regional to a Multi-Agent Approach. *Discrete Dynamics in Nature and Society.* 1999; 3: 149–170.

[111]   Viana J., Ziener V.M., and Holhjem M.S. Optimizing Home Hospital Health Service Delivery in Norway Using a Combined Geographical Information System, Agent Based, Discrete Event Simulation Model. In Chan W.K.V., D'Ambrogio A., Zacharewicz G., Mustafee N., Wainer G., and Page E. (eds.). *Proceedings of the Winter Simulation Conference*, Las Vegas, NV, USA, 3–6 Dec 2017. pp. 1658–1669.

[112]   Onggo B.S. Simulation Modeling in the Social Care Sector: A literature Review. In Laroque C., Himmelspach J., Pasupathy R., Rose O., and Uhrmacher A.M. (eds.). *Proceedings of the Winter Simulation Conference*, Berlin, Germany, 9–12 Dec 2012. pp. 739–750.

[113]   Brailsford S., Silverman E., Rossiter S., *et al.* Complex Systems Modeling for Supply and Demand in Health and Social care. In Jain S., Creasey R.R., Himmelspach J., White K.P., and Fu M. (eds.). *Proceedings of the Winter Simulation Conference*, Phoenix, AZ, USA, 11–14 Dec 2011. pp. 1125–1136.

[114]   Seck M.D., and Honig H.J. Multi-perspective Modelling of Complex Phenomena. *Computational and Mathematical Organization Theory.* 2012; 18: 128–144.

[115]   Davis P.K., and Hillestad R. 1993. Families of Models that Cross Levels of Resolution: Issues for Design, Calibration and Management. In Evans G.W., Mollaghasemi M., Russell E.C., and Biles W.E. (eds.). *Proceedings of the Winter Simulation Conference*, Piscataway, NJ, USA, 12–15 Dec 1993. pp. 1003–1012.

[116]   Reynolds P.F.J., Natrajan A., and Srinivasan S. Consistency Maintenance in Multiresolution Simulations. *ACM Transactions on Computer Modeling and Simulation.* 1997; 7(3): 368–392.

[117]   Davis P.K., and Bigelow J.H. *Experiments in Multiresolution Modeling (MRM).* RAND Research Report MR-1004-DARPA, 1998.

[118]   Davis P.K. *Aggregation, Disaggregation, and the 3:1 Rule in Ground Combat.* Santa Monica, CA, USA: RAND; 1995.

[119]   Duboz R., Ramat E., and Preux P. Scale Transfer Modeling: Using Emergent Computation for Coupling Ordinary Differential Equation System with a Reactive Agent Model. *Systems Analysis Modelling Simulation.* 2003; 43(6): 793–814.

[120]   Muzy A., and Zeigler B.P. Specification of Dynamic Structure Discrete Event Systems Using Single Point Encapsulated Functions. *International Journal of Modeling, Simulation and Scientific Computing.* 2014; 5(3): 1450012 (20 pages).

[121]   Weng S.J., Tsai B.S., Wang L.M., Chang C.Y., and Gotcher D. Using Simulation and Data Envelopment Analysis in Optimal Healthcare Efficiency Allocations. In Jain S., Creasey R.R., Himmelspach J., White K.P., and Fu M. (eds.). *Proceedings of the Winter Simulation Conference*, Phoenix, AZ, USA, 11–14 Dec 2011. pp. 1295–1305.

[122]   Jeffers R.F. *A Model of Population Movement, Disease Epidemic, and Communication for Health Security Investment.* United States: National Nuclear Security Administration; 2014.

Chapter 3

Formalization of the multi-perspective architecture

We present in this chapter a formalization of the framework presented in the previous chapter. This formalization is based on the Discrete Event System Specification (DEVS) formalism. Since its introduction in [1], DEVS has spawned an approach to M&S that has taken root in academia and is emerging into common research and industrial use. A DEVS model is defined as a mathematical and logical object which serves as a way of specifying a dynamic system.

The following briefly summarize the elements of a DEVS model:

- *States*. A state can either be a "hold state" or a "passive state." A hold state is one that the model will stay in for a certain amount of time before automatically changing to another state (via an internal transition). A passive state is one that the model will remain in indefinitely (or until it receives a message that triggers an external transition).
- *Time advance*. Every state has a time advance value which specifies the amount of time that expires before it automatically changes to another state (via an internal transition). The time advance for a "hold" state is a finite real value. The time advance for a "passive" state is infinity.
- *Initial states*. One state in the model must be designated as the initial state from which all interaction with the external word commences.
- *Internal transitions*. Every hold state in the model has one internal transition defined in order to specify the state to which the model should transition after the specified amount of time.
- *Output*. Any state that has an internal transition can also have one output message that is generated before that internal transition occurs.
- *External transitions*. Any state can have one or more external transitions defined. An external transition defines an input message that the model might receive when in a given state and the state to which the model should transition in reaction to that input message.

We show in this chapter how DEVS provides a sound theoretic foundation to the multi-perspective modeling and holistic simulation framework. Conceptually, two key entities stemming from the DEVS paradigm form the building blocks of our framework; they are model and experimental frame (EF). We first present these concepts and we show how they are used in the context of multi-perspective modeling and holistic simulation. Then, we use their formal expression and we bring the concept of parameter at the forefront to capture the key constructs of our framework. An illustrative case is presented to bridge the abstract presentation with a more concrete application.

3.1   Conceptual foundations

Zeigler [1] describes four basic entities to the M&S enterprise, as exhibited by Figure 3.1: the system under study, the model, the EF, and the simulator. The system under study is represented as a source of behavioral data. The EF is the set of conditions under which the system is being observed and is operationally formalized to capture the objectives of the study. The model is a set of rules or mathematical equations that give an abstract representation of the system, which is used to replicate its behavior. The simulator is the automaton that is able to execute the model's instructions.

More theoretical aspects on EF can be found in [2], and Zeigler *et al.* [3] give a thorough description of various algorithms for the simulator, including sequential, as well as parallel and distributed variants. Elaborating on concepts developed in [2], Traoré and Muzy [4] suggested that during modeling activities, models always come with the specification of associated EFs. The specified EF is a model component to be coupled with the model to produce the data of interest under specified conditions, as suggested by Figure 3.2.

These basic entities and the approach form the conceptual foundations of our framework. As depicted by Figure 3.3, when studying a healthcare system, we are interested in addressing different levels of explanation, which we express as much EFs. Therefore, perspective-specific EFs (i.e., HD/RA/PD/IB related) are elaborated to provide answers to questions of orthogonal nature about the same system.

Consequently, they are coupled with models elaborated from the corresponding perspectives (such as models presented in the previous chapter) to derive results of interest in those perspectives.

A summary of such questions and their relation to models presented in the previous chapter is given by Table 3.1.

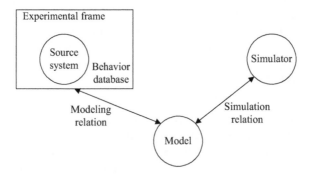

Figure 3.1   Basic entities in M&S and their relationships [3]

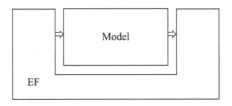

Figure 3.2   Model and EF as interacting components

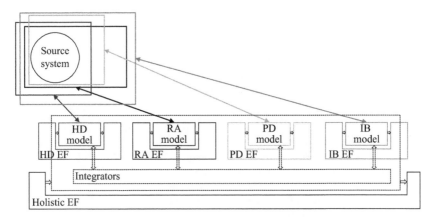

*Figure 3.3   System-theoretic approach to multi-perspective modeling and holistic
simulation*

When the perspective-specific models are holistically integrated (as explained in the previous chapter and presented in Figure 3.3), a holistic EF can be elaborated and questions that are transversal to different perspectives can then be addressed. This holistic EF is to be coupled with the holistic model formed by the perspective-specific models that well-defined integrators glue together using the output-to-parameter scheme previously described.

Also, questions addressed by each perspective-specific EF can be revisited with the corresponding models receiving live feedbacks from the other perspective-specific models, in order to increase the accuracy of answers obtained.

Models, EFs and integrators are all DEVS-specifiable entities. However, due to the key role of parameters in these components, as highlighted in the previous chapter, we focus on how to define a DEVS-based formalization that gives an explicit representation of these parameters so that the holistic integration scheme can be clearly captured.

3.2   Formal foundations

The DEVS framework suggests a specification hierarchy [1–3] to capture the knowledge specific to systems *behavior* and *structure*. Hence, dynamical systems

Table 3.1   EF relation to perspective-specific models

Classes of model	Typical questions formulated through EF	Subclasses of models
Health diffusion models	How to reduce infection spread?	Functional diffusion models (targeted population as a whole) Spatial diffusion models
Resource allocation models	Scheduling and planning problems:	Provision (how health services are delivered)
	• What is the number of beds needed in a surgical unit?	
	• How to reduce average waiting time in an emergency department?	
	• How to improve scheduling of healthcare activities?	
	How to assess value-based healthcare: insurance coverage, premium costs, claims, payments	Provider (who delivers health services)
Individual behavior models	• How do sociocultural decisions affect health status of individuals?	Entity models (e.g., multi-agent systems)
	• How do life strategies depend on the context of competition/selection and scarcity of resources?	
	What is the impact of patient flow optimization on healthcare quality and cost?	Flow models (flow of activities of an individual during pathway of care)
Population dynamics models	• How to forecast a population distribution by gender, social status, or age range?	Functional dynamics models (population growth, death) Spatial dynamics models (includes migration in space)
	• What is the impact of a species strategy on the encapsulating ecosystem?	

can be broken down in knowledge levels and specified in a useful way to understand how they behave. Each of the levels has an associated set-theoretic structure (n-tuple) that allows to describe a system. Going up the hierarchy (from behavior to structure) adds more elements to the n-tuple, since we know more about the system as levels increase. There are corresponding morphisms at each level, i.e., how to tell whether two descriptions of the same system at a level are equivalent or related at that level. Also, the morphisms at one level are consistent with those below, i.e., if two descriptions are equivalent at a higher level, then they are also equivalent at every lower level. Going down the levels is computationally done by simulation, while going up the levels (also known as structural inference) is much harder and can be realized under justifying conditions. On top of the hierarchy is the coupled network (CN) level, below which is the Input–Output System (IOS) level.

Models expressed at the CN level are called coupled models, while the ones expressed at the IOS level are called atomic models. Simulation modelers usually describe their models at those levels using code equivalents (depending on the programming language) of the set-theoretic specifications given below. These levels are often the most convenient to describe the *structure* of the system under study, while the well-defined DEVS simulation algorithms generate the *behavior* of these models, which is described at the lower levels of the hierarchy. Perspective-specific models and EFs mentioned earlier, as well as integrators and EFs, are to be specified as DEVS atomic and coupled models.

3.2.1   DEVS specification of atomic and coupled models

DEVS formalism was initially known as classic DEVS while presenting some limitations to perform parallel implementation. Some of its limitations comprise a tie-breaking function that handles simultaneous occurring internal transitions of the components of a coupled model and the fact that it ignores an internal transition function while occurring at the same time with an external input event (collision), in which case the external transition function always takes place. Chow and Zeigler [5] introduced Parallel DEVS (PDEVS) to alleviate this drawback. For the rest of what will be presented in this chapter, we will be referring to PDEVS as DEVS.

An atomic model is defined by the *n*-tuple:

$$\langle X, Y, S, \delta_{int}, \delta_{ext}, \delta_{conf}, \lambda, \text{ta} \rangle$$

where

- X, Y, and S are, respectively, the input set, output set, and state set (at any time, the system modeled is in one of the possible states);
- ta: $S \rightarrow \Re_0^{+\infty}$ is the time advance function (i.e., it gives the lifespan of each state), with $\Re_0^{+\infty}$ designating the set of nonnegative real numbers, including $+\infty$;
- δ_{int}: $S \rightarrow S$ is the internal transition function (i.e., it is triggered only when the elapsed time in the system's current state s_{curr} has reached ta(s_{curr}) without the system being disturbed by any receipt of input);
- λ: $S \rightarrow Y$ is the output function (i.e., it computes the output of the system, each time an internal transition is occurring);
- δ_{ext}: $Q \times X \rightarrow S$ is the external transition function (i.e., it is triggered only when the system receives an input, while the elapsed time in the system's current state s_{curr} has not reached ta(s_{curr})), and $Q = \{(s,e) \mid s \in S, 0 \leq e < \text{ta}(s)\}$ is called the total state; and
- δ_{conf}: $S \times X \rightarrow S$ is the confluent transition function (i.e., it is triggered only when the system receives an input at exactly the time that the elapsed time in the system's current state s_{curr} has reached ta(s_{curr})).

The operational semantics of an atomic model is informally described as follows: at the start, the systems is in an initial state and remains there until the time specified by ta is exhausted or until input event is received. In the former case, an internal transition function occurs, then the system switches to another state after sending output event as defined by the output function λ. In the latter case, if input event is received before the specified time, then the external transition function is applied. When a collision occurs, i.e., an external event is received concurrently with the elapsed time equal to the time specified by the time advance function, the confluent function is applied in such a way that the system sends output value and changes to a new state.

A coupled DEVS model, CM is a structure:

$$\langle X_{\text{self}}, Y_{\text{self}}, \{M_d\}_{d\in D}, \{I_d\}_{d\in D}, \{Z_{i,j}\}_{i\in D\cup\{\text{self}\}, j\in Ii} \rangle,$$

where

- X_{self} and Y_{self} are defined the same way X and Y are for atomic models (self being here a reference to the coupled model, while component models are referred to using indices such as i, j, or d);
- D is the set of component references (thus, not including self);
- M_d is the component model referenced by d, an atomic or a coupled model, with X_d and Y_d as, respectively, its input and output set;
- I_d is the influence set of component model d, i.e., all other models sending input to d;
- $Z_{\text{self},d\in I\text{self}}: X_{\text{self}} \to X_d$ are the external input transfer functions, which determine how inputs received by self are translated into inputs to component models influenced by self;
- $Z_{d/\text{self},\in Id,\text{self}}: Y_d \to Y_{\text{self}}$ are the external output transfer functions, which determine how outputs sent by component models influencing self are translated into outputs of self; and
- $Z_{i\in D, j\in D-\{i\}}: Y_i \to X_j$ are the internal transfer functions, which determine how outputs sent by component models are translated into inputs to component models they influence.

3.2.2   Parameterized DEVS

If an atomic model is parameterized, its parameters are disjoint from its state variables. Parameters are constant values the model will refer to when triggering its transition functions or when computing its outputs, or even when determining its time advance. Therefore, any change of value of a parameter results in a change of the model's internal rules (and not a state transition). Intuitively, the parameterized model can be seen as a kind of embedding structure for a core atomic model. It distinguishes inputs that impact on the core model's state from inputs that only modify the values of parameters. A variable σ is defined to memorize the remaining time in any current state of the core model (i.e., time before the lifespan expires). Hence, this variable gives the time advance function of the parameterized model. The internal transition of the parameterized model changes the state of the core model according to its internal transition function but does not affect the parameters. The output sent at that time is the one computed by the core model. When only new values for parameters are received by the parameterized model, the state of the core model is kept unchanged, and only the remaining time is updated. When only input values impacting the core model's state are received (without input for modification of parameters), the new situation is defined by the core model's external transition and time advance function. When both input values impacting the core model's state and input for modification of parameters are received, the new situation is defined by the core model's external transition and time advance function; the new state of the core model is computed based on the current values of parameters, but the lifespan of this new state is computed using the new values of parameters. The same rules apply for confluent transition.

Formally, such a model is a DEVS model:

$$\langle X^P, Y^P, S^P, \delta_{\text{int}}{}^P, \delta_{\text{ext}}{}^P, \delta_{\text{conf}}{}^P, \lambda^P, \text{ta}^P \rangle,$$

where

- P is the parameter set;
- $\langle X, Y, S, \delta_{\text{int}P}, \delta_{\text{ext}P}, \delta_{\text{conf}P}, \lambda p, \text{ta}_P \rangle$ is the core model, i.e., the DEVS model obtained with a specific value p of P;
- $X^P = P \times X$;
- $Y^P = Y$;
- $S^P = P \times S \times \Re_0^{+\infty}$;
- $\text{ta}^P : P \times S \times \Re_0^{+\infty} \to \Re_0^{+\infty}$
  $\text{ta}^P(p,s,\sigma) = \sigma$;
- $\delta_{\text{int}}{}^P : P \times S \times \Re_0^{+\infty} \to P \times S \times \Re_0^{+\infty}$
  $\delta_{\text{int}}{}^P(p,s,\sigma) = (p, \delta_{\text{int}P}(S), \sigma)$;
- $\lambda^P : P \times S \times \Re_0^{+\infty} \to Y$
  $\lambda^P(p,s,\sigma) = \lambda_P(S)$;
- $\delta_{\text{ext}}{}^P : Q \times X^P \to S^P$, with $Q^P = \{(p, s, \sigma, e)/(p,s,\sigma) \in S^P, 0 \leq e < \sigma\}$
  $\delta_{\text{ext}}{}^P(p,s,\sigma,e,q,\emptyset) = (q,s,\sigma - e)$
  $\delta_{\text{ext}}{}^P(p,s,\sigma,e,\emptyset,x) = (p, \delta_{\text{ext}P}(s,e,x), \text{ta}_P(\delta_{\text{ext}P}(s,e,x)))$
  $\delta_{\text{ext}}{}^P(p,s,\sigma,e,q,x) = (q, \delta_{\text{ext}P}(s,e,x), \text{ta}_P(\delta_{\text{ext}P}(s,e,x)))$; and
- $\delta_{\text{conf}}{}^P : S^P \times X^P \to S^P$
  $\delta_{\text{conf}}{}^P(p,s,\sigma,\emptyset,x) = (p, \delta_{\text{conf}}(s,x), \text{ta}_P(\delta_{\text{conf}}(s,x)))$
  $\delta_{\text{conf}}{}^P(p,s,\sigma,q,x) = (q, \delta_{\text{conf}}(s,x), \text{ta}_q(\delta_{\text{conf}P}(s,x)))$.

The parameterized model has the following characteristics:

- It distinguishes inputs that impact on the core model's state from inputs that only modify the values of parameters.
- The σ variable memorizes the remaining time in any current state of the core model (i.e., time before the lifespan expires). It defines the time advance function in the parameterized model (while ta_P defines the time advance function in the core model).
- An input $(q, \emptyset)$ changes the current value of the parameter (and σ is updated) but does not affect the current state of the core model. The output of the parameterized model is then the one of the core model.

- An input $(\emptyset, x)$ changes the state of the core model according to its external transition function and time advance function but does not affect the parameters.
- An input (q, x) simultaneously affects parameters and the current state of the model. The new situation is defined by the core model's external transition and time advance function; the new state of the core model is computed based on the old values of parameters, but the lifespan of this new state is computed using the new values of parameters.
- The same rules apply for confluent transition.

In a review of dynamic structure DEVS, Muzy and Zeigler [6] propose a generic framework in which this formalization can be translated, with specific choices for the generic functions defined.

3.2.3   *DEVS-specified multi-perspective healthcare system model*

From what has been presented, we see a significant difference between the coupling of two models (or a model and an EF) within a perspective and the connection of an integrator as defined in our framework with a model. The semantics of the coupling is that stimuli coming from the model's environment provoke a change of the model's internal state, and this change is governed by the model's transition rules (external transition, in the case of DEVS). The semantics of parameter modification through an integrator is that knowledge revealed from another reality (or perspective) of the system of interest provokes a change of the model's internal rules (instead of its state).

With such a formalization, a multi-perspective model (i.e., resulting from the holistic approach previously presented) can be given at the CN level of the DEVS specification hierarchy. Such a coupled model is defined by

$$\left\langle X_{\text{self}}^{P}, Y_{\text{self}}^{P}, D^{P}, \left\{ M_d^{Pd} \right\}_{d \in D}, \left\{ I_d^{Pd} \right\}_{d \in D}, \left\{ Z_{k,j}^{P} \right\}_{k \in D \cup \{\text{self}\}, j \in Ik} \right\rangle$$

where

- $X_{\text{self}}^{P} = (\times P_d)_{d \in D} \times X_{\text{self}}$,
- $Y_{\text{self}}^{P} = Y_{\text{self}}$,
- $D^{P} = D$,
- M_d^{Pd} is a parameterized DEVS model if $P_d \neq \emptyset$ (with $X_d^{Pd} = X_d \times P_d$ as its input set), and a "regular" DEVS model if $P_d = \emptyset$ (with $X_d^{Pd} = X_d$ as its input set),
- I_d^{P} includes all components models sending input to d, whether for parameter modification or internal state change,
- $Z_{\text{self},d}^{P}: X_{\text{self}}^{P} \to P_d \times X_d$
  $Z_{\text{self},d}^{P}(p,x) = (p_d, Z_{\text{self},d}(x))$,
- $Z_{d,\text{self}}^{P} = Z_{d,\text{self}}$, and
- $Z_{i \in D, j \in D - \{i\}}^{P}: Y_i \to P_j \times X_j$
  $Z_{i \in D, j \in D - \{i\}}^{P}(y) = (\emptyset, x)$ for a "regular" coupling
  $Z_{i \in D, j \in D - \{i\}}^{P}(y) = (p_j, \emptyset)$ for an "integration" (or a bridging).

3.3 Illustrative case

On July 20, 2014, the contagious Ebola virus disease (EVD) was imported into Nigeria from a Liberian traveler who, after contracting the virus in his country, flew to the Lagos International Airport [7]. He died 5 days later in a Lagos hospital where he was admitted but after having wreaked havoc by infecting healthcare providers at the hospital. Within the first days of Ebola case diagnose, nine healthcare workers were infected and 898 contacts were generated through the country. The urgent need to control the epidemic prompted the Federal Ministry of Health to declare a national Ebola emergency, and the World Health Organization declared it a public health emergency of international concern. An intervention plan was swiftly developed, with about USD 11.5 million allocated to establish coordination offices and operation centers, along with massive campaign of awareness of Ebola to the public.

Several factors made the control of the pandemic difficult, including the following:

- The transmission vectors of the disease include any contact with sweat, saliva, vomit, and other bodily fluids of an infected person, even when dead. As a result, care providers, women, and children are among the most vulnerable. The former are in direct contact with patients. The others live in a great promiscuity within rural communities.
- With a population of about 14 millions, Lagos, ranked Africa's largest city, is an attractive business area for day laborers, including poor people living in rural areas and slums.
- Cultural practices in some places mean that dead people are transported from one place to another to be buried near their ancestors, while putting carriers, gravediggers, and neighboring places at high risk of infection.

We argue that studying in isolation the Ebola outbreak, without integrating characteristics of the Nigerian healthcare system that can be seen from other perspectives, may fail to provide all the necessary levels of explanation for policy makers to design efficient decisions. This includes the allocation of scarce health resources, the vibrant population dynamics, and sociocultural behavior of individuals. Indeed, it would be interesting to have a platform that allows us to study how the problems in one perspective evolve with respect to those in the others. For example, how do the dynamics of the population and/or the resource allocation strategy deployed by the government affect the evolution of the disease, and vice versa?

While the simulation from one perspective abstracts all realities concerning the rest of the perspectives, connecting different perspectives simultaneously takes into consideration all realities.

We are not showing here an exhaustive study that addresses all the concerns mentioned and the whole Nigerian healthcare system. However, we show how the multi-perspective modeling and holistic simulation framework can be applied and what its key pragmatic aspects are.

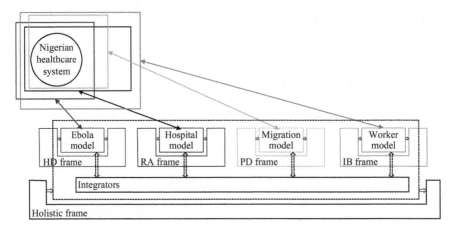

Figure 3.4   Multi-perspective modeling and holistic simulation of Nigerian healthcare system

As presented by Figure 3.4, let us consider the following models in each of the perspectives identified:

• A model of the Ebola outbreak and its EF, from the HD perspective, where we are interested in the distribution of health statuses in the population over a period of time, and the sensitivity of the disease spread to variations of parameters;

• A model of hospital resource allocation in Lagos and its EF, from the RA perspective, where we are interested in the utilization of room beds;

• A model of migrations between Nigerian states and its EF, from the PD perspective, where we are interested in predicting the distribution of population in the Nigerian states over a period of time; and

• A model of daily worker and its EF, from the IB perspective, where we are interested in exploring the impact of various strategies of a certain category of individual in the population on his/her job status and his/her migrations.

Each model is independently developed and studied in isolation coupled with its corresponding EF, and results are derived to answer the perspective-specific questions. All models are then integrated together to produce a holistic view of the situation, with which cross-perspective questions are addressed.

3.3.1   Ebola spreading model

The EVD spreading can be modeled by a compartmental model as proposed by Althaus *et al.* [8]. Taken alone, this perspective provides a level of explanation for the epidemiological understanding of the disease outbreak, where the influence of all other possible perspectives is approximated by parameters. For example, population dynamics, as well as cultural behavior, are all considered together and assumed to lead to transmission rates, without distinction between urban/rural areas or between health workers and the remaining part of the population.

Here, we extended the work presented in [8] to take into account the possibility of infection by dead individuals, leading to the following equations:

$$\frac{dS}{dt} = -\beta SI - \alpha SD$$

$$\frac{dE}{dt} = \beta SI + \alpha SD - \sigma E$$

$$\frac{dI}{dt} = \omega E - \gamma I$$

$$\frac{dR}{dt} = (1-f)\gamma I$$

$$\frac{dD}{dt} = f\gamma I$$

where

- S, E, I, R, and D are, respectively, the number of susceptible, exposed, infected, recovered, and dead individuals in the population;
- β and α are, respectively, the transmission rate with infected and dead individuals; and
- ω, γ, and f are, respectively, the incubation, "recovery or death," and case fatality rate.

Their parameterized DEVS counterpart is specified as follows, accordingly with the formalization scheme proposed earlier:

$$M_{\text{Ebola Spread}} = \langle X^P, Y^P, S^P, \delta_{\text{int}}^{\ P}, \delta_{\text{ext}}^{\ P}, \delta_{\text{conf}}^{\ P}, \lambda^P, \text{ta}^P \rangle$$

where

- $P = \{(\alpha, \beta, \gamma, \omega, f) \in \Re_0^{+\infty} \times \Re_0^{+\infty} \times \Re_0^{+\infty} \times \Re_0^{+\infty}\}$;
- $X^P = \{(p,v), p \in \{\text{set } \alpha, \text{set } \beta, \text{set } \gamma, \text{set } \omega, \text{set} f\}, v \in \Re_0^{+\infty}\}$;
- $Y^P = \{(p,v), p \in \{\#S, \#E, \#I, \#R, \#D\}, v \in \Re_0 +^\infty\}$;
- $S^P = \Re_0^{+\infty} \times \Re_0^{+\infty} \times \Re_0^{+\infty} \times \Re_0^{+\infty} \times \Re_0^{+\infty} \times \{\text{current}\} \times \Re_0^{+\infty}$;
- $\text{ta}^P : S^P \rightarrow \Re_0^{+\infty}$;
  $\text{ta}^P (p, \text{current}, \sigma) = \sigma,$
- $\delta_{\text{int}}^{\ P} : S^P \rightarrow S^P$
  $\delta_{\text{int}}^{\ P}(\alpha, \beta, \gamma, \omega, f, \text{current}, \sigma) = (\alpha, \beta, \gamma, \omega, f, \text{current}, 1 \text{ day});$
- $\lambda^P : S^P \rightarrow Y^P$
  $\lambda^P (\alpha, \beta, \gamma, \omega, f, \text{current}, \sigma) = \{(\#S, S), (\#E, E), (\#I, I), (\#R, R), (\#D, D)\};$ and
- $\delta_{\text{ext}}^{\ P} : Q^P \times X^P \rightarrow S^Y$, with $Q^P = \{(p, s, \sigma\ e)/(p, s, \sigma) \in S^P, 0 \leq e < \sigma\}$
  $\delta_{\text{ext}}^{\ P}(\alpha, \beta, \gamma, \omega, f, \text{current}, \sigma, e, (\text{set } \alpha, v)) = (v, \beta, \gamma, \omega, f, \text{current}, \sigma - e)$
  $\delta_{\text{ext}}^{\ P}(\alpha, \beta, \gamma, \omega, f, \text{current}, \sigma, e, (\text{set } \beta, v)) = (\alpha, v, \gamma, \omega, f, \text{current}, \sigma - e)$
  $\delta_{\text{ext}}^{\ P}(\alpha, \beta, \gamma, \omega, f, \text{current}, \sigma, e, (\text{set } \gamma, v)) = (\alpha, \beta, v, \omega, f, \text{current}, \sigma - e)$

$$\delta_{ext}^{P}(\alpha, \beta, \gamma, \omega, f, \text{current}, \sigma, e, (\text{set}\,\omega, v)) = (\alpha, \beta, \gamma, v, f, \text{current}, \sigma - e)$$
$$\delta_{ext}^{P}(\alpha, \beta, \gamma, \omega, f, \text{current}, \sigma, e, (\text{set}\,f, v)) = (\alpha, \beta, \gamma, \omega, v, \text{current}, \sigma - e).$$

Figure 3.5 shows how the respective numbers of susceptible, exposed, infected, recovered, and dead evolve over a period of 100 days. Initial conditions are 1,000,000 susceptible individuals; only 1 infected person; and no exposed, recovered, or dead individual. Parameters $\beta, \alpha, \sigma, \gamma$, and f are, respectively, set to 1.22e−06, 0, 0.33, 0.71, and 0.42, as calibrated in [8], which model of spreading without control measures coincides with our model for $\alpha = 0$.

Because of the scarcity of reliable data in the Nigerian healthcare management system, validation is a major issue (e.g., a good estimate of the population size of Nigerian states or cities is frequently disputed by national agencies). However, understanding the dynamics of the disease diffusion as regards to the variation of parameters is paramount to getting the exact figures for each health status at a given time.

Figure 3.6 shows such an exploration, with a focus on the level of disease penetration at one hand (variation of I/S, the ratio between initial numbers of infected and susceptible individuals), and on the other hand, the impact of some sociocultural dimension (variation of α).

We considered four levels of infectious situations: disease appearance stage (i.e., only one infection over a million of individuals, something comparable to what happened in big and medium cities in Nigeria, but also in Liberia, Guinea, and Sierra Leone), state of emergency level (i.e., a thousand infections over a million of individuals, a level at which countries often activate very special measures), catastrophe level (i.e., 10% of the population infected), and chaos level (i.e., 50% of the population infected).

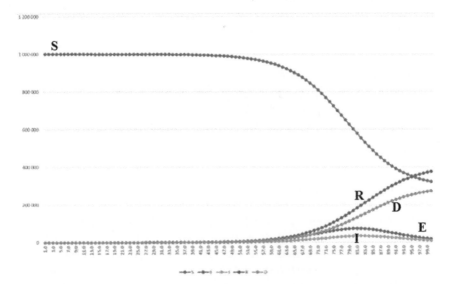

Figure 3.5   Ebola spreading in a period of 100 days, with calibrated parameters

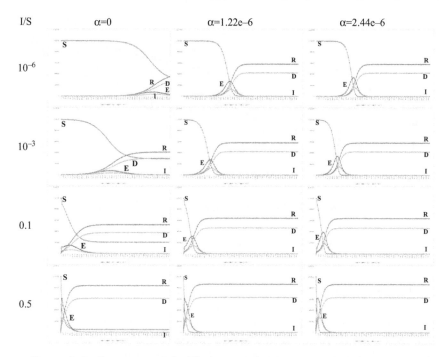

Figure 3.6   Sensitivity of the Ebola spreading to variations of parameters

We also considered three levels of social interaction: safe burial level (i.e., dead persons are buried with the maximum of caution, not allowing any direct contact with any living individuals), classic burial level (i.e., burial ceremonies are making interactions with dead persons as intensive as with living persons), and feasting burial (i.e., burial ceremonies take many days and go at many places, with direct contacts between dead and living individuals).

The top-down reading of Figure 3.6 shows that there is a drastic change of trajectories when burial-based sociocultural interactions come into play compared to safe burial situations, but their intensity does not have a very significant impact above a certain limit. The left-to-right reading of the same figure shows that above a threshold, the infection penetration is out of control, regardless of variations in the sociocultural interactions.

Obviously, much more questions can be explored to get a full level of explanation of this HD perspective-oriented issue.

3.3.2   Hospital beds allocation model

Healthcare affordability is a topic of immense interest to both individuals and national policymakers. An accurate depiction of healthcare affordability requires adequate consideration of the way resources can be allocated to meet the healthcare demand. As identified in the O4HCS ontology previously presented, such resources

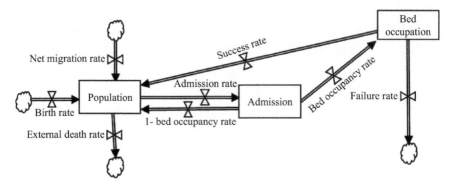

Figure 3.7   Model of demand for hospital services

can be human (doctors, nurses, etc.), physical (beds, rooms, vaccines, drugs, etc.), financial (funds, taxes, out-of-pocket payments, etc.), or information (health records, training, adverts, etc.).

The model developed here and presented by Figure 3.7, using Forrester's system dynamics [9], is meant to help policy-makers understand and anticipate on beds acquisition and management in a Lagos hospital. System dynamics is a popular modeling approach in healthcare systems M&S. A general survey of system dynamics in healthcare systems studies can be found in [10].

The demand for hospital services is modeled by the admission stock, which derives from an admission rate applied to the population stock:

- The population demographics change dynamically due to births, mortality, and migration.
- The mortality rate is disaggregated in the model into external death rate (i.e., deaths caused independently from the hospital intervention) and failure rate (i.e., deaths caused within the hospital).
- The net migration rate aggregates inflow and outflow migrants.
- While the admission rate subtracts quantities from population, the success rate reinjects into the population those hospitalized patients who do not die at the hospital.
- Also, nonhospitalized patients return to the population stock.
- The bed occupancy rate (i.e., the ratio of beds daily occupied by patients over the number of beds available) controls the bed occupation stock, the latter being an indicator of resource need for policy-makers.

The DEVS counterpart of this model is an atomic model which defines a state variable to represent each stock of the system dynamics model and which internal transitions modify the values of these variables according to the rates given as parameters. The time advance is always equal to 1 day. The specification is done according to the parameterized model formalization scheme we presented previously. Therefore, it is not reproduced here.

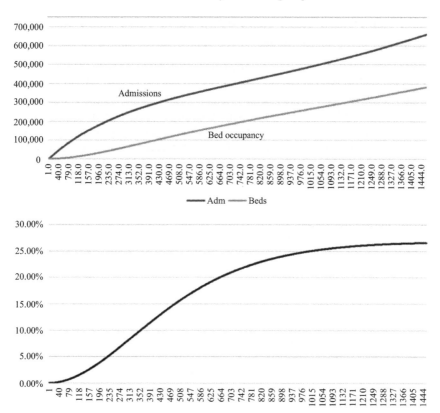

Figure 3.8 Evolution of health demand and supply indicators

Figure 3.8 displays the results obtained with the RA-specific EF used for the study. On top of the figure is shown the daily evolution over 1,460 days (i.e., 4 years) of, respectively, the number of admissions and the number of beds occupied. At the bottom of the figure is shown the ratio of bed occupancy in proportion of the population. The model has been calibrated using 2010 data from the annual report of the National Bureau of Statistics [11], in which hospital-specific data are averaged over major hospitals and health centers of Lagos:

- birth rate = 19.854 per thousand annually,
- net migration rate = −0.40 per thousand annually,
- external death rate = 7.95 per thousand annually,
- admission rate = 0.45 annually,
- bed occupancy rate = 0.73 annually,
- success rate = 850 per thousand annually,
- failure rate = 150 per thousand annually.

3.3.3 Interstate migrations model

Here, we develop a model inspired by [12], but with the following specificities:

- We consider the Nigerian population at states level. Nigeria is a federal country with 36 states (each having its capital city) and a Federal Capital Territory (FCT).
- A CA is used for modeling interstate migrations. The neighborhood of a cell includes all other cells of the CA. Each cell is defined by an index (0 for FCT, and 1–36 for states) and is assigned a geographical position (i.e., latitude and longitude of its capital city). The state of a cell at a given time is the population of the corresponding federal state at that time.

The general rule of the CA is expressed by the following equation:

$$n_i(t+1) = g_i n_i(t) + \sum_{i \neq j} (a_i - a_j) |n_i - n_j| e^{-\tau d_{ij}}$$

where

- $n_i(t)$ is the population of state i at time t;
- g_i is the net growth rate (i.e., birth – death +/– migrations from/toward outside the country) of state i;
- a_i is the relative attractivity of state i (i.e., the GDP (gross domestic product) per capita of state i over the GDP per capita of the country); the GDP is a monetary measure of the market value of all final goods and services produced in a period of time, which is commonly used to determine the economic performance of a country or region;
- d_{ij} is the distance between capital cities of states i and j; and
- τ is a constant positive number.

The equation is inspired by [12], in that the rate of migration between any pair of federal states depends on the population distribution. But, while Sheppard [12] considers the attractivity of a place grows with its size, and eventually declines as it approaches its capacity, we address this aspect in a different way, as follows:

- The interstate migrations for each federal state are addressed in the second member of the equation by its second term.
- $a_i - a_j$ expresses that between any pair of federal states, the more attractive one "wins." One can notice that the number of migrants leaving a source state ($a_i - a_j < 0$) is the same entering the target state ($a_i - a_j > 0$).
- $|n_i - n_j|$ expresses that states with nearly the same size have few attraction to each other. The greater the difference of size is, higher is the attraction (in favor of the more attractive state).
- $e^{-\tau d_{ij}}$ expresses that attractivity between any pair of states is amplified or reduced by the distance between them. Closer states have more attractivity to each other (the extreme case is $d_{ij} = 0$, which gives $e^{-\tau d_{ij}} = 1$), very distant states have a low attractivity to each other (the extreme case is $d_{ij} = +\infty$, which gives $e^{-\tau d_{ij}} = 0$).

The DEVS counterpart of this CA model is a parameterized atomic model which has the CA grid as its state variable, and which applies the CA rules during each of its internal transitions. Time advance is always equal to 1 day.

Figure 3.9 shows how the respective states evolve over a period of 1,460 days (i.e., 4 years period). Calibrating data are taken from [11]:

- The initial distribution of population considers the figures from the 2006 census.
- Net growth rates are calculated for the period of time from 2006 to 2010.
- Attractivity rates are calculated for year 2010.
- We use the Euclidian distance and $\tau = 0.01$.

Figure 3.9 displays snapshots at respective times 1, 183, 364, 545, 726, 907, 1,088, 1,261, and 1,442 (top-down and left-to-right), i.e., every semester, where each state is colored according to the range in which falls the daily growth of its population.

We also compared the evolution curves obtained from the CA simulation, with data available for the period from 2008 to 2011.

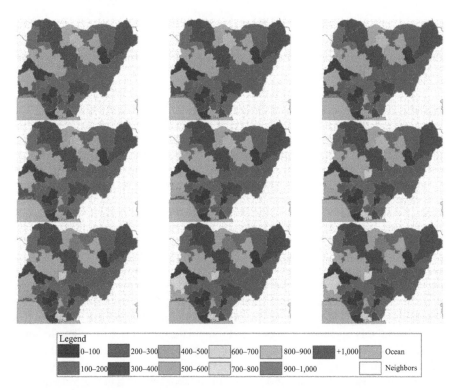

Figure 3.9   Snapshots of population dynamics simulation (daily growth) in Nigerian states

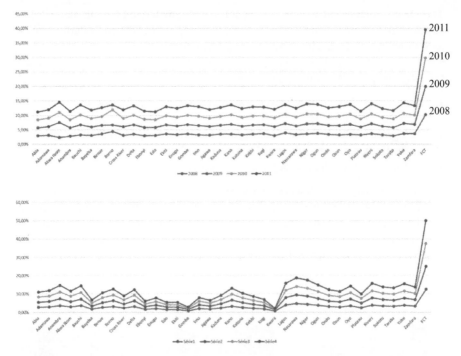

Figure 3.10   Real data versus simulation results (cumulative population growth rate per state)

Figure 3.10 shows (on top) how cumulative real data evolve for all states (horizontal axis) and for 4 years (vertical axis cumulating annual rates). It also shows (at bottom) how cumulative simulation results evolve, using the same layout.

Differences are in the interval of confidence of 95% for all states, except Gombe state and Kwara state which have lesser annual growth rates with simulation than in reality.

3.3.4   Daily worker model

As models from the IB perspective of our framework capture the microlevel of explanation (i.e., at individual level) of phenomena that are often described at the macro level (i.e., at population level) in healthcare systems simulation, let us focus here on the agent-based model of daily workers in the Nigerian population. They constitute a very significant part of intrastate and interstate migration flows. The objective of this agent-based model is to simulate the impact of a simple social strategy in the working condition of a daily worker. The model generates the result of scenarios depicting decisions by a daily worker to move from a working area to another one, based on the situation of the local labor market and the consequential effect on his working rate (i.e., the average number of worked days, hence the worker's earning).

In this study, local labor market refers to a combination of labor parameters such as the probability r for a primo entering to get a job daily, the probability p for a worker to keep the same job for the next day, and the probability q for a jobless to find a new job. These parameters affect the behavior of the daily worker in a way described by Figure 3.9.

Arriving in a new place as a primo entering, it takes 3 days to establish and understand how the local market works. This time represents for the daily worker the cost of moving from one area to another, since the corresponding days are lost in terms of earnings. The transition diagram of Figure 3.11 shows that the primo entering individual gets a job with probability r and is jobless with probability $1 - r$. A job is kept with probability p and lost with probability $1 - p$. A jobless individual will daily seek for a new opportunity, with a level of patience of x days. If he does not get any new job after this deadline, he will move to another working area (a counter n is used to know at each time the number of jobless days). We assume he will not go back to a place he formerly visited, and that the national labor market is uniform (therefore, probabilities do not change from one local labor market to another one). This may look contradictory, since the daily worker would probably move to a new place with higher probabilities. However, in reality, daily workers randomly change their areas of research since they do not have a clear visibility of the labor markets map. Their strategy relies solely on the choice of the value of x. Indeed, r being greater than q, any new relocation increases the potential for a jobless to get a new job, at the cost of the time lost in relocating.

This agent-based model is easily described by a parameterized DEVS atomic model. Each node of the transition diagram given in Figure 3.11 is a state of the DEVS model. Transitions are all internal transitions in the DEVS model. Time advance is 1 day for JOB and JOBLESS states, while it is 3 days for PRIMO state. Internal transitions are triggered depending on probabilities, except for the case of the worker moving to a new place.

The perspective-specific DEVS-based EF built to experiment with the model, explores for various values of x the trajectories of two variables: (1) the percentage of worked days and (2) the frequency of moves. For each value of x, 1,000 experiments are run, each for 1,460 days (4 years).

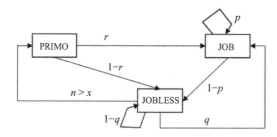

Figure 3.11   Individual behavior model of a daily worker

Figure 3.12 shows the impact of the workers decision (frequency of relocations) on his job performances (percentage of worked days), respectively for $x = 6$,
$x = 3$ and $x = 1$ (top-down).

Values of r, p, and q are, respectively, 0.45, 0.85, and 0.05. The strategy of
highest mobility, although showing high uncertainty of performances at the
beginning, is the most rewarding for the daily worker (highest average number of

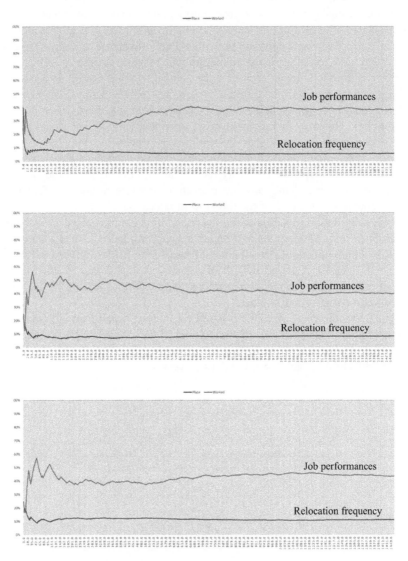

Figure 3.12   Relocations frequency versus job performances

worked days). This result echoes the reality on ground of day labor (low segment of the labor market) and its resulting migration flows.

3.3.5 Integrators

As previously explained, integrators are models that transfer the impact of the outputs of some of the models to the parameters of others. These models, each described as a "regular" DEVS atomic model, allow to integrate together all the models given in the different perspectives of our framework. The integrator model, in each case, has only two states: a waiting state, in which time advance is $+\infty$, and a generating state, in which time advance is 0. Only an external transition is possible from the waiting state to the generating state (which corresponds to the receipt of new outputs from the feeding model). In the generating state, the transfer model computes new values for parameters of its target model, then calls the target model to change the values of its parameters, and then executes an internal transition to go back to its waiting state.

Considering that the hospital of interest is located in a popular area of Lagos, the flow of patients depends on what is going on in the direct environment. Therefore, the individual behavior of the majority of inhabitants (i.e., day workers), as well as the population dynamics of the federal state and the impact of the outbreak of Ebola greatly influence the admission rate and the bed occupancy rate as well. On contrary, performances of the hospital (i.e., cure and death frequencies) impact on the relative attractivity of the area as well as the spreading of the disease. These are relationships that the integrators must express.

3.3.6 Holistic model

Figure 3.13 top shows the four perspective-specific models with rough indications of their parameters (red arrows) and state variables (blue arrows). Each of these models can be given default parameters which remain constant and generate dynamics of their state variables. These can be taken as characterizing normal endogenous activity unperturbed by an exogenous event such as an Ebola outbreak. Figure 3.13 bottom gives more detail in the form of a causal loop diagram of the influence of state variables on parameters. The four vertical layers that are apparent in the figure correspond respectively to the RA, HD, PD, and IB models. Outputs are influencing variables, and parameters are influenced ones. A positive feedback (e.g., from number of infectious individual to admission rate) indicates that an increase (respectively a decrease) of the influencing variable results in an increase (respectively a decrease) of the influenced variable. A negative feedback indicates that both variables evolve in the opposite direction. For example, the transmission rate of Ebola virus is negatively impacted by more hospital admissions and positively increased as the population of a state or locality is increased. The holistic EF built to experiment with the resulting holistic model allows us to see how all models impact on each other simultaneously, and in various scenarios of influence. Experiments are run for 100 days, and each model is initialized to coincide with the outbreak of the EBV period.

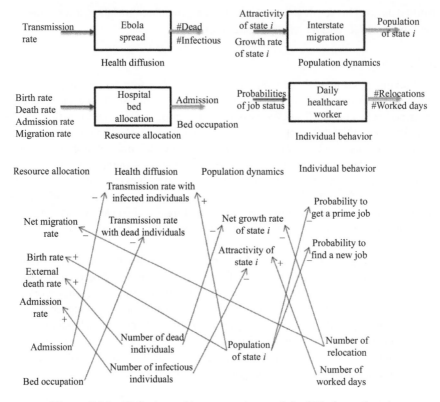

Figure 3.13   Holistic multi-perspective model of Ebola outbreak

Figure 3.14 shows results for the case where the influence relations of Figure 3.13 are treated as linear functions mapping from state variable values to parameter values.

- Top left curves show the distribution of population over a period of 100 days, depending on the health status of individuals (dark blue curve for susceptible individuals, red curve for exposed individuals, green curve for infected individuals, purple curve for recovered individuals, and light blue curve for dead individuals).
- Top right curves show, during the same period of time, the impact of daily workers decision on their job performances (the blue curve indicates the frequency of relocations of the worker, from a working area to another one, while the red curve indicates the ratio of worked days over the total number of days spent).
- The bottom left curve shows the daily evolution of the population in Lagos state at the time of the outbreak.
- The bottom right curve shows the ratio of bed occupancy in proportion of the population in the focused Lagos hospital, at the same time.

The example has illustrated how the framework can address multiple levels of explanation. EFs from the different perspectives focus on perspective-related

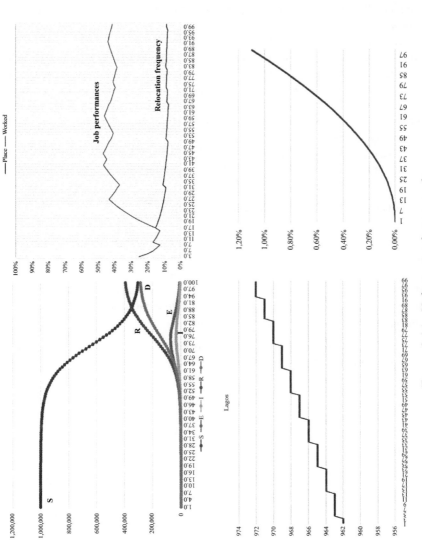

Figure 3.14 Holistic simulation results

questions. Models being abstractions and approximations by essence, a model developed within any perspective will necessarily use parameters to represent the aggregated dynamics of all influencing factors from other perspectives, ceteris paribus. The disaggregation of parameters binds representations from different perspectives to each other. Therefore, models in each perspective are sources of explanation of the hidden influencing processes of models in the other perspectives. That is why, the resulting global model allows deriving results that could not be accurately addressed in any of the perspective taken alone.

If the healthcare system we studied was taken at the scale of the country, each cell of the PD model (i.e., each federal state) would have been associated to a disease spreading model, many hospital models (as much as the number of health centers of the state) and many individual behavior models (for categories of workers). Such a fine-grained holistic model, thought computationally more expensive than simple models, provide a more accurate understanding of the national healthcare system. This is of tremendous interest for decision-makers and has a huge impact on cost, access and affordability concerns.

3.4   Introduction to pathways-based coordination

Interestingly, although not illustrated here, the movement of healthcare workers between locations which is guided by their perception of available jobs may not result in optimal assignments. Such results of holistic modeling point to aspects where coordination as supported by pathways [13] may result in improved performance. Figure 3.15 shows the possible modular integration of a pathway-based coordination model, with the following options:

● The pathway model is coupled with the worker model, and the resulting model is studied in isolation of the other perspectives.

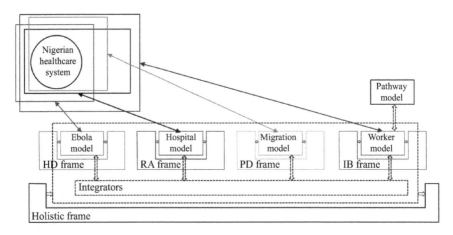

Figure 3.15   Pathway-based coordination of the holistic simulation model

- The worker model is immersed in the holistic model and, therefore, has a dynamic structure nature that the coordination model must take into account.
- The path model can be coupled with the whole holistic model to have a direct link with the different perspectives and to integrate more subtle information into the coordination scheme than that provided by the worker model alone.

To underline the importance of such a coordination, let us quote the Health Foundation [14].

> Quality problems are often treated as if they are one-off events, rather than the inevitable consequence of random combinations of constantly occurring errors and delays in multi-task processes. A typical response therefore is to add more 'checking' tasks to spot and correct errors. However, the most reliable and sustainable way to improve both quality and cost is to systematically redesign processes of care.

Therefore, there is a strong relationship among flow, quality, and cost. Therefore, coordination appears to be a key instrument for addressing the "iron triangle" issue mentioned in the introduction to this book. This aspect is dealt with in the second part of the book.

3.5 Summary

In the previous chapter, we have proposed a holistic approach to healthcare systems modeling and simulation that combines the different perspectives more often developed in isolation in the literature. A common practice is to consider simulation parameters constant throughout the different studies scenarios. However, we have argued that the values of these parameters can change in concurrent simulation studies. Consequently, we have proposed an integration of the underling perspectives where independent simulation processes of disparate concerns in healthcare systems exchange live updates of their influences on one another to bring results closer to the reality. Dedicated EFs can be designed to answer perspective-specific questions, while a global EF can be used to derive answers from the resulting global model, that could not be accurately addressed in any of the perspective taken alone.

In this chapter, we have formalized such a framework using the system-theoretic paradigm provided by the DEVS formalism. The developed integration approach that allows linking up the isolated perspectives has been formalized through a concept based upon DEVS formalism called parameterized DEVS, whereby concurrent simulation processes cause live update through output-to-parameter integration. This sets up the basis for development of DEVS-based holistic models for efficient management of healthcare systems. Results of real case-based simulations are presented to show how holistic models provide new insights not heretofore attainable with unit and facility specific models.

A noteworthy aspect is that the integration approach proposed by the framework allows to link models that have not been initially designed for this purpose. It is a significant difference with the classic "simulation in the large" approach where outputs of existing models are connected to input of other ones, provided the connecting ports were designed to serve that purpose at the time of the construction of these models, and that the ports fit each other. The approach proposed here in the context of healthcare systems M&S can be generalized beyond that limit to integrate models from other domains in a holistic study.

Some previous works are close to our effort to propose a stratification of levels of abstraction and their integration into a holistic framework, though not identical.

Charfeddine and Montreuil [15] introduced a conceptual agent-based framework for modeling and simulation of distributed healthcare delivery systems, which is structured into a three-level categorization, with a simulation engine as the integration platform. The first layer includes agents, objects, environment, and experience. In the second and the third layers, each component is broken down into two or more subcomponents with more details.

Seck and Honig [16] introduced a multi-perspective modeling approach that can be applied to any domain (and not in healthcare M&S only). Therefore, perspectives are not specifically identified, but a generic conceptual framework is proposed and formalized by adding to the DEVS systems specification hierarchy, a top layer to represent multi-perspective models.

An integration approach, very similar to ours, is proposed by Jeffers [17] and though not formalized. The common denominator chosen in [17] to specify models is Forrester's System Dynamics.

A key issue in developing multi-perspective models is the validity of the bridging components, i.e., the way parameters of a model are disaggregated using outputs of other models. Duboz *et al.* [18] asserted the need of awareness for such a legitimacy issue.

More generally, the consideration of different perspectives means that as the modeler attempts to span a wider range of perspectives, there is a need to incorporate the effects of an increasing number of processes. It is therefore of equal importance to identify these perspectives and to manage the phenomenology and mathematics of the interactions between them.

References

[1]   Zeigler B.P. *Theory of Modeling and Simulation.* New York, NY: John Wiley & Sons; 1976.

[2]   Zeigler B.P. *Multifacetted Modelling and Discrete Event Simulation.* London: Academic Press Inc.; 1984.

[3]   Zeigler B.P., Praehofer H., and Kim T.G. *Theory of Modeling and Simulation: Integrating Discrete Event and Continuous Complex Dynamic Systems.* 2nd Edition. New York, NY: Academic Press; 2000.

[4]   Traoré M.K., and Muzy A. Capturing the Dual Relationship between Simulation Models and Their Context. *Simulation Modelling Practice and Theory.* 2006; 14(2): 126–142.

[5] Chow A.C.H., and Zeigler B.P. Parallel DEVS: A Parallel, Hierarchical, Modular, Modeling Formalism. *Proceedings of Winter Simulation Conference*, Orlando, FL, USA, 11–14 Dec 1994. pp. 716–722.

[6] Muzy A., and Zeigler B.P. Specification of Dynamic Structure Discrete Event Systems Using Single Point Encapsulated Functions. *International Journal of Modeling, Simulation and Scientific Computing*. 2014; 5(3): 1450012 (20 pages).

[7] World Health Organization. *Ebola Virus Disease*. Fact Sheet No. 103; 2014. Available at http://www.who.int/mediacentre/factsheets/fs103/en/ [Accessed on 24 Oct 2017].

[8] Althaus C.L., Low N., Musa E.O., Shuaib F., and Gsteiger S. Ebola Virus Disease Outbreak in Nigeria: Transmission Dynamics and Rapid Control. *Epidemics*. 2015; 11: 80–84.

[9] Forrester J. Counterintuitive Behavior of Social Systems. *Technology Review*. 1971; 73(3): 52–68.

[10] Homer J., and Hirsch G. System Dynamics Modeling for Public Health: Background and Opportunities. *American Journal of Public Health*. 2006; 96(3): 452–458.

[11] National Bureau of Statistics. *2012 Annual Abstract of Statistics*. Available at www.nigerianstat.gov.ng [Accessed on 14 May 2018].

[12] Sheppard E. Urban System Population Dynamics: Incorporating Non-linearities. *Geographical Analysis Columbus*. 1985; 17(1): 47–73.

[13] Zeigler B.P., Carter E.L. Redding S.A., and Leath B.A. Pathways Community HUB: A Model for Coordination of Community Health Care. *Population Health Management*. 2014; 17(4): 199–201.

[14] The Health Foundation. *Improving Patient Flow*. Learning Report, London, UK; April 2013.

[15] Charfeddine M., and Montreuil B. Integrated Agent-oriented Modeling and Simulation of Population and Healthcare Delivery Network: Application to COPD Chronic Disease in a Canadian Region. *Proceedings of the Winter Simulation Conference*, Baltimore, MD, USA, 5–8 Dec 2010. pp. 2327–2339.

[16] Seck M.D., and Honig H.J. Multi-perspective Modelling of Complex Phenomena. *Computational and Mathematical Organization Theory*. 2012; 18: 128–144.

[17] Jeffers R.F. *A Model of Population Movement, Disease Epidemic, and Communication for Health Security Investment*. United States: National Nuclear Security Administration; 2014.

[18] Duboz R., Ramat E., and Preux P. Scale Transfer Modeling: Using Emergent Computation for Coupling Ordinary Differential Equation System with a Reactive Agent Model. *Systems Analysis Modelling Simulation*. 2003; 43(6): 793–814.

Chapter 4

Methodological elements for simulation-based healthcare management

Due to the fact that healthcare systems M&S is a complex enterprise, it requires a sound methodology to address it. As a modern complex system, a healthcare system requires multiple levels of explanation be provided to achieve its various objectives, while keeping a holistic understanding of the behavioral pattern of the overall system and its interaction with the surrounding environment. The multi-perspective modeling and holistic simulation framework introduced in this part of the book is meant to achieve this goal. Moreover, a DEVS-based formalization has been provided, allowing to capture in a clear and unambiguous way the key constructs of this framework. However, this does not remove from the shoulder of the modeler the required efforts to identify and highlight all useful abstractions, which a complete study entails.

Modeling complex systems where different aspects of the system are captured by different views have been reported as hard [1]. Usually, the required knowledge to do so is distributed among various stakeholders. Bringing them together is not an obvious operation. In this chapter, we examine some salient aspects of that issue.

Three concerns are scrutinized:

- The heterogeneous nature of healthcare systems M&S, in terms of spatial and temporal scales involved, which calls for a disciplined modeling approach; we do not provide a full methodology to drive such an approach, but we emphasize the importance of scale-driven modeling and the support that some elements of our framework can provide.
- The importance of a visual language to capture all the modeling knowledge while interacting with experts from various domains (as multi-perspective modeling crosses multiple domains of expertise, including social science, epidemiology, and management science); we introduce the High Level Language for System specification (HiLLS), a system modeling language for constructing multi-analysis system models, which can be seen as a visual language for DEVS.
- The hybrid nature of healthcare systems M&S, where hybridization occurs both at conceptual and operational levels; we suggest multi-paradigm modeling as a way to address this aspect. We also have a look on participatory modeling (PM) as a way to involve nonexperts at the conceptual level of the modeling process in order to increase the set of available knowledge for making useful decisions.

The current chapter does not provide a complete methodology as suggested by its title, but it rather examines some elements, among many others, that could be taken into account in a more accomplished methodological approach.

4.1   Scale-driven modeling

A characteristic feature of healthcare systems is the occurrence of interactions between heterogeneous components at different spatiotemporal scales [2–4]. The hierarchy theory [5,6] provides a guideline to model such complex systems, by emphasizing on the fact that at a given level of resolution, a system is composed of interacting lower level components and is itself a component of a higher level component [7]. As such, it opens the way to scale-driven modeling methodologies [8,9], with various interpretations of the notion of scale [5,10–13], and a major concern about scale-transfer processes where inter-scale interactions must be properly described, as emphasized in [14–16].

Scale-driven modeling exhibits a hierarchical organization from differences in temporal and spatial scales between the phenomena of interest. Thresholds between scales are critical points along the scale continuum where a shift in the importance of variables influencing a process occurs. Traditional generic distinctions between scales include the partition into the so-called macro, meso, and microlevels [17], or into strategic, operational, and tactical levels [18], or even into $n-1$, n, and $n+1$ levels [9]. All these efforts agree on the fact that there is a minimum requirement for a triadic view of causalities [19,20].

While the hierarchy theory mainly focuses on a descriptive form, the concrete translation of its derivative scale-driven modeling methodologies into formalized computational models can be achieved within the framework defined in this book. Indeed, under this framework, scale-driven modeling provides a vertical stratification within each perspective, while multi-perspective modeling provides a horizontal stratification within the holistic analysis of the entire system, as shown by Figure 4.1. Consequently, scale-transfer issues (as described in Chapter 2) are addressed both vertically and horizontally.

As part of the plan–generate–evaluate process evocated in Chapter 1, candidate models can be retrieved from the model base for healthcare systems M&S (called MB4HCS) derived from O4HCS (ontology for healthcare systems simulation), to serve the scale-driven modeling purposes.

The architecture of MB4HCS is described by Figure 4.2. Pruning of the O4HCS along a specific perspective results in a top model, probably a multi-scale coupled model that is built by directly connecting component models which belong to the same perspective while addressing intra-scale transfer issues. Global pruning of O4HCS results in a holistic model where component models are connected through well-defined integrators (i.e., the ones that address inter-perspective scale transfer issues).

However, DEVS models specification, including the detailed specification of gray models and integrators, can be challenging for domain experts with few DEVS-based M&S background. Visual modeling, i.e., modeling by the use of semantically rich, graphical, and textual design notations, is known to allow the level of abstraction to be raised, while maintaining rigorous syntax and semantics.

To achieve this goal and help healthcare domain stakeholders (i.e., clinicians, decision-makers, etc.) participate in a rigorous way to the elaboration of the DEVS-based holistic simulation model, we suggest the use of the HiLLS as a pivotal visual language.

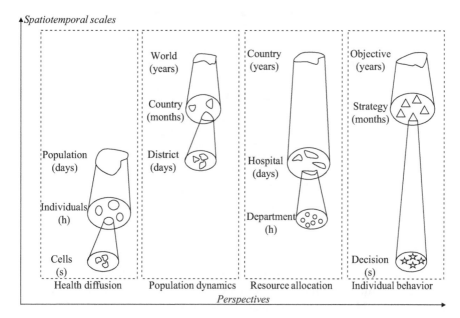

Figure 4.1   Example of scale-driven modeling in multi-perspective and holistic simulation

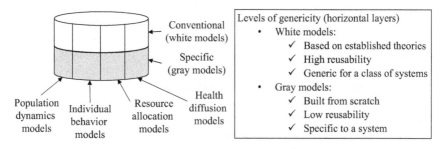

Figure 4.2   Architecture of MB4HCS

4.2   HiLLS modeling

The HiLLS is a specification language for complex systems, based on system-theoretic concepts from DEVS, software engineering concepts from UML, and logic concepts from Object-Z [21], to specify models that are amenable to analysis by both discrete-event simulation, formal methods, and enactment [22]. As such, HiLLS can be seen as a visual language for DEVS, with specific features for formal analysis and direct prototyping. This is the point of view adopted here to present it, although any of the two other formalisms and their underlying paradigms could be used as the entry point. In doing so, we avoid providing unnecessary details, and we refer readers to [22,23], for more on aspects related to enactment and formal analysis.

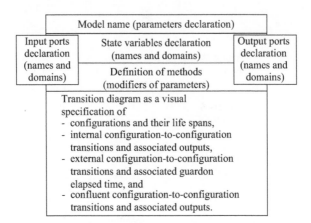

Figure 4.3　Template for HiLLS representation of a DEVS model

A template of how HiLLS represents a DEVS model is shown by Figure 4.3. A HiLLS-specified system is represented by a HSystem, which is denoted by a box similar to the UML class with an additional horizontal compartment and two vertical compartments. The left (respectively right) hand side vertical compartment has input (respectively output) ports attached to it. The concept of port is defined as in DEVS. All declarations in HiLLS (whether ports or any other variables or functions) are done in first-order logic, using the Z declaration schema approach.

The top horizontal compartment contains the name of the model and the declaration of its parameters. The immediate compartment below contains the declaration of state variables. The third compartment from top contains the definitions of operation schemas that use and manipulate all variables, including parameters and ports. Therefore, while a message received on some given input port causes a change of the internal state of the model, a call to some given modifier operation causes a change of value of some given parameters. Remember how this translates into DEVS as described in Chapter 2. The bottom compartment contains the system's behavior described by the configuration transition diagram, a graph the vertices of which represent the configuration set and the edges of which represent configuration-to-configuration transitions that can occur in the system. A configuration corresponds to an assignment of specific values or constraints to each of the state variables. One can notice that the assignment of a specific value to each of the state variables gives a state in the sense of DEVS (i.e., a particular configuration), while the assignment of constraints (rather than specific values) to some or all of the state variables gives a configuration that corresponds in DEVS to a family of states (instead of a single one). As such, a configuration depicts a set of properties that several states share. The main reason of introducing configurations is based on the strong assumption that a modeler may always be able to represent DEVS state-to-state transition functions (internal, external, and confluent transition functions all together) as a finite configuration-to-configuration transition diagram, whether the corresponding DEVS model has a finite or an infinite state set.

Table 4.1   Major visual elements of HiLLS

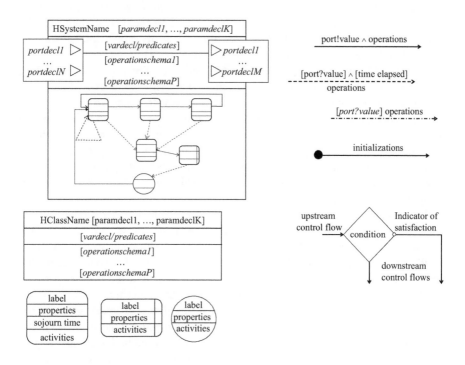

Configurations are a way to cluster any DEVS state set into a finite partition. The assumption that this is always possible has no formal proof. However, experience in modeling discrete-event systems shows that most often, raising the level of abstraction results in lowering the number of atomic properties, and therefore reaching eventually a level where the system's behavior can be captured by a collection of transitions between configurations in a finite set.

Table 4.1 shows how all these elements are graphically captured.

An HClass denotes a software component that does not represent the model of a dynamic system, as opposed to an HSystem. As such, it is a simple resource manipulated by model components and corresponds to a "regular" class in UML with attributes and methods specified by logical declarations and predicates. As an example, think of a queue that is used to store and retrieve entities during the lifetime of a model. An HClass can also be parameterized. HSystem, and HClass components can be related by traditional object-oriented relationships, such as aggregation, composition, generalization, and reference, with cardinalities attached to, as described by the UML metamodel (indeed HClass and Hsystem are specializations of the UML classifier mother class). Consequently, a HSystem can be composed of other HSystem (such a description corresponds to a DEVS coupled model).

As already explained, configurations in HiLLS are what would be high-level states in DEVS (in other words, a HiLLS model represents a finite state set-based

DEVS model which some infinite state set-based DEVS models can be reduced to). A configuration, as shown by Table 4.1, has three possible visual representations. A finite configuration is a four-compartment box, which respectively contains (1) the label of the configuration, (2) the logic specification of its properties, such as the assignment of values and constraints to state variables, (3) its sojourn time, which corresponds in DEVS to the value of the time advance function at states falling within this configuration, and (4) the description of activities to be carried out when the system is in this configuration, which has no equivalence in DEVS but serves for the purpose of enactment. An infinite configuration is a configuration for which sojourn time is $+\infty$; therefore, its visual representation is reduced to a three-compartment box, the compartment related to sojourn time being replaced by a double line at the right hand edge of the box. A transient configuration is the one the sojourn time of which is 0; therefore, its visual representation is reduced to a three-compartment circle.

The three kinds of configuration transitions are denoted by different labeled arrows with the operations accompanying the transitions (for update of state variables, when needed) as part of the labels. For internal transition, the value sent on the output port is also part of the label. For external transition, the label includes the triggering condition on the receipt of a value on a port and the time elapsed by the model in its current configuration. The label for confluent transition is similar, except that there is no condition on the elapsed time (since it is known to be the sojourn time in this case). A black circle allows to make reference to the initial configuration of the model. Also, decision nodes can be used to define various possible routes during a transition, depending on conditions to be met by state variables.

Figure 4.4 displays an example of HiLLS modeling, with some of the concerns going beyond what is needed for only simulation. The system described is a citizen who can contract disease by receipt of a virus. Among external factors that play a role in his/her overall health situation is his/her wealth (such as earnings and

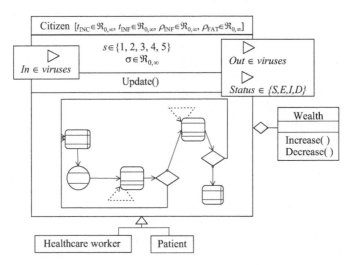

Figure 4.4   Example of HiLLS representation of a parameterized citizen system

various assets). Specializations of this component include patients and healthcare workers (both being subject to potential infection, but having some fundamental differences in the way they proceed in their life). The citizen is a discrete-event system (think of an agent for example), who has an input port (in) by which viruses can be inoculated to him/her (let us assume the term viruses denotes the set of all possible viruses that can be inoculated to a citizen), an output port (out) by which the citizen can inoculate in his/her turn the virus inside him/her to other citizens, another output port to inform the environment (a physician for example) of the health condition of the citizen, two state variables (s for different health statuses of the citizen and σ for the memory of time flowing), an update method to modify the value of his/her parameters, and a configuration transition diagram. Parameters of the citizen model are respectively the incubation duration (t_{INC}: time spent being exposed, i.e., infected but not infectious), the infectious period (t_{INF}: time spent being infectious), the infection rate (ρ_{INF}), and the fatality rate (ρ_{FAT}).

The configuration transition diagram of this system is detailed in Figure 4.5. Five configurations appear, each of which defined by conditions on the state variables. These configurations correspond to health conditions of the citizen when he/she is respectively susceptible, reached (a transient configuration for the very moment when a virus has been inoculated to a susceptible citizen), exposed, infected, and dead (one should notice that the example is not a precise description of the infection process of a given disease). The memory state variable ρ is used to ensure that when the citizen is exposed or infected, the inoculation of a new virus has no effect on the course of the current health process (in such case, the model operates an external transition from the current configuration to itself and just updates the memory that represents the time remaining for being in the current

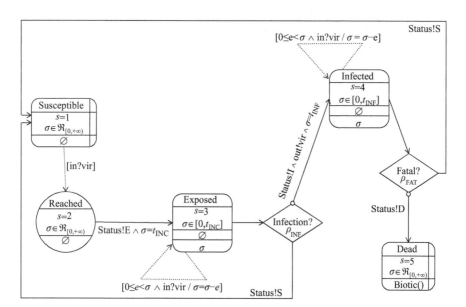

Figure 4.5   Transition diagram of the parameterized citizen

health condition). At the end of the incubation time, the infection rate of the virus helps determine whether the citizen gets infected or recover (in the latter case, we assume the citizen returns to being susceptible, and therefore not immunized against new infections). Similarly, at the end of the infection period, the fatality rate helps determine whether the citizen recovers or dies. Dead bodies start biotic decomposition activities (a knowledge not needed for simulation, but useful if the model has to be enacted).

Interestingly, DEVS atomic and coupled models are visually described in HiLLS the same way. Moreover, while a traditional coupled model will have in HiLLS a single configuration specifying the coupling information between its sub-components, a dynamic structure coupled model will have a configuration transition diagram with more configurations, each of them specifying a given architecture of the coupled model, and the transitions specifying the rules for the dynamic change of structure.

Figure 4.6 shows two examples of HiLLS model, one describing a static structure coupled model (left hand side) and the other a dynamic structure coupled model (right hand side). The left hand side model (permanent control system) defines a permanent coupling between its two components named mnt (a health monitor) and ctz (a citizen). This is captured by the single passive configuration appearing in its transition diagram, which states that whenever an output is sent on the status port of the citizen, this is treated as in input for the monitor.

The dynamic structure model (steady control system) defines a system where the monitor component interacts with the citizen components only 8 time units over 24 (e.g., let the time unit be hour to reflect the fact that the citizen is under daily controls for only some time in a day).

We find HiLLS convenient to visually capture DEVS specifications. A formalized presentation of HiLLS can be found in [23] with premises given by Maïga [24], as well as mapping rules from HiLLS to DEVS. Here, we are particularly interested in the fact that integrators can easily be described as HiLLS entities, and the holistic integration scheme expressed through method calls within activities. That way, the modeling process undergoes gradual refinements, while at the higher

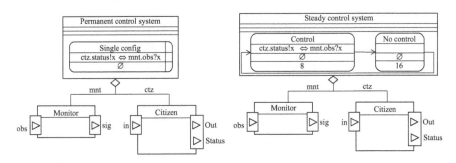

Figure 4.6   Static structure versus dynamic structure HiLLS models

level of abstraction, the way this integration is realized in DEVS is not described, and at a lower level, we know from Chapter 3 how it can be formalized.

4.3 Multi-paradigm modeling

While the HiLLS visual modeling approach is still rooted in the DEVS philosophy, exploring the potential of using other formalisms is of high interest. As argued by Fishwick [25], one formalism cannot effectively model all aspects of a complex system considering their diversity. He then proposed a multi-formalism modeling approach, i.e., the use of a mixture of appropriate formalisms (e.g., Petri nets, differential equations, etc.) to model the different components of a system. An illustration of such a need has already been given with the Nigerian healthcare systems case study presented in Chapter 3, where different formalisms have been used to initially express models developed in the various perspectives, before a DEVS counterpart be built for each.

While multi-formalism enables the use of suitable formalisms to model different perspectives of a system, generating the overall simulation code can be challenging. To achieve this goal, each of the formalism-specific components can be simulated with its corresponding formalism-specific simulator, and interactions due to coupling are resolved at the trajectory level [26]. This approach, known as co-simulation, discards a variety of useful information [27] and involves speed and numerical accuracy problems [28].

Another approach is to transform all the different formalism-specific components into one single target formalism, from which the final simulation code is derived. To automate these transformations, all formalisms involved need to be represented by their metamodels (a metamodel is a model of a formalism), and rules be defined to map metamodels onto each other. That way, for any component specified in a formalism A, its counterpart specification in a formalism B is obtained by applying the rules that map the metamodel of A onto the one of B. Model-driven engineering (MDE) provides the methodology to capture these concepts and organize the systematic application of model transformation [29].

Considering the complexity of healthcare systems and the diversity of its various components, a multi-formalism modeling approach is needed to effectively capture the concerns of the various stakeholders, while accommodating the diverse familiarities of experts with modeling formalisms, reuse of existing models, and easiness to capture some realities in some specific formalisms and other realities in other formalisms.

As depicted by Figure 4.7, the idea is to enable analysts use the most suitable formalisms to model the different perspectives of healthcare systems and systematically generate the final simulation code from the disparate models, through a DEVS-based M&S framework. To achieve this goal, a dedicated model transformation from each of the possible modeling formalisms to DEVS is required. Vangheluwe [30] showed how DEVS can be used as a common denominator to

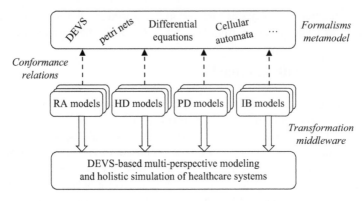

Figure 4.7   MDE-based approach to holistic M&S of healthcare systems

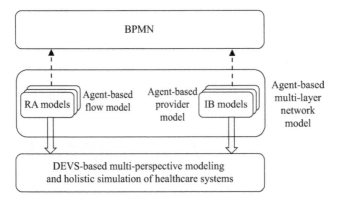

Figure 4.8   Example of multiformalism approach to holistic M&S of healthcare systems

translate that way different component models expressed in different formalisms into their DEVS counterparts.

Figure 4.7 presents the overall conceptual scheme. The topmost layer contains the metamodels of prospective formalisms for modeling the different perspectives of healthcare systems. Note that the list in the figure is not exhaustive. Hence, each of the perspective models in the third layer must conform to the metamodel of the formalism chosen to create it. The conformance relation is the relationship between a model and the meta-model that defines the syntax rules of the language in which this model is written [31]. Therefore, a given model conforms to a given metamodel if the specification of the model respects the syntax rules and constraints specified in the metamodel. Tools like the ATL (Atlas Transformation Language) technology [32] can be used to transform the disparate perspective models into DEVS simulation models at the bottom layer.

The case study presented in the third part of this book is an instantiation of the generic scheme depicted by Figure 4.7, where BPMN is at the domain experts' level, and models are built from the RA and IB perspectives, and then translated into DEVS for simulation. Figure 4.8 shows how this specialization is done.

The RA model is an ABM-based provider model, each healthcare practitioner being represented by an agent. The IB model is a BPMN-based flow model representing a patient in the healthcare system. These models are glued together by encapsulating the flow model in an agent-based entity model. That way, a multi-layer network structure that appears due to the relationships between agents, which derive from the BPMN specifications.

The feedback live between models of both perspectives are ensured by communications triggered between agents. The resulting DEVS model is coupled with an EF that focuses on performance criteria such as the time for a patient to complete his/her care pathway, and the distance covered by a patient during this process.

More details are given in Part 3 of the book.

4.4   Participatory modeling

So far, we have assumed that experts involved in the modeling activities are familiar with at least one modeling formalism. Participatory approaches emerged from the acknowledgment that the inclusion of nonexpert stakeholders and a variety of perspectives are required to improve our understanding of complex systems and problems. The inclusion of nonscientific/expert stakeholders is motivated by the presumed benefit of collaborative problem-solving, a central paradigm in citizen science. This paradigm aims at gaining insights that would otherwise not be available with models constructed by traditional experts/scientists alone. Therefore, simulation modeling also became participatory [33–36].

PM has several definitions. Let us adopt the one that defines it as a "purposeful learning process for action that engages the implicit and explicit knowledge of stakeholders to create formalized and shared representation(s) of reality" [37].

Stakeholders who are not experts in modeling, but have the knowledge and experience (of which expert modelers may lack) of the system of interest, must be included in the modeling activities in an iterative and adaptive way. As such, PM shares common principles with agile methodologies for software development. In software engineering, domain knowledge is critical to the development of any useful and usable application. The close and frequent interactions between the client and software engineers are assumed to increase mutual understanding and improve the global quality of the final product. PM objectives go beyond those of systems design. Its purpose is not only to produce software for a particular task but also to produce a holistic model of the system of interest. Such a model requires knowledge which cannot be held by one stakeholder. It represents a shared and commonly built and agreed description of reality by a set of heterogeneous stakeholders, which is useful for decision support and innovation. As we have seen, such a model has components that can be as diverse as humans, infrastructures, ecosystems, information systems, etc. In the specific case of healthcare, stakeholders can be citizens, health practitioners and managers, participants from pharmaceutical and insurances sectors, etc. By participating in the simulation modeling activities, they share their visions and knowledge (which are as valuable as that derivable from experts).

Nevertheless, PM faces a number of questions about the social dynamics related to the participatory philosophy, such as the following:

• Who, among experts, scientists, and the civil society, initiates the process and defines the questions to be answered?
• How do the different stakeholders interact and how does this influence the resulting knowledge?

Moreover, there are methodological elements that are required that can lead to action-oriented outcomes. These include ways of engaging stakeholders in a modeling process, methods for integrating heterogeneous outcomes, and approaches to embedding modeling into decision-making.

Further, there is also a need for computer interfaces and tools (such as those for visualization, analytics, documentation, and recording) that can assist in linking mental models with systems models. Indeed, computerized assistance for PM may benefit from experience with electronically supported collaboration using group decision support systems [38]. Although there is still a great deal to learn, there are fundamental principles of collaboration technology that apply and difficult problems to solve to achieve fully supportive computer assistance. Such technology can improve communication, structure, and support deliberation and provide access to information but also requires the right kind of leadership for success.

4.4.1   M&S knowledge

PM approaches address the *how, when*, and *by whom*, as it relates to the knowledge collected in PM sessions. This is based on the assumption that such mechanics will improve the acquisition of relevant knowledge in an organized way, namely, the *what*—see [39] for a collective work, with focus on *companion modeling*, an iterative PM approach.

We rather focus here on the *what* to be brought, so that the application of any efficient PM method would be framed within the context of MPM&HS, and the PM method would help better organize obtaining pieces of this framed knowledge from various stakeholders.

What kind of knowledge do we need to apply MPM&HS of healthcare systems? Remember from Chapter 3, that the system-theory-based DEVS framework suggests a specification hierarchy to capture the knowledge specific to systems behavior and structure. We recall hereafter the levels of this hierarchy (each of which being associated to an n-uple, such that going from the lower to the upper levels adds more elements to the n-uple), as detailed in [40] and captured by Figure 4.9.

• At level 0 (observation frame), the system is stimulated with input variables and observed over time to measure the output variables. Each input (I)

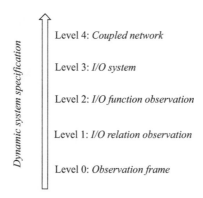

Figure 4.9   DEVS hierarchy of system specification

trajectory can then be paired with the output (O) trajectories that may be observed in experimentation or simulation.

- At level 1 (input/output relation observation), the collection of all I/O pairs observed are gathered and constitutes the I/O behavior of the system.
- At level 2 (input/output function observation level), a system is observed with specified initial states (also called initial conditions) and the I/O pairs at level 1 are segregated into state-based collections. This allows predicting a unique response to any input because the initial state establishes a functional relationship between inputs and outputs.
- At level 3 (input/output system), we are interested in knowing how the state changes when an input is applied to a system when started in its initial state as described at level 2.
- At level 4 (coupled network), a system is specified by a set of subsystems each having its own state and state transition function while linked with a coupling structure to one another.

4.4.2   *Knowledge management for MPM&HS*

Also remember from Chapter 3 that perspective-specific models and EFs, as well as integrators and multi-perspective EF are to be specified as DEVS atomic and coupled models. Consequently, the required knowledge for the building of these components can be framed by the DEVS specification hierarchy.

Figure 4.10 illustrates how, through proper enabling technologies, various stakeholders can build a PM-based knowledge. Such knowledge can clearly encompass M&S knowledge (acquired through climbing the system specification hierarchy), and can be guided/filtered to produce the multi-perspective and holistic simulation model.

To implement this approach, one leg of the effort concerns *multi-perspective modeling* and that the other leg addresses *holistic integration*:

- When dealing with the multi-perspective modeling leg, each of the perspectives leads to a PM process with a particular group of stakeholders. Both the

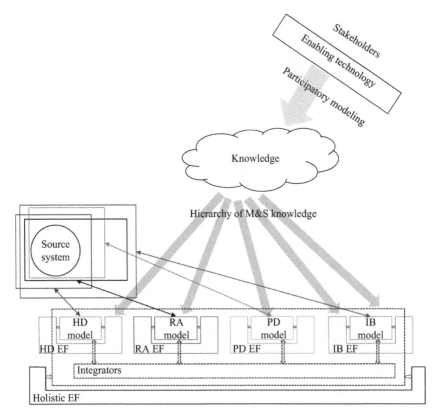

Figure 4.10   PM to MPM&HS of healthcare systems

perspective-specific top model and its associated experimental frame are then derived with the support of the DEVS specification hierarchy.

- When dealing with the holistic integration leg, a PM process is also required from which both the integrators and the holistic experimental frame are derived. The group of stakeholders can be composed by representatives of each perspective or by different stakeholders dealing with a global view of the source system. However, one can see the integrators as model components of the system of interest from a new perspective that assumes interactions exist between phenomena within existing perspectives. As such, they provide model components of the system of interest under objectives, assumptions, and constraints defined by the holistic experimental frame. Consequently, building the integrators falls in a modeling process identical to what prevails during any perspective-specific modeling effort.

When collecting the required M&S knowledge within a given perspective, as well as at the holistic level, it is important to formulate alternatives, specifically at the IOS and CN levels. This results in a family of models as architectural alternatives

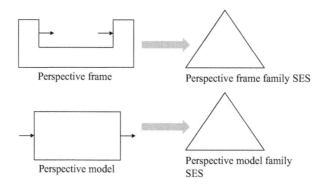

Perspective frame

Perspective frame family SES

Perspective model

Perspective model family
SES

Figure 4.11   Building perspective-specific families of models and frames

for the system of interest, as well as a family of frames as alternatives for the questions to be answered. Both families can be represented by SES tree structures, as shown by Figure 4.11.

The operation of merging SES trees mirrors the operation of composing DEVS models to create a coupled model [41]. As an example, O4HCS is a merging of the sub-trees that correspond to the four perspectives identified (i.e., HD, RA, PD, and IB). As explained in [42],

> merging generalizes the DEVS construction process in which individual models can be developed, tested and then composed to create hierarchical models in stage-wise fashion. This is to say, merging supports hierarchical composition in which families of models are generated and tested via pruning and transforming their SESs in bottom-up manner.

From a practical point of view, an existing SES tree becomes a component of a new larger SES tree. This happens when you terminate the top-down specification of the larger SES at a leaf entity with the same name as the existing SES. Under these circumstances, we say that a component SES will be *merged* into the larger SES. Consequently, a family of holistic models can be produced from merging of perspective-specific model family of SES trees as shown by Figure 4.12.

This can be seen as generalization of the way that a family of holistic studies can be formalized in a merged holistic frame SES tree as illustrated by Figure 4.13.

As a result of merging a holistic frame family SES and a holistic model family SES, the top-level merged SES is used to instantiate specific MPM&HS models such as the holistic Nigerian healthcare system model presented, as depicted by Figure 4.14.

In the context of PM, it is noteworthy that merging at every stage requires stakeholder inputs and agreements to provide refinements and couplings.

In other words, PM for MPM&HS must support knowledge acquisition as formalized by the hierarchy of system specifications and as implemented by merging of SES perspective-specific families of models and frames.

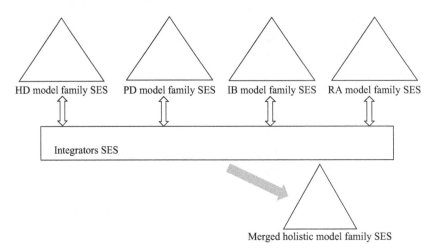

Figure 4.12   Building perspective-specific families of models (and frames)

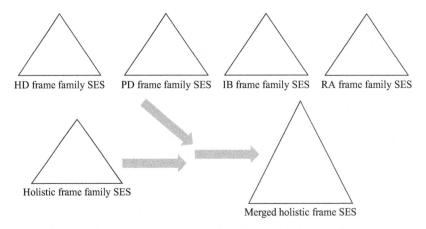

Figure 4.13   Building perspective-specific families of (models and) frames

4.5   Summary

So far, we have adopted a holistic approach to healthcare systems M&S, and we have presented a framework that encompasses common perspectives taken in the research literature but also goes beyond them toward their integration with additional perspectives that are becoming critical in today's environment.

We argue that a viable approach to address the challenging problem of healthcare systems management is a multi-perspective and holistic approach. Multi-perspective modeling allows constructing distinct and separate models from different aspects of healthcare system for a better understanding of its complexity.

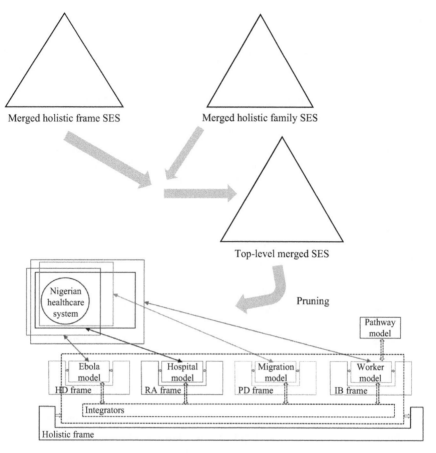

Figure 4.14   Building perspective-specific families of models and frames

Furthermore, an integrative approach is developed to allow the simulation output of a model of a given perspective to update the simulation parameters of another perspective dynamically, while the usual practice is to study problems in each perspective in isolation with some constant parameters representing the influences of other perspectives [43].

Arguably, a closer representation of the real situations can be achieved if these parameters are systematically modified at runtime, since processes of all perspectives run concurrently and influence one another continuously.

To develop the multi-perspective modeling and holistic simulation framework, we have used lessons learnt from literature review of healthcare systems M&S [44–53] to erect an ontology from which we derive a disciplined stratification of concerns where each layer focuses on a particular aspect. A family of conventional models developed through decades of research fall under these perspectives. Consequently, they can populate a model repository (a model base), such that any new specific study can be driven by the systematic process of pruning the ontology and selecting appropriate components in the model base or deriving new ones.

In this particular chapter, we have examined some methodological aspects as regards to the multi-perspective modeling and holistic simulation of healthcare systems.

More specifically, on the one hand, we have emphasized the key role of scale-driven modeling, a concept that appears under various forms in different domains. A recurring challenge is the scale transfer issue, a question to solve in the multi-layered framework proposed, both at intra and inter-perspectives levels.

On the other hand, since we have proposed a DEVS-based formalization of the framework, we have introduced HiLLS, a visual and high-level language that can be used as a graphical concrete syntax for DEVS.

Furthermore, we have addressed the need for extending the capabilities of our framework to multi-formalism in order to allow domain experts bring their knowledge at the conceptual modeling level using languages with which they are more familiar. Recall that multi-formalism allows the possibility that various formalisms can be used to capture the abstractions within and between the different perspectives, according to the objectives of the modeler and the nature of the system components under consideration.

Furthermore, model transformation can turn the abstractions described into their DEVS counterparts, and we have suggested the use of MDE principles [54] to provide such a feature.

Finally, we have discussed the emerging concept of participative modeling and its relation to our framework. However, we have not tackled aspects of the knowledge generated other than construction and elaboration of the simulation model. We recognize that participatory approaches can go farther, extending to the simulation-based decision-making process, including exploratory simulations, diffusion among local stakeholders, training of decision-makers, etc.

This first part of the book has introduced the foundational M&S framework for value-based healthcare systems. Its multi-perspective modeling approach provides multiple levels of explanation, as echoed by other research works [1,55,56], while its holistic integration approach allows deriving results that could not be accurately addressed in any of the perspectives taken alone.

Although this framework has been introduced in the context of healthcare, it is applicable to other complex systems as well, especially when specific features make their understanding amenable to their holistic study, like hybrid, adaptive, or emergent characteristics [4,57–59].

Also, the adoption of DEVS as the formal background is not a constraint. DEVS has been chosen for its universality in discrete-event simulation [40]. Other choices are possible, as long as their expressive power can accommodate the diversity of abstractions revealed in the different perspectives and their composition as well.

In the next part of this book, we expand on the coordination dimension, a key aspect toward M&S for value-based learning healthcare systems.

References

[1] Reineke J., and Tripakis S. Basic Problems in Multi-View Modeling. In Ábrahám E., Havelund K. (eds.). *Tools and Algorithms for the Construction and Analysis of Systems*. Lecture Notes in Computer Science, Vol. 8413. Berlin: Springer; 2014. pp. 217–232.

[2] Cowan G.A., Pines D., and Meltzer D. *Complexity: Metaphors, Models, and Reality*. Reading: Perseus Books; 1994.

[3] Levin S.A. *Fragile Dominions: Complexity and the Commons*. Reading: Perseus Books; 1999.

[4] Patten B.C., Fath B.D., Choi J.S., *et al.* Complex Adaptive Hierarchical Systems, Understanding and Solving Environmental Problems in the 21st Century: Toward a New, Integrated Hard Problem Science. In Costanza R., and Jorgensen S.E. (eds.). *Proceedings of the Second EcoSummit Workshop*, Elsevier, 2002. pp. 41–94.

[5] Allen T.H.F., and Starr T.B. *Hierarchy: Perspectives for Ecological Complexity*. Chicago: The University of Chicago Press; 1982.

[6] O'Neill R.V., DeAngelis D., Waide J., and Allen T.F.H. *A Hierarchical Concept of Ecosystems*. Princeton: Princeton University Press; 1986.

[7] Neill R.V., Johnson A.R., and King A.W. A Hierarchical Framework for the Analysis of Scale. *Landscape Ecology*. 1989; 3: 193–205.

[8] Wu J., and David J.L. A Spatially Explicit Hierarchical Approach to Modeling Complex Ecological Systems: Theory and Applications. *Ecological Modeling*. 2002; 153: 7–26.

[9] Aumann G.A. A Methodology for Developing Simulation Models of Complex Systems. *Ecological Modelling*. 2007; 202: 385–396.

[10] Marceau D. The Scale Issue in Social and Natural Sciences. *Canadian Journal of Remote Sensing*. 1999; 25: 347–356.

[11] Hay G.J., Niemann K.O., and Goodenough D.G. Spatial Thresholds, Image-Objects, and Upscaling: A Multiscale Evaluation. *Remote Sensing of Environment*. 1997; 62: 1–19.

[12] Dungan J.L., Perry J.N., Dale M.R.T., *et al.* A Balanced View of Scale in Spatial Statistical Analysis. *Ecography*. 2002; 25: 626–640.

[13] Ratzé C., Gillet F., Müller J.P., and Stoffel K. Simulation Modelling of Ecological Hierarchies in Constructive Dynamical Systems. *Ecological Complexity*. 2007; 4(1–2):13–25.

[14] Duboz R., Ramat E., and Preux P. Scale Transfer Modeling: Using Emergent Computation for Coupling Ordinary Differential Equation System with a Reactive Agent Model. *Systems Analysis Modelling Simulation*. 2003; 43(6): 793–814.

[15] Jelinski D.E., and Wu J. The Modifiable Area Unit Problem and Implications for Landscape Ecology. *Landscape Ecology*. 1996; 11: 129–140.

[16] Willekens F. Biographic Forecasting: Bridging the Micro–Macro Gap in Population Forecasting. *New Zealand Population Review*. 2005; 31(1): 77–124.

[17]   Blalock H.M. *Social Statistics*. New York: McGraw-Hill; 1979.

[18]   Rainey L.B., Tolk A. *Modeling and Simulation Support for System of Systems Engineering Applications*. New Jersey: Wiley & Sons; 2015.

[19]   Salthe S.N. *Evolving Hierarchical Systems: Their Structure and Representation*. New York: Columbia University Press; 1985.

[20]   Ulanowicz R.E. *Ecology, the Ascendent Perspective*. New York: Columbia University Press; 1997.

[21]   Smith G. *The Object-Z Specification Language*, Vol. 1. New York: Springer Science & Business Media; 2012.

[22]   Aliyu H.O., Maïga O., and Traoré M.K. The High Level Language for System Specification: A Model-Driven Approach to Systems Engineering. *International Journal of Modeling, Simulation, and Scientific Computing*. 2016; 7(1): 1641003.

[23]   Aliyu H.O. *An Integrative Framework for Model-Driven Systems Engineering: Towards the Co-Evolution of Simulation, Formal Analysis and Enactment Methodologies for Discrete Event Systems*. PhD Thesis, Université Blaise Pascal – Clermont Ferrand 2, France, Dec 2016.

[24]   Maïga O. *An Integrated Language for the Specification, Simulation, Formal Analysis and Enactment of Discrete Event Systems*. PhD Thesis, Université Blaise Pascal – Clermont Ferrand 2, France, Dec 2015.

[25]   Fishwick P.A. *Simulation Model Design and Execution: Building Digital Worlds*. New Jersey: Prentice Hall; 1995.

[26]   Sarjoughian H, and Huang D. A Multi-formalism Modeling Composability Framework: Agent and Discrete-Event Models. *Proceedings of the 9th IEEE International Symposium on Distributed Simulation and Real-Time Applications*, Montreal, Quebec, Canada, 10–12 Oct 2005. pp. 249–256.

[27]   Vangheluwe H., De Lara J., and Mosterman P.J. An Introduction to Multiparadigm Modelling and Simulation. In Barros F., and Giambiasi N. (eds.). *Proceedings of the AIS Conference*, Lisboa, Portugal, 7–10 Apr 2002. pp. 9–20.

[28]   Foster L., and Yelmgren K. Accuracy in DoD High Level Architecture Federations. *Proceedings of the Summer Computer Simulation Conference*, Arlington, VA, USA, 13–17 Jul 1997. pp. 451–460.

[29]   Cetinkaya D., Verbraeck A., and Seck M. Model Transformation from BPMN to DEVS in the MDD4MS Framework. *Proceedings of the Symposium on Theory of Modeling and Simulation – DEVS Integrative M&S*, Orlando, FL, USA, 26–29 Mar 2012. pp. 28:1–28:6.

[30]   Vangheluwe H.L. DEVS as a Common Denominator for Multi-formalism Hybrid Systems Modelling. *IEEE International Symposium on Computer-Aided Control System Design*, Anchorage, AK, USA, 25–27 Sep 2000. pp. 129–134.

[31]   Favre J.M. Megamodeling and Etymology – A Story of Words: From MED to MDE via MODEL in Five Millenniums. *Dagstuhl Seminar on Transformation Techniques in Software Engineering*, 2005. #05161 in DROPS 04101. IFBI.

[32] Jouault F., Allilaire F., Bézivin J., Kurtev I., and Valduriez P. ATL: A QVT-like Transformation Language. *Companion to the 21st ACM SIGPLAN Symposium on Object-Oriented Programming Systems, Languages, and Applications*, Portland, Oregon, USA, 22–26 Oct 2006. pp. 719–720.

[33] Pahl-Wostl C. Participative and Stakeholder-Based Policy Design, Evaluation and Modeling Processes. *Integrated Assessment*. 2002; 3(1): 3–14.

[34] Bousquet F., Barreteau O., D'Aquino P., *et al*. Multi-Agent Systems and Role games: Collective Learning Processes for Ecosystem Management. In Janssen M.A. (ed.). *Complexity and Ecosystem Management: The Theory and Practice of Multi-Agent Systems*. Cheltenham: E. Elgar Publishers; 2002. pp. 248–285.

[35] Antunes P, Santos R, and Videira N. Participatory Decision Making for Sustainable Development – The Use of Mediated Modelling Techniques. *Land Use Policy*. 2006; 23: 44–52.

[36] Binot A., Duboz R., Promburom P., *et al*. A Framework to Promote Collective Action within the One Health Community of Practice: Using Participatory Modelling to Enable Interdisciplinary, Cross-Sectoral and Multi-Level Integration. *One Health*. 2015; 1: 44–48.

[37] Participatory Modeling Web portal. https://ParticipatoryModeling.org [Accessed on 15 May 2018].

[38] Nunamaker J.F.Jr., Briggs R.O., Mittleman D.D., Vogel D.R., and Balthazard A.P. Lessons from a Dozen Years of Group Support Systems Research: A Discussion of Lab and Field Findings. *Journal of Management Information Systems*. 1996; 13(3): 163–207.

[39] Etienne M. *Companion Modelling – A Participatory Approach Supporting Sustainable Development*. Versailles: Quae (Collection Update Sciences & Technologies); 2011.

[40] Zeigler B.P. *Theory of Modeling and Simulation*. New York: John Wiley & Sons; 1976.

[41] Zeigler B.P, and Hammonds P. *Modeling & Simulation-Based Data Engineering: Introducing Pragmatics into Ontologies for Net-Centric Information Exchange*. Boston: Academic Press; 2007.

[42] Zeigler B.P. Creating Suites of Models with System Entity Structure: Global Warming Example. *Proceedings of the Symposium on Theory of Modeling & Simulation – DEVS Integrative M&S Symposium*, Society for Computer Simulation International, San Diego, California, 7–10 Apr 2014. Article 32.

[43] Djitog I., Aliyu H.O., and Traoré M.K. Multi-Perspective Modeling of Healthcare Systems. *Privacy and Health Information Management*. 2017; 5(2): 1–20.

[44] Almagooshi S. Simulation Modelling in Healthcare: Challenges and Trends. *Procedia Manufacturing*. 2015; 3: 301–307.

[45] Barjis J. Healthcare Simulation and its Potential Areas and Future Trends. *SCS M&S Magazine*. 2011; 2(5): 1–6.

[46] Bountourelis T., Ulukus M.Y., Kharoufeh J.P., and Nabors S.G. The Modeling Analysis and Management of Intensive Care Units. In *Handbook of Healthcare Operations Management*. New York: Springer; 2011, pp. 153–182.

[47] Brailsford S.C. Advances and Challenges in Healthcare Simulation Model-
      ing: Tutorial. *Proceedings of the 39th Winter Simulation Conference:
      40 Years! The Best is Yet to Come*, Washington, DC, USA, 9–12 Dec 2007.
      pp. 1436–1448.

[48] Günal M.M., and Pidd M. Discrete Event Simulation for Performance Mod-
      elling in Health Care: a Review of the Literature. *Journal of Simulation*. 2010;
      4(1): 42–51.

[49] Katsaliaki K., and Mustafee N. Applications of Simulation within the
      Healthcare Context. *Journal of the Operational Research Society*. 2011;
      62(8): 1431–1451.

[50] Onggo B.S. Simulation Modeling in the Social Care Sector: A literature
      Review. In Laroque C, Himmelspach J., Pasupathy R., Rose O., Uhrmacher
      A.M. (eds.). *Proceedings of the Winter Simulation Conference*, Berlin,
      Germany, 9–12 Dec 2012. pp. 739–750.

[51] Powell J.H., and Mustafee N. Widening Requirements Capture with Soft
      Methods: An Investigation of Hybrid M&S Studies in Healthcare. *Journal of
      the Operational Research Society*. 2016; 68(10): 1211–1222.

[52] Roberts S.D. Tutorial on the Simulation of Healthcare Systems. In Jain S.,
      Creasey R.R., Himmelspach J., White K.P., and Fu M. (eds.). *Proceedings of
      the Winter Simulation Conference*, Phoenix, AZ, USA, 11–14 Dec 2011.
      pp. 1408–1419.

[53] Thorwarth M., and Arisha A. *Application of Discrete-Event Simulation in
      Healthcare: A Review*. Dublin Institute of Technology Reports 3, 2009.

[54] Da Silva A.R. Model-Driven Engineering: A Survey Supported by the
      Unified Conceptual Model. *Computer Languages, Systems & Structures*.
      2015; 43: 139–155.

[55] Tekinay Ç., Seck M., Fumarola M., and Verbraeck A. 2010. A Context-
      Based Multi-Perspective Modeling and Simulation Framework. *Proceedings
      of the Winter Simulation Conference*, Baltimore, MD, USA, 5–8 Dec 2010.
      pp. 479–489.

[56] Seck M.D., and Honig H.J. Multi-perspective Modelling of Complex
      Phenomena. *Computational and Mathematical Organization Theory*. 2012;
      18: 128–144.

[57] Tolk A., Diallo S.Y., and Mittal S. Complex Systems Engineering and the
      Challenge of Emergence. In Mittal S., Diallo S., Tolk A. (eds.). *Emergent
      Behavior in Complex Systems Engineering – A M&S Approach*. Hoboken:
      John Wiley; 2018. pp. 79–97.

[58] Mittal S., Durak U., and Oren T. (eds.). *Guide to Simulation-based Dis-
      ciplines: Advancing our Computational Future*. Herndon: Springer; 2017.

[59] Mitchell B., and Yilmaz L. Symbiotic Adaptive Multisimulation: An Auto-
      nomic Simulation Framework for Real-time Decision Support under
      Uncertainty. *ACM Transactions on Modeling and Simulation*. 2008; 19(1):
      (31 pages). Article 2.

Part II

Modeling and simulation of pathways-based care coordination

Overview—A healthcare delivery system of systems is made of humans and technology where for the foreseeable future, self-improvement will be primarily based on human understanding rather than machine learning. Therefore, for such a system to continually self-improve, it must provide the right data and models to support human decisions on selection of alternatives likely to improve the quality of its services. Our focus in this part is to show how modeling and simulation can help design service infrastructures that introduce coordination and bring into play the conditions for learning and continuous improvement. To do this, we discuss the application of the discrete event system specification formalism within system of systems engineering to develop coordination models for transactions that involve multiple disparate activities of component systems that need to be selectively sequenced to implement patient-centered coordinated care interventions. We show how such coordination concepts provide a layer to support a proposed information technology for continuous improvement of healthcare as a learning collaborative system of systems.

Chapter 5

DEVS methodology for coordination modeling

5.1 Introduction

So far we have presented a framework for multi-perspective modeling and holistic simulation of healthcare systems. In this framework, components of the healthcare system are modeled as autonomous systems that can influence and be influenced by their environments. The resulting global model can be coupled with multiple experimental frames to more realistically represent the system individually in multiple perspectives as well as in holistic ones. In this part, we expand on the coordination dimension toward modeling and simulation (M&S) for value-based learning healthcare systems.

M&S has been applied to a variety of levels of analysis in medicine and healthcare [1,2]. However, M&S has had little impact at the level of reform involving radical restructuring of the ways in which multiple systems interact to deliver healthcare [3]. Indeed, healthcare delivery can be regarded as a *service system* that comprises service providers and clients working together to coproduce value in complex value chains.

Following Spohrer *et al.* [4], we raise the question: under what conditions does a healthcare delivery system of systems (HDSoS) improve itself, and how can we design such a system to improve in this manner?

Roughly our argument is as follows: an HDSoS is made of humans and technology where for the foreseeable future, self-improvement will be primarily based on human understanding rather than machine learning, artificial intelligence, and cognitive computing, such as IBM's WatsonPaths [5], will be increasingly better at generating and evaluating hypotheses about improved treatments and other interventions. However, humans must make decisions about protocols, processes, and procedures to actually put in place to improve healthcare delivery. Therefore, in order for an HDSoS to continually self-improve, it must provide the right data and models to support human selection of alternatives likely to improve the quality of its services.

As in Figure 5.1, it follows that

1. there must be working definitions of *quality of service*, for example, Porter's Healthcare Value, defined as outcome divided by cost [6],

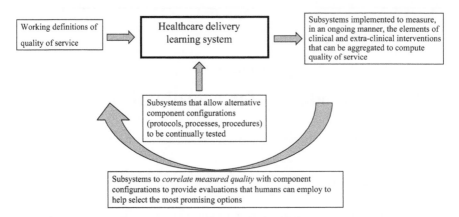

Figure 5.1　Prerequisites for healthcare delivery learning system

2. there must be *systems implemented to measure*, in an ongoing manner, the elements of clinical and extra-clinical interventions that can be aggregated to compute quality of service as defined,
3. likewise, there must be implemented systems that allow *alternative component configurations* (protocols, processes, procedures) to be continually tested [7], and
4. there must be systems to *correlate measured quality* with component configurations to provide evaluations that humans can employ to help select the most promising options.

A prerequisite for such conditions to prevail in an HDSoS is that sufficient organization and infrastructure exists to support their implementation. Currently, most national healthcare systems do not meet this prerequisite. At the high end is healthcare delivery in the United States. Although the most costly in the world, it focuses on medical services and fails to include social services that are equally important in achieving good health outcomes [8]. US healthcare has been diagnosed as consisting of loosely coupled, fragmented systems that are not sufficiently integrated or coordinated to provide high quality of service [9] or to enable self-learning [10]. At the low end, some national healthcare infrastructures are both underdeveloped and uncoordinated so that leap-frogging into twenty-first century learning systems is critical to meeting the challenges of burgeoning populations (Chapter 3). In the middle, nationalized systems are better organized from the top-down but still lack the infrastructure to experiment, measure, and evaluate on the large scales required to implement self-improving HDSoSs.

Our focus in this book is to show how M&S can help design service infrastructures that introduce coordination and bring into play the conditions (a)–(d) for a self-improving HDSoS. To do this, we discuss the application of the discrete event system specification (DEVS) formalism [11] to design of self-improving healthcare service systems. Systems theory, especially as formulated by Wymore [12–14], provides a conceptual basis for formulating the coordination problem of interest here. In particular, we discuss a concept of coordination models for trans-actions that involve multiple activities of component systems and coordination mechanisms implementable in the DEVS formalism. We show how system of systems engineering (SoSE) concepts [15,16] enable formal representation that combines Porter's [6] value-based care concepts with Pathway Community HUB care coordination [17,18] to enable implementation of criteria for measurement of outcome and cost. This leads to a pathways-based approach to coordination of HDSoSs and to a proposed mechanism for continuous improvement of HDSoSs as learning collaborative system of systems (SoS).

The framework will be expressed at the fine-grained level in which individual patients are explicitly represented, because this is level of analysis at which the metrics of quality and cost are fundamentally measured. However, means for aggregation of data to more abstract levels of analysis are also included to support generalization and quality improvement. In the sequel, we will point out the tech-nical benefit that DEVS offers over other approaches. Finally, we will suggest how the book offers a fertile framework for healthcare service SoSE that can stimulate further research in both the underlying theory and its application.

5.2   Overview of DEVS methodology for coordination modeling

As illustrated in Figure 5.2, a SoS, which is a system composed of multiple com-plex systems, can be abstracted to a simulation model [11]. *It is a DEVS model obtained by coupling together component models representing the systems com-ponents and derived from the multi-perspective modeling framework presented in Part 1.*

The representative SoS model can be used to test mechanisms developed to coordinate transactions that cut across multiple component systems. Such compo-nent models can be derived by abstracting the features (activities, services, etc.) of

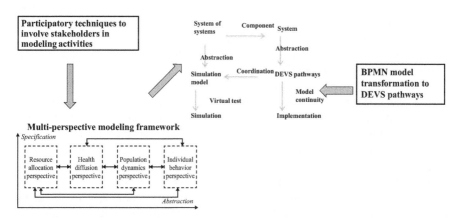

Figure 5.2   Abstraction of healthcare delivery SoS into DEVS simulation model using the modeling framework of part I

the component systems that are relevant to defining coordination mechanisms for cross-system transactions of interest [19]. After virtual testing in the SoS simulation, the same models can be implemented in net-centric information technology (IT) using the model-continuity properties of the DEVS framework. Such model-continuity allows simulation models to be executed in real-time as software or hardware by replacing the underlying simulator engine [20,21].

Several M&S environments based on DEVS support design, testing, and implementation of coordination mechanisms in a SoS engineering approach [22,23]. *Here we discuss automated approaches to transforming models expressed in Business Process Model Notation (BPMN) to DEVS pathways which implement the coordination mechanisms to be discussed.* In this part, we also discuss new techniques being developed that are centered on the multi-perspective modeling framework introduced in Part 1 and that directly involve the participation of all stakeholders (from population served to clinical service personnel) for codesigning new forms of healthcare service systems.

5.3   Healthcare as a learning collaborative system of systems

On the US national level, a learning health system is being envisioned that lays out several requirements, including one for a stable, certifiable, adaptable, and self-improving system. A workshop on the topic raised such questions as: what is the relationship between healthcare delivery innovations, such as team practice and patient engagement, and the extent and quality of learning in such a system? [10].

Porter and Teisberg [24] advocate radical reform of healthcare that requires that physicians reorganize themselves into integrated practice units (IPUs) moving away from care that is currently based on specialties with associated hospital departments. An IPU is centered on a medical condition defined as an interrelated set of patient medical circumstances best addressed in an integrated way. Porter's formulation for an IPU emphasizes knowledge acquisition and continuous improvement based on measurement of outcomes but does not provide a mechanism to do this. We will discuss a generalization of Porter's IPU concept based on DEVS and SoS concepts that lays the groundwork for application of DEVS to continuous improvement of HDSoS.

First, we summarize and review some of the basic concepts required to formulate the continuous improvement problem for collaborative HDSoS [25]. Table 5.1 defines features characteristic of continuous improvement and exemplifies them for multidisciplinary physician teams characteristic of IPUs.

Component systems have variants that are interchangeable in the slot represented by the component. This is referred to as specialization. The variants are available for substitution alternatives in the component slots. Outcome variety will be created due to the composite effect at the SoS level of an assignment of alternatives to component slots. It is the global behavior of the enclosing system that is the healthcare outcome to be evaluated. The analogy with genetic evolution is noted and discussed in Muzy and Zeigler [25]. A decision must be made on what

Table 5.1 Exemplifying features characteristic of continuous improvement in multidisciplinary physician teams

Characteristic feature	Definition	Multidisciplinary physician team manifestation
Enclosing system	SoS for which goal requires collaboration, may enclose components of more than one identified system	Integrated practice unit
Component system	A component that participates in the enclosing system	Physician
Collaboration requirement	Description of the goal that requires collaboration	Each physician must provide his/her service to the assure successful treatments
Modularity	Component system has well-defined interfaces and its own contained state	Physicians within the same discipline can be interchanged to play the same role in a team
Specialization	Component systems have variants that are interchangeable in the slot represented by the component. Typically, the variants represent the behavior characterized by the component in specialized manners	Physicians specialize via disciplines to play specific roles in a team
Variety at component level	The variants available for substitution alternatives in the component slot	Physicians schedules and participation in multiple teams provide variety in components
Outcome quality of service	Composite effect at the SoS level of an assignment of alternatives to component slots. It is the global behavior of the enclosing system that is healthcare outcome to be evaluated	Some physicians work well with others, some do not. So selecting the best team composition for a given full cycle of patient care is a challenge
What constitutes a single trial	The time interval during which activity of components, global outcome, and their correlation are evaluated as a single instance	Time spent by physician in full cycle of care rendered to a patient
Evaluation of trial	The evaluation of activity of components, global outcome, and their correlation for a trial instance	Healthcare value (outcome per unit cost as will be defined in text)

constitutes a single trial—this is the time interval during which the activity of system components and the global outcome are measured. At the end of a trial, the correlation of component activity and global outcome in such time series is computed as the result of the trial (how evaluation of a trial occurs will be explained in detail below).

While the table shows the elements needed for implementing continuous improvement strategies, it does not show how to employ these elements in a manner to implement such a strategy. We now turn to a proposal for such a strategy and its implementation.

In the traditional formulation of the coordination problem, each system has a goal and often the goal of the SoS conflicts in part with those of the components. Coordination is then conceived as a mechanism to achieve optimal alignment of component goals to the overall goal [26]. In contrast, as mentioned above, our concern here is the organization of activities among individual clients and service providers to coordinate the appropriate delivery of services. Although salient in healthcare, this concept of coordination is applicable to many situations where multiple providers offer multiple services to multiple clients.

In the sequel, we show how the DEVS formalism provides a clear and precise way to define and implement coordination mechanisms in systems of systems.

5.4   Care coordination

Care coordination does not have a universally accepted definition. One view is that is the process of identifying at-risk individuals and connecting them to the care and services. This view emphasizes concentration on at-risk individuals, subpopulations that have little to no access to needed services and notes that the work products purchased often have no confirmed benefit to the individual served, efforts may be duplicated or inefficient, leading to payments well above the "market value" of the service provided, and there is no requirement or incentive to focus on those at greatest risk.

Consequently, care coordination should be a process which identifies those at greatest risk and ensures that they receive needed evidence-based health and social services (e.g., prenatal care, immunizations, chronic disease management, parenting education, housing, food, and clothing).

Equally important, such coordination must measure, track, document, and evaluate benchmarks and final outcomes. Craig *et al.* [27] present a care coordination framework aimed at improving care at lower cost for people with multiple health and social needs. Although such a framework provides a starting point, it does not afford a rigorous predictive model that takes account of emerging health information networks and electronic health records (EHRs).

The Pathways Community HUB Model is a delivery system for care coordination services provided in a community setting [28]. The model is designed to identify the most at-risk individuals in a community, connect them to evidence-based interventions, and measure the results [17]. Community care coordination works at the SoS level to coordinate care of individuals in the community to help address health disparities, including the social barriers to health.

In this book, we delve into the Pathways Community HUB model in some depth as a basis for a more extended framework for care coordination. As outlined in Figure 5.3, this framework elaborates on the basic approach of Figure 5.2 in which the multi-perspective modeling framework is employed to provide models for the HDSoS, which are of use in developing simulation-testable coordination mechanisms for the systems and services involved.

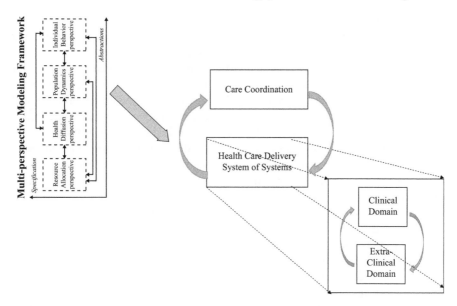

Figure 5.3   Extended framework for care coordination of health delivery SoS

Furthermore, we emphasize the recognition that the HDSoS is a complex assemblage of elements that involves both medical services in clinical settings (hospitals, labs, clinics, etc. where primary care is concentrated) as well as social and other services beyond primary care carried out in extra-clinical settings such as patient homes, pharmacies, education centers, and so on.

Later in this chapter, we provide an example relating to HIV–AIDS, which illustrates the need for incorporating both clinical and extra-clinical elements into consideration.

Moreover, in subsequent chapters of this book, we explore in more depth the implications of such an expanded consideration as well as the associated IT requirements and their interaction with our M&S framework.

5.5   Pathways-based care coordination

The Pathways Community HUB model is a construct that enforces threaded distributed tracking of individual clients experiencing certain pathways of intervention, thereby supporting coordination of care and fee-for-performance based on end-to-end outcomes [28]. As an essential by-product, the pathway concept also opens up possibilities for system level metrics that enable more coherent transparency of behavior than previously possible, therefore greater process control and improvement reengineering.

The pathways model [18,29,30] is a coordination structure that offers a potential for application to at-risk populations but also more generally where services exist but are not efficiently nor cost effectively deployed. This model is derived from industrial metaphors and provides a structure for production of outcome-based deliverables via formalized pathways that support accountability and can be used as standards for healthcare that promote positive outcomes.

In the pathways model, the steps of a pathway are designed to provide for problem resolution. Pathways enable coordination of healthcare components that is likely to increase positive outcomes while minimizing expenditures of additional resources. In short, a well-designed and implemented pathways model can improve patient health and social service outcomes while reducing total cost of care.

As a coordination tool, pathways can be used within an agency, a community, or statewide to focus on and track outcomes. A pathway simply defines the problem to be addressed, the desired positive outcome, and the key intervention steps required to achieve the outcome. By reporting on common variables—the pathways—it is possible to compare different employees, agencies, or communities to learn and share successful strategies.

In Chapter 6, we go into more detail on a *coordination model* that abstracts essential features of the Pathways Community HUB Model so that the kind of coordination it offers can be understood and employed, in a general SoS context. This allows development of the M&S framework discussed here to design, test, and implement such coordination models in a variety of SoS settings, exemplified by healthcare, that present the issues that such coordination models address. Formalization provides a firm basis for capitalizing on the transparency that is afforded by the Pathways Community HUB Model [31]. Such pathways are represented as DEVS atomic models and can become components of coupled models, thereby enabling activation of successors and sharing of information.

As we will see in more depth in Chapter 6, such pathway models represent steps in a pathway as states that can constrain steps to follow each other in proper succession with limited branching as required; external input can represent the effect of a transition from one step to next due to data entry. Moreover, temporal aspects of the pathways, including allowable duration of steps, can be directly represented by the DEVS atomic model's assignment of residence times in states.

5.5.1   Coordination of cross-system transactions

In the kind of coordination considered here, there are multiple service providers (component systems) whose activities must be brought together in different ways to serve different clients.

In the as-is situation, a client is, to a large extent, responsible for selecting, sequencing, and scheduling encounters with providers. Since multiple activities are located in different component systems, the client needs to traverse several activities across different systems to complete a *cross-system transaction*.

Thus, an adequate coordination model is characterized by the following requirements:

- Coordination design must define cross-system transactions and criteria for their successful completion.
- One or more cross-system transactions may be assigned to a client.

- A coordination agent must aim to assure that clients will successfully complete their assigned transactions.
- Coordination tracks the completion state and provides accountability for success/ failure of the client and coordination agent in completing assigned transactions.
- Coordination allows computing the costs of sets of cross-system transactions by accumulating the costs of activities involved in such sets.

5.5.2 Pathways as coordination models

Viewed as coordination models as just defined, coordination pathways provide concrete means to

- define steps in terms of goals and subgoals along paths to complete cross-system transactions;
- test for achievement and confirmation of pathway goals and subgoals;
- track, and measure progress of, clients along the pathways they are following; and
- maintain accountability of the compliance/adherence of the individual and responsible coordination agent.

An IT implementation of such pathways can provide abilities to

- query for the state of a client on a pathway,
- query for population statistics based on aggregation of pathway states for individuals, and
- support time-driven activity-based costing (TDABC) [32] based on pathway steps and their completion times.

5.5.3 Atomic pathways models

Three aspects of atomic pathway models to note are as follows:

- *Their primary role* is to request and receive data about a main goal and benchmarks (or subgoals) accomplishment—we will call these Questions and Answers.
- *Bounded times* are given for answers to be received.
- *Accomplishment of the main goal is decidable after a finite time* in the sense that the model is guaranteed to wind up (and remains) in one of three classes of states: *known success, known failure, or incomplete.* In the last type, the model explicitly reports that it is unknown whether the goal has been achieved or not.

In the following, we illustrate how atomic pathway models are formally defined as a class of DEVS models.

An atomic pathways model is a DEVS:

$$AtomicPathway = (X, Y, S, \delta_{ext}, \delta_{int}, \lambda, ta)$$

where

   X is the set of inputs;
   Y is the set of outputs;

Table 5.2   Definition of the sets and functions in atomic pathway model

Set and functions	Explanation
$X = Answers \cup \{Activate\}$ $Y = Queries \cup \{Activate\}$	Inputs are answers received by sending out queries plus the ability to send and receive an activation signal
$S = \{s_0, s_1, s_2, s_3, \dots, s_N\} \cup$ $\{Success, Failure,$ $Incomplete, End\}$	The states form a sequence starting with subscript 0 and ending with subscript N where N is an even integer. In addition, there are states for successful and unsuccessful completion, as well an incomplete state (see text)
$ta(S_0) = \infty$	The starting state is a passive state (waits for input)
$\delta_{ext} (S0.,_e, Activate) = S_1$	Upon receiving an activation signal, the initial state goes to the first indexed state
$\delta_{int}(S_i) = S_{i+1}$ $ta(S_i) = 0$ $\lambda(S_i) \in Queries$	The first indexed state and all odd indexed states immediately output queries and transition to the next even indexed state
$\delta_{ext}(S_{i+1,e,ans} = S_{i+2})$ *for ans* $\in Answers$ $ta(S_{i+1}) = T_{i+1}$ $\delta_{int}(S_{i+1}) =$ $Incomplete$	An even indexed state waits for a specified time interval (parameter of the model); if it receives an expected answer within that time, it transitions to the next odd indexed state; otherwise (a timeout situation) it transits to the incomplete state
$\delta_{ext}(SN.,e,ans) \in$ $\{Success, Failure\}$ $ta(S_N) = T_N$ $\delta_{int}(S_N) =$ $Incomplete$	In the last state of the sequence and answer indicates either success or failure. Timeout is again to the incomplete state
$ta(Success) = 0$ $\lambda(Success) =$ $Activate$ $\delta_{int}(Success) = End$ $ta(End) = \infty$ $ta(Failure) = \infty$ $ta(Incomplete) = \infty$	Success outputs an activation signal and transitions to the passive end state. Failure and incomplete states are passive. None of these states accept input

S is the set of *sequential* states;
$\delta_{ext}: Q \times X \to S$ is the *external state transition function*;
$\delta_{int}: S \to S$ is the *internal state transition function*;
$\lambda: S \to Y$ is the output function;
$ta: S \to \Re_0^+ \cup \infty$ is the *time advance function*;
with $Q = \{(s, e)|s \in S., 0 \le e \le ta(s)\}$ is the set of *total states*.

Table 5.2 gives the definition of the sets and functions in the specification, and an example of an atomic model representing a pathway with one goal is given in Figure 5.4.

The model starts in state WA (for waitForActivate) which is passive (its time advance, ta is infinity). When an Activate is received (input ports are noted by "?", output ports by "!"), the model transitions to the Initialization state, I which is a transient state ($ta = 0$). This state immediately outputs the question, GoalReached and transitions to the state WG (waitForGoal). In this state, the model can receive

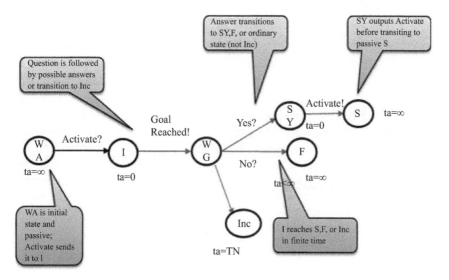

Figure 5.4 DEVS model of pathway with one goal

answers "Yes" or "No" and eventually enter passive states S (Success) and F (Failed) resp. (S is entered after an Activate output is generated from state SY.) However, WG has a finite time advance, TN, so that it transitions to states Inc (incomplete) if it does not receive one of the "Yes" or "No" answers within this interval. Since Inc is a passive state, it is easy to see that, as required, this simple model always winds up (and remains) in one of the three states S, F, or Inc.

Following the methodology presented in Figure 5.2, such a specification can be automatically transformed into a component of a simulation model and tested through simulation as well as implemented in actual software to function in an actual HDSoS setting.

5.5.4 Coupled pathways models

Coupling atomic pathway models to compose coupled models enables us to coordinate the behavior of multiple concurrent pathways. For simplicity in exposition, coupling will be limited to activations by one pathway of one or more others. The DEVS formalism's closure under coupling will assure that the resultant is a DEVS model. More than that, we can show that the resultant is also expressible as an atomic pathway model, establishing closure under coupling when restricted to the subset of DEVS defined as pathway models. The following property is essential to such closure:

Finite termination property:

- For any atomic pathway model, there is a finite time *T*, such that the model reaches, and passivates, in any one the three types of states: Success, Failed, or Incomplete within time *T* after initialization.

> • For any coupled pathway model, there is a finite time *T*, such that all the components of the model reach, and passivates, in any one the three types of states: Success, Failed, or Incomplete within time *T* after initialization.

Zeigler [19] proved the finite termination property and the closure of under coupling of pathway models. Examples of coupled pathway models are presented in the upcoming discussion. Many of the features discussed above are common to both coordinated care and clinical pathways commonly employed in hospital settings [29,33]. However, coordinated care pathways are focused on accomplishment of steps, with associated accountability and payment schemes. Consequently, they specify tests for accomplishment and time bounds within which such tests must be satisfied [18].

5.6   Quality of service measurement

As indicated, we must have working definitions of *quality of service*, and there must be *systems implemented to measure*, in an ongoing manner, the elements of clinical and extra-clinical interventions that can be aggregated to compute quality as defined. We turn to formalizing quality as the quotient of health value delivered divided by cost to deliver it.

5.6.1   *Porter's integrated practice unit*

As indicated, Porter and Teisberg [24] advocate radical reform of healthcare that requires that physicians reorganize themselves into IPUs moving away from care that is currently based on specialties with associated hospital departments (geriatrics, obstetrics, etc.). As formulated by Porter and Lee [34,35], an IPU is centered on a medical condition defined as an interrelated set of medical circumstances best addressed in an integrated way. Examples of IPUs are those centered on asthma, diabetes, congestive heart failure, and so on. These target a cluster of related adverse health conditions that includes the most common co-occurring complications. As such, the IPU may bring together a host of specialists and services needed to treat the target in an integral manner—as a team rather than as a collection of individual entities. This assemblage of individual independent entities into a single collaborative organization fits the pattern of SoS and motivates research to provide a firm basis for such integration. The IPU delivers all the services needed for the target condition, which are organized into an *end-to-end interaction* with the patient called a full cycle of care covering a care delivery value chain (CDVC). Here, "value" is defined as health outcome achieved per dollar of cost. The critical requirement is that such a metric be quantifiable, so that it can be compared by the patient—or surrogate payer such as insurance company—to the equivalent number offered by the competition. Much like the increase in value in a manufacturing process, the value chain is a linked set of activities that increase value (i.e., contribute to the outcome) from the initiation of the care cycle to its termination.

Table 5.3   Template for instantiating a care delivery value chain

Knowledge development ment					
Informing					
Measuring					
Accessing					
Monitoring					
Preventing	Diagnosing	Preparing	Intervening	Recovering/Re habilitating	Managing

5.6.2   Formal representation for care delivery value chain

The CDVC is formulated as a chain of activities that constitutes the architectural blueprint of the integrated team-based practice unit. From a SoS perspective, the CDVC specifies the organization of its components and their coupling. Adopting this perspective allows us to generalize the application of the CDVC concept to HDSoSs beyond the IPU. To do this, we interpret Porter's discussion of requirements for a properly constituted CDVC as follows:

- The set and sequence of activities are aligned with value—generally value should increase and cannot decrease, later activities cannot have lesser value than precursors. Taken together, the activities must achieve the desired outcomes.
- The activities have the right scopes to cover the target medical cluster of conditions and to minimally overlap.
- The activities form a coherent whole with seamless handoffs from one to the other—this will ultimately minimize process delays and "dropping the baton" [36].

Porter provides examples of CDVCs for particular targeted medical conditions, following the template below. The main value-producing activities are shown along the bottom row. Supporting activities are shown in the first column. In principle, any of the supporting activities can be paired with the main activities. Further, supporting activities can also operate across the full care cycle, e.g., knowledge development can concern the interrelationship of the main activities (Table 5.3).

The template will be illustrated in relation to the design of a coordinated care HDSoS for HIV–AIDS.

5.6.3   Pathways-based outcome measurement

As indicated, the CDVC enables computing the numerator in the value definition by defining how outcomes are produced. As shown in the template, the CDVC also includes measurement and other activities that cut across the outcome producing activities and that are capable of observing the behavior of the outcome producing activities. The key guiding principle is that "whatever is measured tends to improve" [35]. The denominator in the value quotient is the cost attributable to the activities that produced an outcome. It requires that the activities are sufficiently granular to support activity-based cost analysis.

We now consider how both numerator and denominator are formalized in our DEVS-based approach.

Porter's outcome measurement hierarchy [34,35] reproduced in Appendix D3 provides a comprehensive basis for the measurement system. The hierarchy has three tiers relating to health status, process of recovery, and sustainability of health. Each tier is has two parts. Tier 1 concerns survival and degree of health or recovery; Tier 2 concerns time to recovery and disutility of care or treatment process; and Tier 3 concerns sustainability of health or recovery including nature of recurrences and long-term consequences of therapy. Measuring the full set of outcomes that matter is indispensable to better meeting patients' needs and a powerful vehicle for lowering healthcare costs.

As illustrated in Figure 5.5, the form of the System Entity Structure (SES) shows the HealthCareSystem composed of three components: HealthStatusAchieved, ProcessOfRecovery, and SustainabilityOfHealth. Following Porter's approach, each of these is decomposed into the two types of measures illustrated in Figure 5.5. The basic event-based pathways models implement specific measures into the six slots to flesh out the full measuring system.

We employ the DEVS pathway representation for the measurement system along the lines of Porter's outcome hierarchy design approach. We can define a comprehensive set of outcome dimensions, and specific measures based on the event-based experimental frame methods implementable using DEVS.

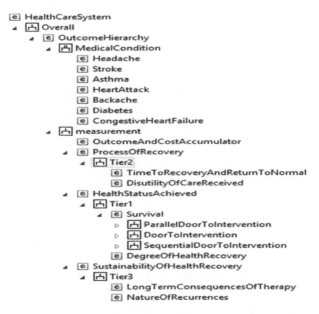

Figure 5.5   SES for outcome measurement hierarchy

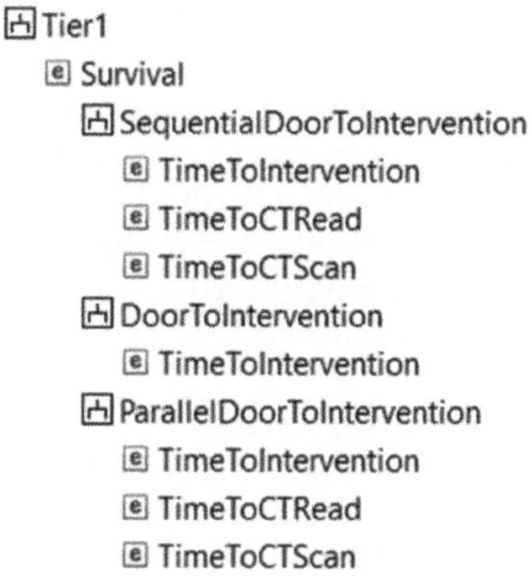

⊞ Tier1
   ⓔ Survival
      ⊞ SequentialDoorToIntervention
         ⓔ TimeToIntervention
         ⓔ TimeToCTRead
         ⓔ TimeToCTScan
      ⊞ DoorToIntervention
         ⓔ TimeToIntervention
      ⊞ ParallelDoorToIntervention
         ⓔ TimeToIntervention
         ⓔ TimeToCTRead
         ⓔ TimeToCTScan

Figure 5.6   Expansion of the SES of Figure 5.5

Following the Pathways Coordination Model allows tracking patients through the full cycle of care to accumulate actual costs of care (not how they are charged, currently often done in arbitrary fashion).

Figure 5.6 shows how the SES expands the Survival entity with alternative architectures for measurement of the time taken from entry of the patient to an intended intervention. Such alternatives and corresponding DEVS models are discussed in Section 8.

5.7 Pathways-based cost measurement for value-based purchasing

Pathways and activity-based evaluation of components provide a basis for aligning payment incentives for subgoal completion and replacement of fee-for-service with fee-for-performance (value-based payment).

Qualitatively, an activity is a label assigned to a state trajectory over an interval. Events that start and end of such activity cause discrete changes in the state of the system when formulated in discrete event terms. A quantitative measure of activity was provided by the framework presented [37]. In this approach, the *activity* of a DEVS model is simply measured by the count of its state transitions. Thus, as a DEVS model, activity of a pathway over a time interval is measured by the number of state transitions that occurred in the interval.

More generally, we stress the importance of activity-based costing—ABC [32] for the design of healthcare systems. Fixing a cost to the resources and services that are consumed by a particular activity is a main concern. To deal with this issue, costing systems are designed to support decision-making in business management.

Among the available accounting methods, ABC uses the concept of activity to estimate direct and indirect costs in the production of goods or services. ABC belongs to the set a cost estimation methods. It establishes a relationship between costs, products and services by identifying the cost drivers in a set of inter-dependent activities. ABC was revealed as complex to implement due to the dynamical nature of processes and resources spending.

Furthermore, the activities can change over time. Therefore, an improved version of ABC was proposed, the TDABC [32]. TDABC has been largely experimented in enterprises, and its effectiveness has been discussed for particular sectors, such as logistics. Recently, TDABC has successfully been tested in healthcare with an emphasis on value-based healthcare delivery at the scale of an entire pediatric hospital [38].

We can use the main ideas supporting the TDABC, and the weighted activity formulation proposed by Hu and Zeigler [37] in the DEVS framework. Duboz [39] illustrated this connection in the field of surveillance and control in animal epide-miology. In a discrete event system, the weighted activity is the sum of weighted state transitions over a time interval.

Making an analogy between TDABC and this definition, we consider the weight of the activity to be a measure of the cost of the activity. In addition, we consider the following two variables of the TDABC:

- The estimation of the costs per time unit to supply resources to the activities (the total overhead expenditure of a component divided by the total employee time available) and
- The amount of time necessary to carry out one unit of each kind of activity (as estimated or observed).

The second variable is computed by the DEVS time advance function. It depends on the state of the model or it can be a constant value. Similarly, the first variable of the TDABC can be a constant value changing from one simulation to another or a function depending on the state of the model. For both variables, the DEVS for-mulation adds some potential to the expressiveness of the TDABC model. The formalization of the weighting transition functions should be elaborate with the decision maker. We first estimate the costs per time unit, $c_r \in s$, for all the activities modeled as state transition function in the DEVS model. The weighting transition functions, w_t, can be specified as follows: $w_t : S \rightarrow R$.

From here, three different cases can be identified for this function. First, some transitions (activities) are performed at no cost, then $w_t(s) = 0$. Second, the cost can depend on the duration of the activity, i.e., $w_t(s) = f(s,e)$, with $c_r \in s$ the set of state, $f()$ a weighting function, and e the elapsed time in the activity. Third, the cost can be invariant in time, i.e., stationary, and depend only on the model state, then w_t $(s) = g(s)$, another weighting function. Finally, we can defined the total cost c of the activities in the system at time $t + e$ as follow: $c(t + e) = f(s, e) + g(s) + c(t)$. It is a strictly positive monotonic function. It is then straightforward to compute the cost growth rate α in any time interval $[t_1, t_2]$, for any components, with the following formulation: $\alpha = [c(t_2) - c(t_1)]/(t_2 - t_1)$.

The total cost c and the rate α can be used to compare the cost of activities in different components or DEVS models using simple ratios. The activity of the overall system is estimated by the aggregation of all individual pathway activities.

When activity is aggregated over all individuals that traversed a component, we get an estimate of the component's activity. These measures can be subindexed by pathway to rank the overall system activity from most active to least active pathway, thereby providing insight into how the system is being utilized.

Further subindexing by factors such as condition treated, patient attributes, source of client referral, enable analysis of the variation due to such factors [28]. Pathway activity can be correlated to personnel and resource expenditures to calculate costs using TDABC. Distributions of such activity can be used to inform continuous improvement as will be discussed soon.

5.8 Alternative component configurations

As indicated, there must be implemented systems that allow *alternative component configurations* (protocols, processes, and procedures) to be continually tested. The SES outcome hierarchy of Figure 5.5 offers an example of alternative architectures for "door-to-critical-interventions." These are shown as specializations for Survival in Figure 5.6 that can be selected as appropriate for different medical conditions. For example, a heart attack implementation [40] might use only a single atomic pathway model to measure door-to-balloon times and survival rates. In contrast, a stroke implementation might employ one of the sequential or parallel alternative architectures for its time-lost-is-brain-lost interventions. The SES supports automated generation of the SoS model once all selections have been made from component models in a model base.

Let us now provide more detail on such DEVS models, by considering more on the model of "time lost is brain lost."

The approach starts with the simplest, yet meaningful, example of DEVS measurement system. In the application to stroke, the activation event is the arrival of a patient at the door, and the goal is that the infusion of the blood clot breakdown agent, IV tPA starts before 60 min have elapsed since the patient's arrival. The measurement system must support detecting the events of patient arrival and infusion occurrence and increment the count of goal success if, and when, it does. The output is the measure of success—percent of patients receiving the injection in time.

Figure 5.7 shows a coupled model in which the atomic pathway model for "time lost is brain lost" sends Success or Failed outputs to an Accumulator model, which counts patients and percent of successful infusion of IV tPA before 60 min. Figure 5.7 shows the Accumulator as an atomic model in the state design graphical form supported by the simulation environment, MS4 Me [22].

Subgoals for the door-to-IV tPA process are formulated based on critical points in the process in which a CAT Scan is taken, read, and interpreted. As illustrated on the timeline in Figure 5.8, the subgoals CT Scan and CT Read can be established with benchmarks of 25 and 45 min expiration from arrival at the door. Note that the scan and read measures are process measures not outcomes of direct interest to the patient but that can help the organization meet its overall goal. Figure 5.8 also depicts an atomic pathway model with states for CT scan and CT read subgoals

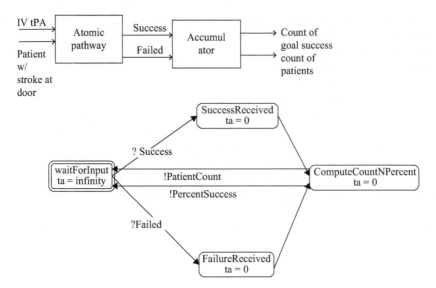

Figure 5.7   Measurement of pathway outcomes

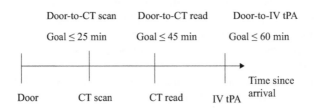

Figure 5.8   Subgoals for the door-to-IV tPA process

where the scan must happen within 25 min and the read must take place within 20 min later for success. Extending the model with a state for the goal of IV tPA infusion completes it.

The single atomic model, and its variants, is possible realizations of measurement for the door-to-IV tPA process. In addition, there can be coupled model alternatives, and some are sketched in Figure 5.9. Here, the components are pathway models for the individual subgoals and goal: CT scan, CT read, and IV tPA, respectively. These components can be coupled in series or in parallel as illustrated. The atomic model and coupled model alternatives for realization of the measurement can be considered as alternative architectures available to explore as design options for implementation.

In the parallel case, each of the times is measured from the arrival event at the door. In contrast, in the sequential case, these times are relative to the time of the earlier stage completion. Thus, the former (parallel) architecture allows independent, less error prone, measurement while the latter (sequential) one may be more natural to implement since it conforms to the care delivery pathways.

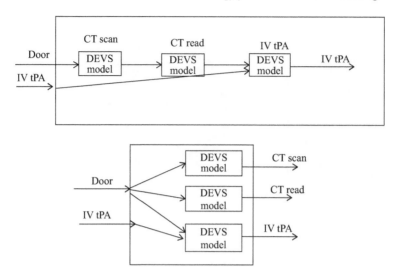

Figure 5.9   Alternative sequential and parallel coupled models for multiple pathways

Bradley *et al.* [40] researched strategies to reduce door-to-balloon time in acute myocardial infarction (AMI), including a time based goal of expecting staff to arrive in the catheterization laboratory within 20 min after being paged and the catheterization laboratory to use real-time data feedback. They found that despite the effectiveness of these strategies, only a minority of hospitals surveyed were using them—perhaps indicating the need for further automation such as the current approach can provide.

The three alternative architectures for what might be called door-to-critical-intervention can be represented as specializations in the SES of Figure 5.5 and can be selected as appropriate for different medical conditions. For example, a heart attack (AMI) implementation might use only a single atomic pathway model to measure door-to-balloon times and survival rates. In contrast, a stoke implementation might employ one of the sequential or parallel alternative architectures for its time-lost-is-brain-lost interventions.

5.9   Example: coordinated HIV–AIDS care system model

The continuity spectrum of HIV–AIDS intervention spans HIV diagnosis, full engagement in care, receipt of antiretroviral therapy, and achievement of complete viral suppression (Figure 5.10). However, Gardner *et al.* [41] estimate that only 19% of HIV-infected individuals in the United States have been treated to the point where their virus is undetectable. This occurs because achievement of an undetectable viral load is dependent on overcoming the barriers posed by patients "falling through the cracks" in traversing each of the sequential stages shown in Figure 5.10. The authors conclude that recognition of the "pipeline" and support for successful handoff of

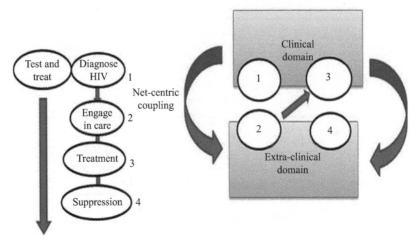

Figure 5.10   HIV–AIDS continuity of care pathways model

patients from stage to stage is necessary to achieve a substantial increase in successfully treated HIV population. Figure 5.10 depicts the stages of care continuity roughly assigned to both clinical and extra-clinical domains and that they alternate between the two domains (shown cycling from 1 to 4).

Here, we consider the approach of formulating the DEVS pathways discussed above for stages 1 and 3 to form an IPU. Also, DEVS pathways are proposed for stages 2 and 4 which are similar to those of the Pathways Community HUB. Using a DEVS coupled model, the clinical domain pathways are interfaced to the extra-clinical ones. The objective is that patients are handed-off from one DEVS pathway to the next without being dropped from care. Such cross-organization care pathways require sufficient EHR system and health IT networking support to track and monitor patients as they traverse the treatment pipeline [30,36]. Recall that this will require definition of goals and subgoals along paths to complete cross-system transactions, testing for achievement and confirmation of pathway goals and subgoals, tracking, and measuring progress of, patients along the pathways they are following, and maintaining accountability of compliance and adherence. The implementation of such IT can then provide a "dashboard" for viewing the overall disposition of patients through the complete cycle of continuity of care required for successful HIV–AIDS treatment.

Although formulated for IPUs, the criteria for a well-specified CDVC apply generally to healthcare SoS. Table 5.4 applies the criteria to provide a basis for achieving a CDVC for the HIV–AIDS example in Figure 5.10. The main value chain activities appear in this example as diagnosis, engagement, treatment, and suppression. They are organized in a sequence and must satisfy the criteria given for well-defined CDVCs in order to support value-based healthcare.

The generalization of value chain concepts from IPUs to HDSoS in general allows us to achieve a synthesis that applies to HDSoS with both clinical and

Table 5.4 Illustrating criteria for well-specified CDVC for HIV–AIDS

Criteria for well-specified CVDC	Application to HIV–AIDS
The set and sequence of activities are aligned with value	• Set: Diagnosis, engagement, treatment, suppression ○ Sequence: Shown in Figure 5.6 • Earlier stages must be completed before later stages • All four stages must be completed for positive outcome
The activities have the right scopes to cover the target medical cluster of conditions and to minimally overlap	• Diagnosis determines presence of HIV • HIV presence triggers engagement • Engagement enables treatment • Treatment enables suppression
The activities form a coherent whole with seamless handoffs from one to the other— minimize process delays and "dropping the baton"	• The sequence in Figure 5.6 is minimal connection at pathway level • Must be implemented faithfully with minimal delays at service level • Must assure transfer without dropping patient

extra-clinical aspects. The synthesis combines Porter's value-based concepts of CDVC and outcome hierarchy with pathways concepts that support implementation capabilities such as individual end-to-end goal-based tracking.

5.10 Pathways-based learning system implementation

Returning to the conditions that allow an HDSoS to continually self-improve, we have laid the foundation with a working definition of quality of service, DEVS pathway models for systems implemented to measure and compute quality of service in an ongoing manner, as well as systems that allow alternative component configurations (protocols, processes, and procedures) to be continually tested.

Finally, we noted that there must be systems to *correlate measured quality* with component configurations to provide evaluations that humans can employ to help select the most promising options. In this regard, continuous improvement in healthcare can be productively viewed as a specific kind of adaptation over time of a collaborative system whose components can take on alternative variants [42]. The goal is to keep improving the value (outcome/cost) of the system's CDVC by finding combinations of component variants that produce high value outcomes.

In the following, we present an approach based on the application of credit assignment and activity-based selection of component alternatives to successively increase the level of collaboration needed to produce progressively higher valued outcomes.

In this regard, Muzy *et al.* [43] identify the following three layers of an activity-based adaptive system:

- **Time-driven activity-based costing:** Using a built-in system for measurement of component activity and performance (outcome value).
- **Activity evaluation and storage:** Using the built-in detection mechanisms of level 1, activity can be measured as the fractional time that a component contributes to the outcome. Correlating contribution with outcome, a credit can be attributed to components. Such a measure of performance of components can be memorized in relation to the experimental frame, or context, in which it transpired.
- **Activity awareness:** Feedback of the activity–outcome correlation to inform the selection of combinations of component variants, so as to drive the system toward increased performance.

This kind of adaptive system differs from other simulation-based optimization systems (see e.g., [44]). It also differs from supervised learning (e.g., [45]) in that here learning is based on correlation between the activity of component systems and the behavior achieved at the SoS composition level.

Muzy and Zeigler [25] describe a system that implements these layers in a simulation that involves finding winning combinations of players on a hockey team. A simulation-based stochastic search is achieved based on the performance model-base. Correlation between the activity of a component and corresponding composition outcome is referred to the credit assignment problem. The credit of components is used to bias their selection.

Activity-based credit assignment (ACA) was shown to

1. apply to any level in the hierarchy of components within any experimental frame,
2. converge on good compositions much faster than a repository-based random search, and
3. automatically synthesize an SoS from a model-base, thus enabling reusability of highly rated components in compositions.

Pathway coordination models lend themselves to support critical features of such learning systems. As mentioned before, DEVS pathways enable TDABC based on pathway steps and their completion times. Recall that, as a DEVS model, the activity of a pathway over a time interval is measured by the number of state transitions that occurred in the interval. We can estimate a component's activity by aggregating the pathway activities over all individuals that traversed the component during an interval. Moreover, since pathways include outcome measurement, they enable correlation of activity and performance (CDVC value) for each individual. Aggregation over individual traversals of components yields estimates of activity–outcome correlation for components. Components or variants that do not perform well in this measure are candidates for replacement by other alternatives that can replace them. Such activity-based performance correlation and feedback exhibits the continuous improvement characteristic of an evidence-based learning health-care system advocated by Porter *et al.* (Figure 5.11).

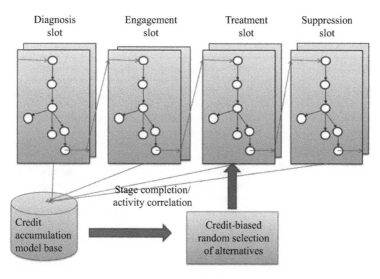

Diagnosis slot   Engagement slot   Treatment slot   Suppression slot

Figure 5.11   Activity-based credit assignment in the HIV–AIDS SoS

5.11   Activity-based credit assignment applied to HIV–AIDS example

Table 5.5, exemplifying features characteristic of continuous improvement in collaborative healthcare, sets up the basis for application of the ACA-based continuous improvement to the case of the HIV–AIDS continuity of care. As indicated the stages of continuity of care (diagnosis, engagement, treatment, and suppression) form a pipeline in that each stage must follow the prior one and set up the next one.

As in Figure 5.6, the stages can be viewed as component systems coupled together in the overall SoS which represents the pipeline. The stages are distinct from each other having different goals with alternative processes for stages being specialized to support the goals of the stage. Not all combinations of pipeline component variations will work to achieve the overall SoS goal of enabling patients to traverse the full pipeline, i.e., to receive the complete intervention required by continuity of care. Each patient constitutes the basis for a trial with the evaluation of a trial being the how many stages the patient successfully traversed.

The overall objective of continuous improvement is to increase the number of successful patient pipeline traversals, ideally to reach 100%, but with an objective of reaching a level of 65% [41]. Muzy and Zeigler [46] consider a family of pipeline coupled models with alternatives selected from an independent identically distributed random process. They proved that for such a pipeline, the activity-based credit assignment converges to an equilibrium distribution in which the best alternative at each stage has a credit that exceeds the others at that stage. This result offers an analytic confirmation to support the simulation results of Muzy and Zeigler [25] and a basis to propose that the implementation of a continuous

Table 5.5   Exemplifying features characteristic of continuous improvement in continuity of care

Characteristic feature	HIV–AIDS continuity of care manifestation
Component system	Stage of continuity of care (diagnosis, engagement, treatment, suppression)
Collaboration requirement	Stages form pipeline, each stage must follow the prior one and set up the next one
Modularity	Stages are distinct from each other having different goals and (conceptually) well-defined interfaces
Specialization	Alternative processes for stages are specialized to support the different goals of the respective stages
Variety at component level	A continuous improvement approach would seek to alter subprocesses and/or internal couplings (information flows) to provide requisite variety
Outcome quality of service	Not all combinations of pipeline component variations will work together
What constitutes a single trial	Time consumed by a stage in patient traversal of the pipeline
Evaluation of trial	The number of stages successfully traversed by a patient

improvement strategy based on ACA such as discussed above will prove successful in real application.

5.12   Relations to current research

An AHRQ/NSF workshop [9] envisioned an ideal healthcare system that is unlike today's fragmented, loosely coupled, and uncoordinated assemblage of component systems. The workshop concluded, "An ideal (optimal) healthcare delivery system will require methods to model large scale distributed complex systems."

Improving the healthcare sector presents a challenge in that the optimization cannot be achieved by suboptimizing the component systems, but must be directed at the entire system itself. On the other hand, healthcare has been compared to manufacturing with the premise that many of the same techniques can be transferred to it.

However, complex patient flows, numerous human resources, dynamic evolution of patient's health state motivated Augusto and Xie [2] to develop Petri-net-based software for modeling, simulation, and activity planning and scheduling of healthcare services. Their goal was to provide a mathematical framework to design models of a wide range of medical units of a hospital in order to model and simulate a wide range of healthcare services and organizations and to support such design with a unified modeling language/business process modeling interface for decision-makers.

In contrast, our concern here is not within the hospital, but at the SoS level where hospitals interact with other components such as physicians, community workers, social services, and health plan payers.

5.13   Summary

At the SoS level, care coordination is the organization of all activities, both clinical and extra-clinical, among the individual patient and providers involved in the patient's care to facilitate the appropriate delivery of healthcare services.

In this part, we expanded upon a SoSE formalization and simulation modeling methodology for a more in-depth application of DEVS coordination Pathways to re-engineer healthcare service systems. In particular, we discussed a concept of coordination models for transactions that involve multiple activities of component systems and coordination pathways implementable in the DEVS formalism. We showed how SoS concepts and pathway coordination models enable a formal representation of complex healthcare systems requiring collaboration, including Porter's integrated practice concepts, to enable implementation of his criteria for measurement of outcome and cost. This leads to a pathways-based approach to coordination of such systems and to the proposed mechanism for continuous improvement of healthcare systems as learning collaborative SoS. The DEVS comprehensive methodology, illustrated in Figure 5.2, underlies the development of pathways-based self-improving HDSoS architecture models. It proceeds along parallel paths of HDSoS simulation model development, coordination and learning sub-models, testing of the sub-models in the simulation model, and implementation of the sub-models within actual healthcare environments. The formalization in terms of DEVS provides enabled temporal analysis that would difficult to undertake with conventional biostatistics [47].

The US President's Council on Science and Technology [48] advocates that the US healthcare industry adopt a systems-engineering approach used in other industries to improve the health data infrastructure and boost overall quality and delivery of care. Our results support the contention that understanding healthcare as an HDSoS and applying SoSE methods based on simulation modeling helps to address these recommendations. The pathways and activity-based evaluation of components provides a basis for aligning payment incentives for subgoal completion. Distributed individual-based tracking enabled by pathways provide a basis for effective design of a health data infrastructure. Increased data supply on the community level enabled by extra-clinical pathway hubs and the analysis supported by M&S will enable better understanding of healthcare delivery as a self-improving service system [49].

References

[1]   Brailsford S.C. Advances and Challenges in Healthcare Simulation Modeling: Tutorial. *Proceedings of the 39th Winter Simulation Conference: 40 years! The Best Is Yet to Come*, Washington, DC, USA, 9–12 Dec 2007. pp. 1436–1448.

[2]   Augusto V., and Xie X. Modelling and Simulation Framework for Health Care Systems. *IEEE Transactions on Systems, Man and Cybernetics, Part A: Systems and Humans*. 2014; 44(1): 30–46.

[3]   Wickragemansighe N., Chalasani S., Boppana R.V., and Madni A.M. In Jamshidi M. (ed.). *Systems of Systems – Innovations for the 21st Century*. 1st Edition. Hoboken: Wiley; 2008. pp. 542–550.

[4]   Spohrer J., Maglio P.P., Bailey J., and Gruhl D. Steps Toward a Science of Service Systems. *Computer*. 2011; 40(1): 71–77.

[5]   Lally A., Bagchi S., Barborak M., *et al.* WatsonPaths: Scenario-Based Question Answering and Inference over Unstructured Information. *AI Magazine*. 2017; 38: 59–76.

[6]   Porter M.E. Value-Based Health Care Delivery. *Annals of Surgery*. 2008; 248(4): 503–509.

[7]   Rippel Foundation. *Can Simulation Modeling Fill Knowledge Gaps About Health Care?* Available at https://www.rethinkhealth.org/the-rethinkers-blog/can-simulation-modeling-fill-knowledge-gaps-about-health-care/ [accessed on 17 May 2018].

[8]   Bradley E.H., Elkins B.R., Herrin J., and Elbel B. 2011, Health and Social Services Expenditures: Associations with Health Outcomes. *BMJ Quality & Safety*. 2011; 20(10): 826–831.

[9]   Valdez R.S., Ramly E., and Brennan P.F. *Industrial and Systems Engineering and Health Care: Critical Areas of Research*. Rockville: AHRQ (Agency for Healthcare Research and Quality) Publication No. 10–0079; 2010.

[10]   NSF Workshop: Learning Health System. *Toward a Science of Learning Systems: The Research Challenges Underlying a National Scale Learning Health System*. Findings from a Multi-Disciplinary Workshop Supported by the National Science Foundation; 2011.

[11]   Zeigler B.P., Praehofer H., and Kim T.G. *Theory of Modeling and Simulation: Integrating Discrete Event and Continuous Complex Dynamic Systems*. 2nd Edition. New York, NY: Academic Press; 2000.

[12]   Ören T.I., and Zeigler B.P. System Theoretic Foundations of Modeling and Simulation: A Historic Perspective and the Legacy of A. Wayne Wymore. *Simulation*. 2012; 88(9): 1033–1046.

[13]   Wymore A.W. *A Mathematical Theory of Systems Engineering: The Elements*. New York, NY: Wiley & Sons; 1967.

[14]   Wymore A.W. *Model Based Systems Engineering*. Boca Raton, FL: CRC Press; 1993.

[15]   Mittal S., Zeigler B.P., Risco-Martin J.L., Sahin F., and Jamshidi M. 2008. Modeling and Simulation for Systems of Systems Engineering. In Jamshidi M. (ed.). *Systems of Systems – Innovations for the 21st Century*. 1st Edition. Hoboken, NJ: Wiley; 2008. pp. 101–149.

[16]   Boardman J., and Sauser B. System of Systems – The Meaning of "of". *IEEE/SMC International Conference on System of Systems Engineering*, Los Angeles, CA, USA, 24–26 Apr 2006. pp. 118–123.

[17]   Redding S., Conrey E., Porter K., Paulson J., Hughes K., and Redding M. Pathways Community Care Coordination in Low Birth Weight Prevention. *Maternal and Child Health Journal*. 2014; 18(6): 1–8.

[18] Zeigler B.P., Carter E.L., Redding S.A., and Leath B.A. Pathways Community HUB: A Model for Coordination of Community Health Care. *Population Health Management.* 2014; 17(4): 199–201.

[19] Zeigler B.P. The Role of Modeling and Simulation in Coordination of Health Care *Proceedings of the International Conference on Simulation and Modeling Methodologies, Technologies and Applications,* Vienna, Austria, 28–30 Aug 2014. pp. IS5–IS16.

[20] Soo K.B., Choi C.B., and Kim T.G. Multifaceted Modeling and Simulation Framework for System of Systems Using HLA/RTI. *Proceedings of the 16th Communications & Networking Symposium,* Society for Computer Simulation International San Diego, CA, USA, 7–10 Apr 2013, Article 4.

[21] Mittal S., and Risco-Martín J.L. *DEVS Net-Centric System of Systems Engineering with DEVS Unified Process.* Boca Raton: CRC Press (Taylor & Francis Series on System of Systems Engineering); 2012.

[22] Zeigler B.P., and Sarjoughian H.S. *Guide to Modeling and Simulation of Systems of Systems,* 2nd edition. Berlin: Springer; 2017.

[23] Seo C., Zeigler B.P., Kim D., and Duncan K. Integrating Web-Based Simulation on IT Systems with Finite Probabilistic DEVS. *Proceedings of the Symposium on Theory of Modeling & Simulation – DEVS Integrative M&S Symposium,* Alexandria, VA, USA, 12–15 Apr 2015. pp. 173–180.

[24] Porter M.E., and Teisberg E.O. *Redefining Health Care: Creating Value-Based Competition on Results,* 1st edition. Boston, MA: Harvard Business Review Press; 2006.

[25] Muzy A., and Zeigler B.P. Activity-Based Credit Assignment Heuristic for Simulation-based Stochastic Search in a Hierarchical Model-base of Systems. *IEEE Systems Journal.* 2014; 11(4): 1–14.

[26] Mesarovic M.D., Macko D., and Takahara Y. (eds.). *Theory of Multi-Level Hierarchical Systems.* New York, NY: Academic Press; 1970.

[27] Craig C., Eby D., and Whittington J. *Care Coordination Model: Better Care at Lower Cost for People with Multiple Health and Social Needs.* IHI Innovation Series white paper, Cambridge, Massachusetts Institute for Healthcare Improvement; 2011.

[28] Zeigler B.P., Carter E.L., Redding S.A., Leath B.A., and Russell C. *Care Coordination: Formalization of Pathways for Standardization and Certification,* Report for Project Health System Modeling and Simulation – Coordinated Care Example, National Science Foundation Grant Award No. CMMI 1235364; 2014.

[29] de Bleser L., Depreitere R., de Waele K., Vanhaecht K., Vlayen J., and Sermeus W. Defining Pathways. *Journal of Nursing Management.* 2006; 14(7): 553–563.

[30] Van Houdt S., Heyrman J., Vanhaecht K., Sermeus W., and De Lepeleire J. Care Pathways Across The Primary-Hospital Care Continuum: Using The Multi-Level Framework In Explaining Care Coordination. *BMC Health Services Research.* 2013; 13:296.

[31] Zeigler B.P., and Redding S.A. 2014. *Formalization of the Pathways Model Facilitates Standards and Certification,* Report for Project Health System

Modeling and Simulation – Coordinated Care Example, National Science Foundation Grant Award No. CMMI 1235364.

[32]   Robert S., Kaplan S., and Anderson R. Time-Driven Activity-Based Costing. *Harvard Business Review*. 2004; 2(11): 131–138.

[33]   Ozcan Y.A., Tànfani E., and Testi A. A Simulation-Based Modeling Framework to Deal with Clinical Pathways. In Jain S., Creasey R.R., Himmelspach J., White K.P., and Fu M. (eds.). *Proceedings of the Winter Simulation Conference*, Phoenix, AZ, USA, 11–14 Dec 2011. pp. 1190–1201.

[34]   Porter M.E. Measuring Health Outcomes. *New England Journal of Medicine*. 2010; 363: 2477–2481.

[35]   Porter M.E., and Lee T.H. The Strategy That Will Fix Health Care. *Harvard Business Review*. Oct 2013; 91(10): 1–19.

[36]   Rudin R., David S., and Bates W. Let the Left Hand Know What the Right is Doing: A Vision for Care Coordination and Electronic Health Records. *Journal of the American Medical Informatics Association*. 2014; 21(1): 13–16.

[37]   Hu X., and Zeigler B.P. Linking Information and Energy-Activity-Based Energy-Aware Information Processing. *Simulation*. 2013; 89(4): 435–450.

[38]   Yu Y.R., Abbas P.I., Smith C.M., *et al.* Time-Driven Activity-Based Costing: A Dynamic Value Assessment Model in Pediatric Appendicitis. *Journal of Pediatric Surgery* 52(6): 1045–1049.

[39]   Duboz R. Weighted Activity and Costing of Surveillance and Control in Animal Epidemiology. *Proceedings of the Activity-Based Modeling & Simulation Conference*, ITM Web of Conferences, Volume 1, 2013. Article 01002.

[40]   Bradley E.H., Herrin J., Wang Y., *et al.* Strategies for Reducing the Door-to-Balloon Time in Acute Myocardial Infarction. *The New England Journal of Medicine*. 2006; 355(22): 2308–2320.

[41]   Gardner E.M., McLees M.P., Steiner J.F., Del Rio C., and Burman W.J. The Spectrum of Engagement in HIV Care and Its Relevance to Test-and-Treat Strategies for Prevention of HIV Infection. *Clinical Infectious Diseases*. 2011; 52(6): 793–800.

[42]   Tao F., Laili Y., Liu Y., *et al.* Concept, Principle and Application of Dynamic Configuration for Intelligent Algorithms. *IEEE Systems Journal*. 2014; 8(1): 28–42.

[43]   Muzy A., Varenne F., Zeigler B.P., *et al.* Refounding of Activity Concept? Towards a Federative Paradigm for Modeling and Simulation. *Simulation*. 2013; 89(2): 156–177.

[44]   Swisher J.R., Hyden P.D., Jacobson S.H., and Schruben L.W. A Survey of Simulation Optimization Techniques and Procedures. *Proceedings of the Winter Simulation Conference*, Orlando, FL, USA, 10–13 Dec 2000. pp. 119–128.

[45]   Sutton R., and Barto A. Reinforcement Learning: An Introduction (Adaptive Computation and Machine Learning). Cambridge: MIT Press; 1998.

[46]   Muzy A., and Zeigler B.P. Conjectures from Simulation Learning on Series and Parallel Connections of Components. *Proceedings of the 26th European*

Modeling & Simulation Symposium, Bordeaux, France, 10–12 Sep 2014. pp. 550–557.

[47]   Seo C., Kang W., Zeigler B.P., and Kim D. Expanding DEVS and SES Applicability: Using M&S Kernels within IT Systems. *Proceedings of the Symposium on Theory of Modeling & Simulation – DEVS Integrative M&S Symposium*, Tampa, Florida, 13–16 Apr 2014, Article 7.

[48]   President's Council of Advisors on Science and Technology. *Better Health Care and Lower Costs: Accelerating Improvement Through Systems Engineering*. PCAST Report To The President, 2014.

[49]   Adler-Milstein J., Embi P.J., Middleton B., Sarkar I.N., and Smith J. Crossing the Health IT Chasm: Considerations and Policy Recommendations to Overcome Current Challenges and Enable Value-Based Care. *Journal of the American Medical Informatics Association*. 2017; 24(5): 1036–1043.

Chapter 6

Care coordination and pathways-based experimental frames

6.1 Introduction

Today's health information technology (IT) infrastructure remains largely a collection of systems that are not designed to support a transition to value-based healthcare. From the Systems of System (SoS) view, US healthcare, the most expensive in the world, has been described as an assemblage of uncoordinated component subsystems embedded in a market economy that promotes price setting by components independently and without reference to the end-to-end quality of care (and related costs) delivered to patients.

Provider organizations pursuing new models of healthcare delivery and payment are finding that their electronic systems lack the capabilities needed to succeed. Efforts are underway to increase the level of IT to improve patient recordkeeping and portability as well as the move to performance-based costing. Yet, such an IT infrastructure by itself will not provide significantly greater coordination since it does not provide transparency into the threads of transactions that represent patient treatments, their outcomes, and total costs.

Care coordination is the organization of care activities among the individual patient and providers involved in the patient's care to facilitate the appropriate delivery of healthcare services. In this book, we focus on the pathways model of care coordination [1–3], a construct that enforces threaded distributed tracking of individual patients experiencing certain pathways of intervention, thereby supporting coordination of care and fee-for-performance based on end-to-end outcomes.

Community care coordination works at the community level to coordinate care of individuals in the community to help address health disparities including the social barriers to health. Leath and Mardon [4] reported on a study that defined performance measures and tested them at several community care coordination sites. The project addressed the lack of performance measures in community care coordination and assessed, to some extent, the usefulness of the measures in helping to inform local quality-improvement activities. The study found that in many cases, documentation on which quality improvement could be based was scarce, and that client adherence to recommended activities was problematic.

The explicit formulation of problem-resolution processes, an essential element of the pathway care coordination concept, opens up possibilities for system-level metrics and analytics. These enable more coherent visualization of system behavior than previously possible, and thereby foster greater process control and improvement reengineering.

In this chapter, we review the pathways model and provide a formalization of pathways that serves as a basis for quality improvements in coordination of care. We explain how this formalization came about, expand on the pathway concept, and its formalization.

We also discuss metrics defined based on the formalization, and how such metrics can support simulation and analysis to provide insights into factors that influence care coordination quality of service, and, therefore, health outcomes. We conclude the chapter with a discussion of how the pathways model and its formalization will enable exploiting the emerging IT infrastructure to afford significantly greater coordination of care and fee-for-performance based on end-to-end outcomes.

6.2   Background: origins of the pathways model

The community pathways model was developed by the Community Health Access Project (CHAP) in Richland County, Ohio, USA, to improve health and preventive care for high-risk mothers and children in difficult-to-serve areas [3]. CHAP was formed to focus on pregnant women at risk of poor birth outcomes, and its pregnancy pathway was enhanced to work as a common outcome measurement tool in the seven community agencies designed to identify and engage pregnant women [2]. In Ohio, low birth weight (LBW) births represent only about 10% of all Medicaid births but account for more than 50% of all Medicaid birth expenditures. The pathways approach reduced the countywide LBW rate from 9.7% in 2005 to 8.0% in 2008 (Ohio Department of Health Data Warehouse). CHAP successfully demonstrated how the Pathways Community HUB Model worked with high-risk pregnant women in Richland County to improve both health outcomes and cost savings.

In the Community HUB Care Framework, the pathways system serves as the documentation and reporting tool that captures (but does not represent in a formal manner) each of a set of guiding principles—finding those at risk, ensuring that they are treated with evidence-based medical and social interventions, and measuring the health outcomes and costs of these efforts. These basic principles are applicable to the full range of coordinated care efforts.

The pathways documentation and reporting system, enhanced to work as a common outcome measurement tool, has been extensively employed by CHAP to identify and engage women with high-risk pregnancies.

6.3   Review: formulating and defining pathways

The pathways model is a tool that can coordinate the activities of otherwise uncoordinated agencies and services by focusing on the progress and outcomes of individual clients as they traverse such care organizations. As the pathways model is deployed to improve community health and social service outcomes, basic regional needs assessments, geo-mapping, and other data are evaluated to determine the areas of greatest need. The information required for these evaluations is readily available and often duplicative in many communities. As a result of these evaluations, pathways can be chosen to specifically address the highest priority health and social outcomes.

6.3.1 Pathway concept

Pathways are unique in that the outcomes are tracked at the level of the individual being served. Each step of the pathway addresses a clearly defined action toward problem resolution. Many steps deal with social and cultural issues, and these steps are just as important as the traditional activities of the health and human service systems.

Pathways have been developed for many issues, including homelessness, pregnancy, medical home, immunizations, lead exposure, childhood behavior issues, just to name a few. One client (or patient) may be assigned to many different pathways depending on the problems identified.

At first glance, pathways may resemble clinical guidelines or protocols. Coordinated care pathways are, however, different from clinical protocols and pathways in two essential dimensions: accountability and basis of payment. In a protocol, accountability is not taken into consideration in a specific sense. If the patient does not show up for follow-up appointments or the medication is not being taking correctly, then the provider is not held accountable as long as he or she followed the protocol.

This is not the case in a pathway. The pathway is not considered complete until an identified problem is successfully resolved. Conversely, at some definitive point, a pathway that has not been successfully completed must be closed in a documented fashion. Moreover, as indicated above, coordinated care pathways are associated with payment for specific benchmarks along the pathway with the highest payment provided for successful outcomes at completion, thereby linking payments to accomplishments.

As we shall see, care coordination pathways attempt to be analogous to skeletons showing paths and benchmarks rather than detailed handbooks of actions to achieve these benchmarks.

Upon enrollment, a client is interviewed with a checklist of questions that determine the set of pathways to be initiated. Pathways are predefined threads of transactions that the client is encouraged to carry out to achieve certain subgoals to improve the prospects of attaining the main goal. Pathways can coexist concurrently as illustrated in Figure 6.1. Pathways may be initiated and terminated as conditions require.

6.3.2 Pathway: initiation, action, completion, and closure steps

The structure of a pathway is summarized in Table 6.1. The following sections discuss each type of step in the structure.

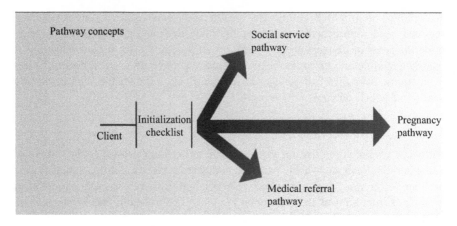

Figure 6.1   Illustrating the pathway concept

Table 6.1   Pathway structure

Initiation step	Defines the problem and target population
Action steps	Examples: High-risk pregnancy, asthma in poor control, lack of medical home, etc.
	• Provide standardized education to the client/family regarding the problem identified. Identify and develop a plan to eliminate identified barriers to receive services related to the problem
	• Assist client/family in identifying qualified provider or agency to resolve identified problem. This may include scheduling appointment, arranging transportation, submitting forms, etc.
Completion step	• Must be measurable outcome
	• Confirm resolution or significant improvement of identified problem (i.e., normal birth weight, improved control of diabetes, immunizations up to date) or
	• Confirm that client is receiving an evidence-based service proven to be effective in resolving or improving the identified problem (i.e., smoking cessation program)
Closure step	Conditions that justify closing a pathway in the absence of successful completion as measure by completion criteria. This may include time-based criteria such as expiration of time allowed time in which to reach successful completion

6.3.2.1   Completion step

Pathways are built from the objective back to the initiation. The completion step is the successful resolution outcome of an identified problem. This outcome must be a variable that can be measured. The completion step is clearly defined, easy to measure, and based on accepted criteria. When an agency or community meets to select pathways, the first task is to prioritize what the desired, measurable outcomes will be.

A pathway is not complete until the problem has been resolved. The completion step documents that the client has achieved all requirements for confirmed resolution or definitive improvement of the problem identified in the initiation step. The completion step clearly defines the desired positive outcome. Particular qualifiers of the completion step may be required.

These qualifiers may be stated as part of the completion step or be further described in the quality assurance manual that supports the pathways process. Examples follow:

- In an employment pathway, the client must remain employed for 1 month before the pathway is documented as completed. In the pregnancy pathway, the baby must be viable and weigh at least 2,500 g at birth.
- Completion steps must result in a defined positive outcome. For example, a client's receipt of a flyer on smoking cessation provides no evidence that this represents any defined positive outcome. The client must achieve some clear decrease in smoking or complete a training/treatment process that has been proven through evidence-based mechanisms to decrease smoking.
- If a client has been given bus tokens and a medical referral, these alone do not define positive outcomes, unless it is confirmed that the client was actually seen by a medical provider.

6.3.2.2 Closure step

It was quickly recognized that not all pathways would be successfully completed, and that for recording purposes, it would be important to officially recognize a negative outcome. Thus, an alternative to the completion step was added with an appropriate closure condition that would enable the tracker to close out the pathway. Clearly, the closure criteria must be well considered. For example, a "time out" condition is based purely on the lapse of a set amount of time without achievement of the objective. Such time outs apply to keeping of a variety of doctor's and other appointments. Such a time out can mark the closure of a pregnancy pathway for a client who has left the system for some reason unknown to the tracker.

On the other hand, the birth of a baby below 2,500 g is a definite event that marks the end of the pregnancy pathway and provides a closure step. Having one of, but not both, a documented completion and closure step is critical to being able to assess the effectiveness of a pathway application because otherwise there is irreducible uncertainty in the proposition of positive outcomes relative to the total. Furthermore, documented closure of pathways supports diagnosis of what might have been responsible for a negative outcome. Further, closure of a pathway in the hands of one agency may then give another agency a chance to complete the same pathway (see discussion of coordination below).

6.3.2.3 Initiation step

Once the completion step is clearly defined, the next pathway step to be built is the initiation step. The specific problem to be addressed, as well as the target population, is identified in this first step. The initiation step must clearly define who meets the criteria for the pathway. It is critical that the initiation step and completion step be carefully defined in order to maintain the accountability and credibility of the pathway. The information included in the initiation step may be further qualified by the quality assurance manual or guidelines. The initiation step must be easy to understand and specific as to the manner of documentation. In some cases, the initiation step is very straightforward, such as the client is unemployed or is pregnant. Some pathways benefit from the utilization of national guidelines or rating scales to define problems, such as out of control diabetes, hypertension, obesity, etc. Resolution of the identified problem will be documented in the completion step and the connection between initiation and completion must be clear.

6.3.2.4   Action steps

The next series of pathway steps are termed action steps. These steps are evidence-based interventions that build upon one another leading to a positive outcome. There may be up to five action steps before reaching the completion step. More than four or five action steps cause the model to lose strength in simplicity and increase the documentation requirements. When significantly more action steps are needed, more than one pathway may be needed. The action steps are ordered by priority. For example, if the first step in getting a child's immunizations up to date is believed to be educating the family about the importance of immunizations, then that should be the first action step. When the pathways coordinator/community health worker (CHW) is working through the pathway steps, the action steps may not be completed in series (one after the other). One of the key features of pathways reporting is finding the steps that took the longest to complete. These rate-limiting steps are the ones that may be delaying or restricting the pathways process. Addressing issues related to the rate-limiting-steps will often improve the outcome production process.

6.3.3   Example: pregnancy pathway

The pregnancy pathway offers an example. As discussed in more detail below, steps are represented symbolically and stand for events in the pregnancy process, as depicted in Table 6.2. Some further explanation is needed. The PREG1 and PREG2 steps must occur exactly once for each enrolled client. PREG1 occurs at, and records, the date of enrollment with the community care coordinator. PREG2 follows PREG1 as a step but may have an earlier date referring to the date at which the client first consulted with a physician concerning a pregnancy. The completion step PREG4 documents a successful outcome with a normal birth weight, while step PREG5 documents an LBW outcome; PREG6 is the closure step which, as described above, indicates the date at which the pathway was closed without a known outcome.

6.3.4   Initialization checklists

In addition to the release of information form, there are data collection forms that are key to deciding on services needed by the client and coordination of the

Table 6.2   Steps in the pregnancy process

PREG1	Pregnancy initiation date
PREG2	First prenatal appointment date
PREG3	Kept prenatal appointment date
PREG4	Delivery date, with weight $\geq 2{,}500$ g
PREG5	Delivery date, with weight $<2{,}500$ g
PREG6	Pregnancy pathway finished incomplete

agencies that provide such services. The Pathway Community HUB model is an extension that supports a higher level of coordination among agencies. As shown in Figure 6.1, an enrollment form developed by collaborative agencies captures the key pieces of information that all agencies will need, such as demographics, the agency enrolling the client, and date such information is submitted to the Community HUB. This form also serves as a request to initiate pathways, with the submitting agency indicating which pathways it would like to initiate. The HUB reviews the community database to determine if another agency is already working with that client on the issues identified.

6.4   Formalization of pathways

We now discuss formalization of the pathway concept, motivated by experienced limitations in current CHAP data infrastructure implementation. Formalization involves representing pathways in a symbolic form that lends itself to manipulation by well-defined logical and mathematical rules as well as simulate or animate them to generate and visualize their behavior.

6.4.1   Atomic model representation

Formalization will proceed by casting pathways as DEVS atomic models. We represent steps in a pathway as states in a DEVS atomic model. Such a representation can constrain steps to follow each other in proper succession with limited branching as required. Moreover, external input can represent the effect of a transition from one step to next due to data entry. Moreover, temporal aspects of the pathways, including allowable duration of steps, can be directly represented by the DEVS atomic model's assignment of residence times in states.

We will use the social service referral pathway (Table 6.3) as an example to illustrate the concepts.

As illustrated in Figure 6.2, the normal progression through the pathway begins with a starting state, Ref1, which is the state in which the atomic model is initialized.

When a social service referral appointment is made, this is considered an external event performed by the CHW and brings the model to state Ref2. When, and if, the appointment is kept by the client, it is also considered an external event and brings the model to the end state Ref3, where the pathway is completed. This is signified by the fact that there are no transitions out of the end state. Dates are

Table 6.3   Social service pathway

SSREF1	Social service referral initiation date
SSREF2	Social service referral scheduled appointment date
SSREF3	Social service referral appointment kept
SSREF4	Social service referral pathway finished incomplete

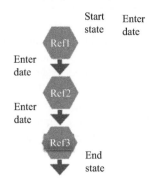

Figure 6.2   DEVS representation

associated with the states Ref1, Ref2, and Ref3, documenting when the pathway was established, the appointment for services was made, and finally kept. In the current implementation, these dates are entered in fields SSREF1, SSREF2, and SSREF3, respectively, of the client's record.

In a future implementation, these entries might be automatically "time-stamped" by the atomic model by pulling the current date from the calendar object associated with the application.

Figure 6.3 shows the atomic model extended to include the end state, Ref4, corresponding to the incomplete step SSREF4. This requires the CHW to close the pathway via an external transition to Ref4 by entering a date in states Ref1 and Ref2. This should occur when it becomes known that the pathway will not continue because an appointment will not be made, or because it has not been kept.

Figure 6.4 illustrates how formalization can employ the temporal properties of the DEVS atomic model to extend the features of pathways making them easier to manage. The dashed arrows signify internal transitions of the model which take states Ref1 and Ref2 to Ref1' and Ref2', respectively. Such an internal transition, or "time out," occurs when the time duration assigned to the state has expired, and there has been no external event to transition it to another state. For example, a time out value of 60 days can be assigned to Ref1, which would cause it to close out the pathway if there is no appointment scheduled within this period. Similarly, the time-out value assigned to Ref2 might be computed as the time until the scheduled appointment and for 10 days after it, which would automatically close the pathway after a reasonable time without having an external event indicating the appointment was kept to complete it.

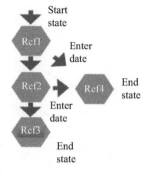

Figure 6.3 Extended DEVS representation

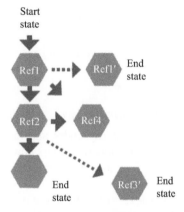

Figure 6.4 Further extended DEVS representation

Social service pathways may repeatedly be reopened to meet different needs, e.g., food, shelter, childcare, etc. In the current implementation, new entries are made with the same field labels but different dates to keep track of such repetitions. In the formalization, we support the creation of new instances of the atomic model to instantiate new pathways as required.

An atomic model representation of the pregnancy pathway is shown in Figure 6.5. It differs from the social service and other pathways in that only one instance is created for each client and to allow scheduling and recording repeated prenatal appointments during the life span of the pathway. This is represented in the model by the external transition taking the PREG3 state (for a completed appointment) back to the state PREG2R in which another appointment can be made. When (and if) this appointment is kept, the model returns to PREG3 from which the cycle can be repeated. The states PREG3 and PREG2R both have transitions that allow recording the outcome (normal, LBW, or unknown) of the pregnancy pathway.

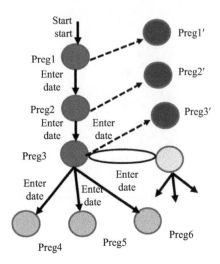

Figure 6.5   DEVS representation of pregnancy pathway

6.4.2   Experimental frames for pathway simulation

As will be discussed, counts of pathway codes and ratios are statistics that can provide client measures of a client's *adherence*, i.e., how well a client has followed the steps required of her. However, in a study of CHAP [1], we found that such counts can often provide inconsistent results.

For example, if a client's record indicates more appointments kept than made, what are we to make of her adherence? Having a formal specification of a pathway enables us to develop consistency requirements for client records resident in a database. Having such requirements, we can filter out records that do not meet such requirements and calculate adherence on the remaining subset. Further, the relative size of the remaining set is a measure of how consistent the data set is, and by implication, how correctly the CHWs are entering data into the database.

Filtering of data for input to experimental frames is illustrated in Figure 6.6. In the following, we discuss metrics for quality of reporting that produces measured pathway data.

6.5   Metrics for quality of reporting

Quality of reporting is measured by consistency and completeness of client records pulled from a database or generated by a simulation. Although simulation generation may not be prone to inconsistency or incompleteness, it may be useful to include filters for such attributes where CHWs who input such data are modeled or such input data is included in the simulation from uncontrolled sources.

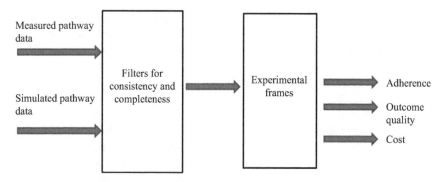

Figure 6.6   Filtering of data for experimental frames

6.5.1   Consistency of reporting

We start with pathways such as social service, medical referral, postpartum, and family planning pathways that have simple structures involving initiation, appointment making, and appointment keeping. Examining the state graph in Figure 6.3 reveals that the only way to get to the appointment kept state, Ref3, is by starting in initial state, Ref1, and traversing through appointment made state, Ref2. Thus, the following relation must be true:

$$\#Ref1 \geq \#Ref2 \geq \#Ref3$$

where, for example, #Ref1 denotes the number of codes of the form Ref1 in a client's record. Applying this relation as a criterion for consistency requires that the CHW to have officially entered a SSREF1 code with date before making an appointment and noting it in an SSREF2, finally entering an SSREF3 to document that the appointment was kept. Examination of the data revealed that often CHWs neglected to enter the SSREF2 but did properly enter the other codes. We realized that the SSREF2 entry is redundant, in that it can be implied by the SSREF3 entry (to keep an appointment implies that it had to be made.) Therefore, to maximize the filtered data while retaining records of interest, we imposed the following relaxed criteria for consistency:

A client's record is *consistent* if

$$\#Ref1 > 0 \quad \text{and} \quad \#Ref1 \geq \#Ref3$$

Relative to such a definition of consistency of a client record, we can define the consistency of a set of records as the percentage of consistent records in the set.

6.5.2   Completeness metrics

In the CHAP [1] study, we found that CHWs are prone not to complete the records they have started, thereby nullifying the dictum to have a definite outcome for each initiated pathway. For example, the pregnancy pathway is more complex than the single appointment pathways in that it has a multiple branching termination that provides information about normal or LBW. In this case, we are concerned that a pathway has been properly opened and closed. Referring to Figure 6.6, we make the definition:

A client's pregnancy pathway record is *OpenAndClosed if*

$$\# PREG1 > 0 \text{ and } \#PREG4 + \#PREG5 + \#PREG6 = 1$$

This requires that the pathway record has been initiated (#PREG1>0) and that at exactly one of the termination steps, PREG4 (normal), PREG5 (LBW), or PREG6 (incomplete) has been taken. The closure percentage of a set of records is then the percentage of pregnancy pathway records that satisfy the OpenAndClosed criterion.

For the pregnancy pathway, we are interested in the records that convey a definite result (normal or LBW). This motivates the definition:

A client's pregnancy pathway record is *complete* if

$$\#PREG4 + \#PREG5 = 1$$

This requires that exactly one entry for either PREG4 or PREG5 has been made.

The *completeness* of a set of records is then the percentage of pregnancy pathway records that satisfy the completeness criterion.

6.5.3   Adherence metrics

With the consistency filter producing a subset of reliable records, we can measure client adherence in the social service and other pathways. This leads to the definition for adherence in such pathways: Adherence of a client in a social service pathway satisfying Figure 6.5 is measured by

$$\text{Adherence} = \begin{cases} \dfrac{\#Ref3}{\#Ref1} & \text{if record is consistent} \\ 0 & \text{otherwise} \end{cases}$$

This defines adherence as the fraction of initiated pathway repetitions that were successful in keeping appointments (this does not use the number of appointments made as the denominator, per previous discussion).

The *adherence* of clients in a set of records is the sum of their adherences (as measured) divided by the number of consistent records. Note that if all clients are fully adherent, then the adherence of the set is unity (or 100%).

6.6   Exemplary data analysis

In the study of CHAP [1] mentioned before, we employed the metrics just discussed to analyze data obtained from the CHAP database over a period of several years. The results illustrate the experimental frames and data analyses that can arise from modeling and simulation of care coordination employing the multi-perspective modeling framework.

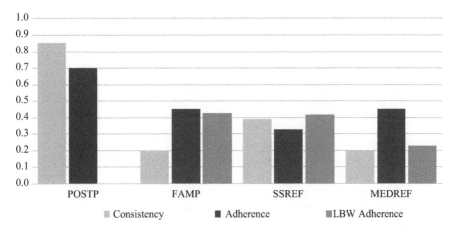

Figure 6.7   Consistency and adherence measurements

6.6.1   Consistency and adherence

The leftmost bars in Figure 6.7 show values obtained by applying the consistency metric to the pathways shown for the data set previously described.

There is a large variance between the consistency of the postpartum pathway at 86% and the others which range from 20% and 40%. The adherence values for the data set as restricted to the consistent subset are shown in the middle bars. These range from 33% to 69%.

Adherence levels are also shown for the subset of LBW outcomes in the rightmost bars (there are no postpartum pathways for LBWs). While there is little difference between the overall and LBW adherences, there are noticeable differences, in opposite directions, between the social service referral and medical referral pathways.

6.6.2   Pregnancy pathway closure and completeness

Table 6.4 shows the overall and LBW subsets' measures of closure and completeness. Although both are relatively high levels, the LBW subset has a noticeably higher level (see Section 6.6).

6.6.3   Reporting competence analysis

The importance of reporting metrics such as consistency in pathway reporting has been mentioned above. To summarize, the consequence for data analysis is the more the data satisfy consistency criteria, the greater are the useful data that remain after elimination of the invalid records. For example, at a consistency level of 20% for the medical referral pathway, only one-fifth of the records can be used for computing client adherence in that pathway. Moreover, discrepancy in consistencies values across pathways can lead to examination of plausible explanations for the observations. For example, the large variance between consistency of reporting for the

Table 6.4   Overall and low birth weight (LBW) subsets measures of closure and completeness

Subsets	OpenAndClosed fraction	Completeness fraction
Overall	0.77	0.80
LBW	0.89	0.89

postpartum pathway and the others in Figure 6.7 called out for an analysis of the CHAP operations that might explain the results. In fact, the Ohio Medicaid Managed Care Plans (MCPs) contract with CHAP to outreach to high risk pregnant members and coordinate their care. The contracts are designed as pay-for-performance with payments tied directly to pathway completion. The MCPs are most interested in contracting for pathways that align with Healthcare Effectiveness Data and Information Set (HEDIS) measures.

The HEDIS is a tool used by health plans to measure performance. For example, the HEDIS measure for postpartum care is the percentage of deliveries that had a postpartum visit on or between 21 and 56 days after delivery. MCPs contract for the postpartum pathway, because it is a measure that is important for their own performance. This graph shows the difference between a pathway that is linked to payment (postpartum) and three pathways that are not directly tied to payments (family planning, social service referral, and medical referral). If a pathway is linked to payment, then there are several reviews done at the agency by both the clinical and financial staff. Pathways need to be confirmed—all steps— prior to invoicing the MCPs.

The results for the pregnancy pathway in Table 6.2 indicate a relatively high degree of reporting competence for that pathway. Since this pathway is central to CHAP's primary objective of reducing LBW outcomes, and its evaluation for this result, the high competence accords with the importance of the pathway. That the LBW subset indicates a higher competence accords with the observation that the CHWs who care for the population at highest risk for LBW are among the most experienced.

Thus, the reporting competence results suggest that pathways may receive different levels of reporting attention as a result of several factors. However, per-haps the most surprising result is that the existence of standard and reimbursable metrics for some pathways may distort the balance of effort to those pathways, draining effort from other pathways that are not standardized but may be equally important to attain the overall objective. Our formalization can contribute to stan-dardizing such pathways by providing a well-defined basis for defining them and the associated metrics for reporting quality. Furthermore, the ability to monitor CHW effort based on such metrics can support the development of incentive schemes for CHWs that encourage higher levels of reporting. Finally, the for-malization in the form of DEVS models as given above supports implementation of more computerized assistance to CHWs in executing their reporting responsibilities.

Many of the mistakes that are possible with the current manual implementation can be obviated with the active calendar implementation that employs atomic model instances to control the possible next steps and provides automatic time outs and reminders on pathway closures.

6.6.4   Client adherence analysis

Leath *et al.* [4] reported on a study which defined performance measures and tested them out at several community care coordination sites including CHAP. The project addressed the lack of performance measures in community care coordination. The study's use of community-based participatory research and other scientific-based approaches to measure development is a major contribution to the field of care coordination and assessed, to some extent, the usefulness of the measures in helping to inform local quality improvement activities. Most of the measures were implemented as survey instruments rather than as measured directly from operational data as done here. The most relevant non-survey metrics are related to developing and maintaining a care-coordination plan, healthcare referral scheduling, and healthcare referral completion. There were insufficient pilot test data to report on the care-coordination measure. For the care-coordination plan, few records were found complying with these activities, and sites that reported care plans under development did not actively document this task—a requirement per measure specifications.

Results are presented for timeliness of referral scheduling and completion. Quoting the report

> More than 40% of referrals (40.8%, n = 1,157) were completed within 14 days and nearly two-thirds were completed within 30 days (Figure 3). Most of these completed referrals were for primary care visits. The completion rates were lower for mental and behavioral health services as well as other types of specialty care. Of the 101 clients who did not complete their referral, more than half (58.4%) did not appear for a scheduled appointment. This measure highlighted a documentation challenge for some of the sites. The measure specification requires confirmation from the healthcare provider to demonstrate the completion of the referral. Yet in practice, some sites relied on client confirmation.

Our results are consistent with these findings as they show that adherence is a problem with even a lower adherence rate. However, more significantly from a methodological standpoint, we provide a formalized approach to obtaining adherence values from primary pathway data with well-defined consistency filtering. This enables adherence measurement to become a standardized feature of care coordination based on pathways and supported by computerized implementation as discussed soon.

6.7   Health outcome metrics

The completeness metric developed for the pregnancy pathway illustrates the need for filtering client records to eliminate certain end states where appropriate.

The tools developed enable outcome analysis of subsets of clients subject to the filtering criteria just discussed and further constrained by various attributes available on client profiles in the data. Such analytics and the results of application to CHAP data are described in the CHAP technical report [1]. These results lead to the conclusion that the percentage of normal births at the output of the pregnancy intervention process appears to be more appropriate than the percentage of low weight births as an effectiveness measure. The normal births percentage takes account of both low weight births and undocumented cases and clients that left before completion of care.

Using this approach, we examined how outcomes vary in client subsets such as those based on race and age, as well as from external sources such as referrals and payer contracts. We found that the use of metrics of CHW and client performance such as activity, consistency, and adherence seem to verify anecdotal observations that different client streams receive differing levels of quality of care. The local minority care coordinators are able to reach out to African-American women and engage them quickly into the program. This contrasts with the administrative delays that reduce the time available for managed care clients to receive proper intervention. This tends to corroborate our conjecture that different risk policies and client-processing times of such sources can negatively influence pregnancy outcomes.

6.8   Pathway-based systems operational metrics

6.8.1   *Pathway activity distribution*

According to our data, about half of the pathways constituted most of those employed in practice in the years 2009–13. The total activity of a pathway is computed as the total number of events recorded for that pathway's overall clients during the period of interest. The mean activity is the total activity divided by the number of clients in that period. As shown in the chart of Figure 6.8, the pregnancy pathway is by far the most active (as is to be expected), while of the four other pathways, the most active are postpartum, family planning, medical referral, and social service referral. These and other results shown in this paper are based on the actual set of clients ($N = 262$) resident in the database from 2009 to 2013.

Pathway activity can be correlated to personnel and resource expenditures to calculate costs using time-driven activity-based costing [5]. Distributions of activity such as in Figure 6.8 (based on our analysis of CHAP data [1]) can be used to inform quality-improvement planning.

6.8.2   *Pathway-duration distribution*

Another measure that is computable is the duration of a pathway instance; this is the difference between the dates of the initiating step and the completion step (either successful or not) or the closure step. Figure 6.9 orders pathways according to the mean durations of pathways (pathway durations averaged over clients

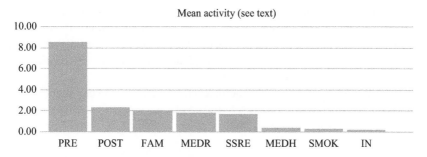

Figure 6.8   Pathway activities ordered from largest to smallest

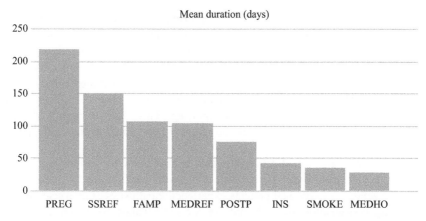

Figure 6.9   Pathway durations ordered from largest to smallest

participating in pathway). Note that the durations of the pregnancy pathway relate to the length of pregnancies but include the effect of the pregnancy stage at which the client enrolled. Deeper analysis of the activity and duration data probes the dependence on client profile and can yield insights into the program operation [5].

The longest duration pathways in Figure 6.8 are the same high activity pathways in Figure 6.9. However, the order in which they are placed reflects the natural processes (e.g., the postpartum pathway is very active as it is required of all clients), but the required appointments take only a month or two to accomplish.

6.8.3   Temporal metrics

As is clear from the formalization given above, the pathway model associates dates with pathway steps. Therefore, it enables metrics, such as the pregnancy pathway duration, to be defined based on differences between dates. Metrics involving time difference between successive dates can answer questions about how long it takes for clients to make and keep appointments, or viewed from the supply side, how long it takes to deliver services of various kinds. Alternatively, dividing the number of successive events within a time interval by the length of the interval gives rates of event occurrence. Rate-limiting steps, mentioned in Section 6.3.1, can be identified by criteria which set thresholds below which rates are considered to be too low. Warnings and alerts can be generated automatically based on the active calendar implementation of the pathway model.

More in-depth analysis of time-indexed events can reveal more in-depth understanding of pathway processes. For example, the number of different pathway types that are concurrent for a client can be computed by observing multiple pathway time series during the same period.

6.9 Summary

This chapter provided a brief review of the original formulation of the Pathways Community HUB model. This review set the context for the multiperspective modeling and simulation framework and its support of pathways-based care coordination. We then discussed the DEVS formalization of pathways in more depth and showed how pathways serve as a basis for quality improvements in coordination of care. The DEVS formalization provides a firm basis for computerized support for monitoring, tracking, and reporting. This leads to improved client adherence to their assigned activities and improved coordination among the payers and agencies involved.

More generally, formalization provides a firm basis for capitalizing on the transparency that is afforded by the pathways model which enforces threaded distributed tracking of individual patients experiencing pathways of intervention, thus supporting coordination of care and fee-for-performance based on end-to-end outcomes.

Based on the formalization and applied to data from an actual HUB, we showed how metrics can be defined that provide insight into the effectiveness of the HUB, the means to measure such effectiveness, and its dependence on various factors of HUB operation.

References

[1]   Zeigler B.P., Carter E.L., Redding S.A., Leath B.A., and Russell C. *Care Coordination: Formalization of Pathways for Standardization and Certification*, Report for Project Health System Modeling and Simulation – Coordinated Care Example, National Science Foundation Grant Award No. CMMI 1235364; 2014.

[2]   Agency for Healthcare Research and Quality. *Pathways Community HUB Manual: A Guide to Identify and Address Risk Factors, Reduce Costs, and Improve Outcomes*. AHRQ Publication No. 15(16)-0070-EF, Rockville, MD, USA, January 2016.

[3]   Pathways Community HUB Institute, Community Care Coordination Learning Network. Connecting Those at Risk to Care: The Quick Start

Guide To Developing Community Care Coordination Pathways – A Companion to the Pathways Community Hub Manual. AHRQ Publication No. 15 (16)-0070–1-EF, Rockville, MD, USA, January 2016.

[4] Leath B., Mardon R., Atkinson D., *et al. Advancing Quality Improvement Efforts Through the use of Standardized Community Care Coordination Performance Measures*. WESTAT Report to AHRQ. National Institutes of Health Grant Award No. 1RC2MD004781–01, 2012.

[5] Robert S., Kaplan S., and Anderson R. Time-Driven Activity-Based Costing. *Harvard Business Review*. 2004; 2(11): 131–138.

Chapter 7

Pathways-based care coordination simulation example

7.1 Introduction

For the foreseeable future, healthcare services quality improvement will depend on implanting health information infrastructures that supports human decision-making about protocols, processes, and procedures that work together to support value-based delivery of services. The formal representation of pathway structures (Chapters 5 and 6) provides a lens to examine the data, obtain insights, and make recommendations for improvements in coordination.

Based on this formalization, this chapter will discuss a simulation framework to guide design, development, and evaluation of architectures for pathways-based coordination of care. A simulation model will be presented that exemplifies this framework and is intended to predict return on investment (ROI) for implementing pathways-based coordination as well its sustainability over the long run.

In order to more fully elaborate the framework to support, we point to needed expansion of the model to include dimensions such as client risk characteristics, the referral source of clients to the coordination program, the effect of incentives on service workers' performance, and alignment of pathways with payments.

7.2 Simulation model framework

Recall from Part I that we view coordination of care as a component of a larger system that includes healthcare system, viewed as an assemblage of medical and related social services. We treat patients as agents traversing the healthcare system interacting with its agencies and services. Our models can represent this interaction in the absence of the coordination of care component to establish a baseline from which the effect of coordination on quality and cost of care can be measured. The models also support inclusion of alternative care coordination architectures that can be simulated and tested against the baseline in the same metrics of quality and cost.

Four primary components to coordination of care are as follows:

- Identification of patients for treatment—often phrased as the "hotspot" problem of identifying the 20% of the population that run 80% of the cost. The outlines of such problematic populations are generally clear [e.g., women at risk for low birth weight (LBW) pregnancies]; however, identification "profiles" that are both outcome and cost effective are more difficult to come up with.

- Compliant scheduling—assuring that patients show up to medical and social service appointment.
- Medication reconciliation—assuring that the patient takes the right medicine at right time—more complex than it appears at first glance, since multiple participants (primary and secondary physicians, pharmacists, service navigators, and patients) may have different lists of medicines for the same patient.
- Care management—management of the patient's health condition—beyond taking their medications are patients following the doctors care plans (diet, exercise, etc.) and action plans (e.g., for asthma) on a continuous basis between episodic assessments (e.g., visits to the doctor).

7.3   Evaluation framework for coordinated care architectures

A framework for modeling of coordinated care architectures aims to include components for such aspects and allows their interactions, efficacies, and costs to be assessed in combination. In particular, we are here concerned with architectures modeled after the pathways model for community care coordination.

Our framework for simulation evaluation is illustrated in Figure 7.1 which refines the overview of coordination of Health Delivery SoS (HDSoS) developed previously.

Here, we conceive of coordination as a layer added to the already existing health information technology that links patients and the HSSoS.

The evaluation is viewed as implemented in one or more experimental frames that implement the following capabilities:

1. Generate target population with appropriate physical, behavioral, and socio-economic characteristics;
2. Track patients through system with and without coordination of care;
3. Evaluate cost of coordination as addition to total cost;
4. Evaluate improvement in outcome and effect on reduction in total cost; and
5. Compute ROI = cost saving per dollar of coordination cost.

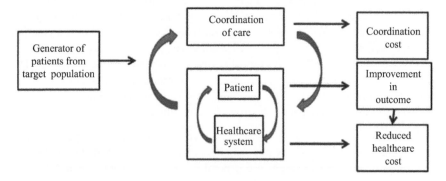

Figure 7.1   Generic framework for evaluation of coordination of care

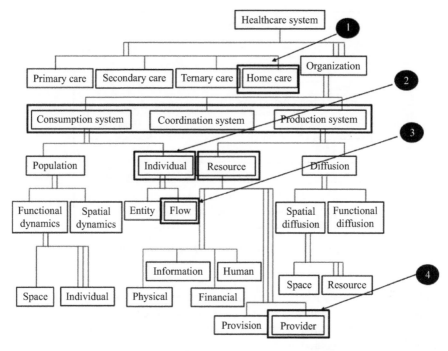

Figure 7.2   Pruning of the SES ontology for healthcare system

From the perspective of the System Entity Structure (SES) ontology and model-base for healthcare of Chapter 1, the just given framework for evaluation of coordination of care is derived from pruning of the SES shown in Figure 7.2.

Home care is selected as the type of care under consideration in step 1. A single form of organization is considered with individual selected from consumption system in step 2, and flow selected from individual in step 3. Multiple providers are selected from resource under production system in step 4. Although not shown, the type of provider under focus is the community healthworker (CHW) and the coordination system under consideration is based on pathways.

The resulting coupled model corresponding to the pruned entity structure of Figure 7.2 is shown in Figure 7.3. It shows that the Discrete Event System Specification (DEVS)-coupled model contains two types of components, resource allocation models and individual based models, populated by agent-based patient flow models and agent-based CHW models, respectively.

7.4   Pathways home care coordination in low birth weight prevention

We now consider a simulation of an HDSoS that accommodates this frame and demonstrates how smart analytics can be derived from it. The Community Health Access Project (CHAP) [1] utilizes CHWs to identify women at risk of having poor

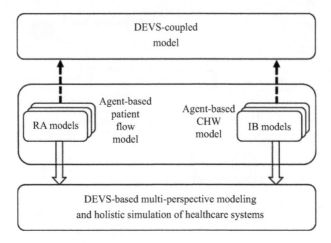

Figure 7.3   Pruned model for evaluation of care

birth outcomes and connects them to health and social services. CHWs are trained individuals from the same highest risk communities. The CHAP pathways model is used to track each maternal health and social service need to resolution, and CHWs are paid based upon outcomes.

Recent studies of CHAP [2] produced evidence on the effectiveness of CHAP's coordination of home visiting in addressing poor birth outcome, including LBW.

The generic capabilities depicted in Figure 7.1 are given concrete form in the following functionality:

1. *Generate* agents representing pregnant women patients from distribution with given characteristics: compute baseline LBW outcomes for optimally matched population from census data;
2. *Track* patients through system with CHW-implemented pathways coordination;
3. *Evaluate* cost of CHWs based on outcome-based compensation;
4. *Evaluate* outcome improvement versus baseline—simulation generated LBWs: evaluate reduction in total cost as savings from reduced LBW pregnancies; and
5. *Compute* ROI = cost saving per dollar of coordination cost: assess sustainability of implementation.

7.4.1   Model description

The model we developed [3] uses hierarchical modeling methodology to evaluate the ROI and sustainability of coordinated pathway-based intervention in LBW pregnancies. The top-level components of the model are detailed in Table 7.1. The appendix to this chapter (Appendix A) provides examples of model and SES specifications in MS4 Me [4].

Table 7.1   Description of top-level components

Component	Description
Patient	Generates patient agents to enroll in CHAP (if it is included) at one of three trimesters—i.e., immediately, after 90 days, or after 180 days. The probability distribution is taken from the study as described in the text
CoordinatedHDSoS	Is itself a coupled model constructed along the lines of its name; further detail is given in Table 7.2
EFCoordinationCost	Computes the cost of coordination by tracking the number of CHWs used and multiplying by the cost per CHW taken from the study as described in the text
EFOutcome	Tracks the outcome of pregnancies as LBW, Normal, or Normal-GivenCHAP (LBW was averted due to participation in CHAP)
EFOutcomeCost	Computes the cost of outcomes using the excess cost of LBW as described in the text
EFROI	Computes the ratio of inputs from EFOutcomeCost and EFCoordinationCost which can be interpreted as return on investment (ROI)

The description of the core model component, CoordinatedHDSOS (medical and social, services within CHAP), is given in Table 7.2.

The basic flow of activity involving coordination in the top-level model is shown in Figure 7.4.

A patient starts by sending ConsumeHealthCare and EnterCHAPTrimester messages to CoordinatedHDSOS. These messages are channeled to, and consumed by, the internally contained MedicalNSocialServices and CHAPEnrollment components (Figure 7.5). The latter notifies the EFCoordination of the CoordinationLevel (related to the trimester of entry into CHAP) as well as notifying the CHW component. This activates a CHW and sends a ConnectToServices message to MedicalNSocialServices. This message notifies the latter that CHAP is providing coordination for the patient with potential to alter the birth outcome.

The outcome of coordinated intervention, cost of coordination, and cost savings are sent to the value computing component, which here computes ROI. The decision flow embedded in the MedicalNSocialServices model is depicted in Figure 7.6.

The arrival of the patient (signaled by the ConsumeHealthCare message) triggers the determination of Normal or LBW outcome according to a random selection mentioned earlier made with probability P_{LBW}. If the patient has been enrolled in CHAP then a second random selection is made with probability $P_{Normal/LBW}$.

The birth outcome probabilities are estimates related to quantities taken from the CHAP studies [5] and summarized in Table 7.3.

7.4.2   Return on investment

The frame component EFROI (Experimental Frame for ROI) computes the ratio of outcome savings due to coordination cost. The coordination cost is defined as the

Table 7.2   Description of components of the core model component

Component	Description
CHAPEnrollment (present when coordination is included)	Generates patient agents to enroll in CHAP (if it is included) at one of three trimesters—i.e., immediately, after 90 days, or after 180 days. The probability distribution is taken from the study as described in the text
CHW (present when coordination is included)	Is itself a coupled model constructed along the lines of its name; further detail is given in Table 7.3
MedicalNSocialServices	Represents the medical and associated social services available to the patient. In the absence of coordination by CHAP, the outcomes of pregnancies are determined as LBW or Normal using a probability taken from the study as described in the text. In the presence of CHAP, the probability that an LBW pregnancy is converted back to NORMAL is taken from the study as described in the text

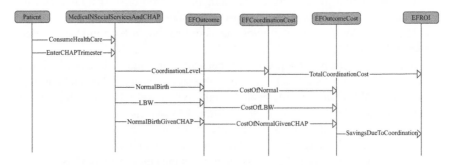

Figure 7.4   Sequence diagram of activity flow in top-level model

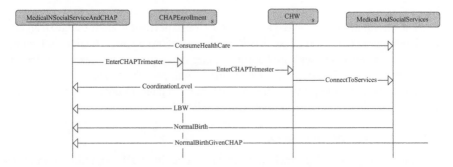

Figure 7.5   Sequence diagram of activity flow in CoordinatedHDSOS model

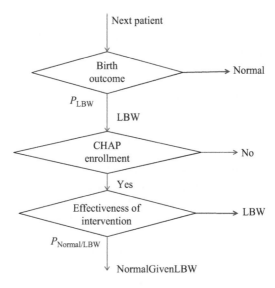

Figure 7.6   Decision flow for birth outcomes

Table 7.3   *Parameter value estimates*

Parameter	Value	Literature source [5]
Trimester of entry	Distribution of trimester of entry: first—0.56%, second—0.20%, and third—0.24%	The greatest cost of the program was time spent by a CHW to provide care coordination, and the amount of time spent by a CHW was primarily driven by trimester of entry into CHAP
CHAP cost per pregnant client	$751/patient	The estimated cost to provide pathways community care coordination
Probability of LBW pregnancy	$P_{LBW} = 0.13$ $P_{Normal/LBW} = 0.47$	We identified 115 CHAP clients and 115 control births. Among the intervention group, there were seven LBW births (6.1%) compared with 15 (13.0%) among non-CHAP clients

product of the number of CHWs in the program and the cost per CHW (this is the largest component of cost). Taking the number of CHWs per patient as fixed, the coordination cost is actually computed as the number of patients enrolled in CHAP times the cost per patient taken as $751. The cost of savings due to coordination is taken as the excess cost incurred due to a LBW birth (relative to a normal birth) multiplied by the number of NormalGivenLBW births (those averted by CHAP intervention). Quoting the study [5], the Institute of Medicine estimates that in the first year of life, excess medical expenses per LBW infant were $29,000 and long term costs (including maternal costs, early intervention, special education and lost household and labor market productivity) were $48,275.

7.4.3   Sustainability

Using the Monte Carlo stochastic simulation capabilities of the DEVS simulator in MS4Me, we obtained the savings-to-coordination cost ratio for 30 replications of simulations runs with 30 different initial seeds. The trajectories are observed to converge to a small region around the value 2.4. A plot (see Figure A3) of the fraction of samples lying below the value 2.0 approximates the probability that an implementation of the CHAP program does not return enough in savings to cover twice its cost. This probability appears to settle at 0.0 after approximately 2,000 patients have been processed. This suggests that to be sustainable, a program has to treat a large enough pool of patients, so that the number of prevented LBW births is predictable and large enough to cover the program costs.

To verify these results, the value to which the ROI converges can be computed by the formula:

$$\text{ROI} = \frac{P_{\text{LBW}} \times P_{\text{Normal/LBW}} \times \text{excess cost of LBW}}{\text{cost per patient}}$$

(multiplying the numerator and denominator by the number of patients gives the estimated cost of avoided LBW births and the coordination cost, respectively) with values given in Table 7.3 and employing 29,000 for excess cost, we obtain that ROI = 2.36, in reasonable agreement with the simulation results.

7.4.4   The model as an exemplar of value-based HDSoS

Simulation models in healthcare have been oriented to operations research in hospital and clinical settings [6,7]. In contrast, the evaluation frame and the model presented here are oriented to healthcare from a more holistic view as a service system. Recognizing its current limitations, the model serves as a kernel from which to grow toward increased validity with respect to the real system and greater generality of representation of value-based coordinated care health systems in general.

To support effective simulation investigation of smart coordination of health services, the core model must be expanded in at least the following directions:

- In addition to client risk characteristics, the referral source of clients to the coordination program must be taken into account. Analysis of CHAP data [2] showed that African-American women who were referred from community-based care sources received more coordinator attention, showed greater adherence to their assigned tasks, were more likely to remain in the program, and had a higher percentage of normal births than the predominantly white women referred by the Medicaid managed care plans.
- The effect of incentives on CHW performance must be included: Our data analysis showed that compensation based in part on outcome-based payments leads to higher rates of pathways completion, compared with contracts without

such incentives. The optimal ratio of fixed compensation to performance-based bonuses can be studied through simulation of the model.

- The alignment of pathways with payments must be modeled. Typically, HUBs receive an initial payment after they engage clients and conduct an initial assessment. The remaining payments are made after a pathway is completed. Payers like this model, because payment is tied to results. However, current health data measures are not the best way to approach payment for pathways, because payers may select only action steps that have such measures to reimburse and may ignore other critical steps. New measures that better align payment with outcome can be developed with the help of the simulation framework.
- The model should be refined to enable quality improvement at the pathway step level. Our analysis identified several steps common to many pathways that were critical to the overall achievement of outcomes and that provided important opportunities for process analysis, measurement and improvement. For example, we developed new measures for care coordinator competency, because we found that many pathway reports were incomplete or inconsistent. We also developed measures for client adherence that document the client's role in completing pathway steps, such as keeping appointments. These measures should be incorporated into the evaluation frame to enable the simulation to support cross-comparison of step times across CHWs and service tasks enabling standards by which to support CHW compensation—critical element in value-based purchasing.

7.5   Conclusions and recommendations

Based on the pathways formalization in Chapters 5 and 6, this chapter discussed a simulation framework to guide design, development, and evaluation of architectures for pathways-based coordination of care. A simulation model was presented that exemplifies this framework and is intended to predict ROI for implementing pathways-based coordination as well its sustainability over the long term.

The model does not include dimensions such as client risk characteristics, the referral source of clients to the coordination program, the effect of incentives on service workers' performance, and alignment of pathways with payments. Including these dimensions would be needed to increase the predictive capability to address questions about pathways that are important for their implementation in value-based healthcare.

Moreover, the development of such a framework is part of a methodology to implement and test pathways-based data architecture in a cloud environment to be discussed in Chapter 8, up next.

References

[1]   Zeigler B.P., Carter E.L. Redding S.A., and Leath B.A. Pathways Community HUB: A Model for Coordination of Community Health Care. *Population Health Management*. 2014; 17(4): 199–201.

[2]   Zeigler B.P., Carter E.L., Redding S.A., Leath B.A., and Russell C. *Care Coordination: Formalization of Pathways for Standardization and Certification, Report for Project Health System Modeling and Simulation – Coordinated Care Example*, National Science Foundation Grant Award No. CMMI 1235364; 2014.

[3]   Zeigler B.P., Carter E.L., Molloy O., and Elbattah M. Using Simulation Modeling to Design Value-Based Healthcare Systems. *OR58 Annual Conference, Portsmouth*, 6–8 Sep 2016. pp. 33–48.

[4]   Zeigler B.P., and H.S. Sarjoughian, *Guide to Modeling and Simulation of Systems of Systems*, New York, NY: Springer; 2017 edition.

[5]   Redding S., Conrey E., Porter K., Paulson J., Hughes K., and Redding M. Pathways Community Care Coordination in Low Birth Weight Prevention. *Maternal and Child Health Journal*. 2014; 18(6): 1–8.

[6]   Gunal M.M. A Guide for Building Hospital Simulation Models. *Health Systems*. 2012; 1(1): 17–25.

[7]   Ozcan Y.A., Tànfani E., and Testi A. A Simulation-Based Modeling Framework to Deal with Clinical Pathways. In Jain S., Creasey R.R., Himmelspach J., White K.P., and Fu M. (eds.). *Proceedings of the Winter Simulation Conference*, Phoenix, AZ, USA, 11–14 Dec 2011. pp. 1190–1201.

Appendix A

A.1 MS4 Me code and graphical displays for the CHAP ROI model

This appendix presents the MS4 Me code that implemented the model discussed in this chapter. Figures A1–2 are views of the models produced by the MS4 Me simulation viewer. Figure A3 is a plot of savings/cost trajectories. Listings A1–A11 are model specification files.

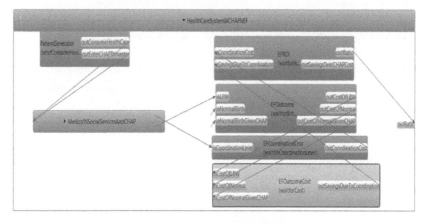

Figure A.1   HealthCareSystemWCHAPNEF-coupled model

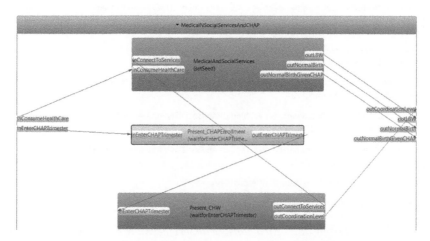

Figure A.2   MedicalNSocialServicesAndCHAP-coupled model

From the HealthCoordANDEF perspective, HealthCareSystemWCHAPNEF is made of PatientGenerator, MedicalNSocialServicesAndCHAP, EFOutcome, EFCoordinationCost, EFOutcomeCost, and EFROI!

From the HealthCoordANDEF perspective, MedicalNSocialServicesAndCHAP sends CoordinationLevel to EFCoordinationCost!

From the HealthCoordANDEF perspective, PatientGenerator sends EnterCHAPTrimester to MedicalNSocialServicesAndCHAP!

From the HealthCoordANDEF perspective, PatientGenerator sends ConsumeHealthCare to MedicalNSocialServicesAndCHAP!

From the HealthCoordANDEF perspective, EFCoordinationCost sends CoordinationCost to EFROI!

From the HealthCoordANDEF perspective, MedicalNSocialServicesAndCHAP sends NormalBirth to EFOutcome!

From the HealthCoordANDEF perspective, EFOutcome sends CostOfNormal to EFOutcomeCost!

From the HealthCoordANDEF perspective, MedicalNSocialServicesAndCHAP sends LBW to EFOutcome!

From the HealthCoordANDEF perspective, EFOutcome sends CostOfLBW to EFOutcomeCost!

From the HealthCoordANDEF perspective, MedicalNSocialServicesAndCHAP sends NormalBirthGivenCHAP to EFOutcome!

From the HealthCoordANDEF perspective, EFOutcome sends CostOfNormalGivenCHAP to EFOutcomeCost!

From the HealthCoordANDEF perspective, EFOutcomeCost sends SavingsDueToCoordination to EFROI!

From the HealthCoordANDEF perspective, EFROI sends Ratio to HealthCareSystemWCHAPNEF!

Listing A.1   HealthCareSystemWCHAPNEF.ses

From the HealthSystemCoord perspective, MedicalNSocialServicesAndCHAP is made of CHAPEnrollment, CHW, and MedicalAndSocialServices!

From the HealthSystemCoord perspective, MedicalNSocialServicesAndCHAP sends ConsumeHealthCare to MedicalAndSocialServices!

From the HealthSystemCoord perspective, MedicalNSocialServicesAndCHAP sends EnterCHAPTrimester to CHAPEnrollment!

From the HealthSystemCoord perspective, CHW sends ConnectToServices to MedicalAndSocialServices!

From the HealthSystemCoord perspective, CHAPEnrollment sends EnterCHAPTrimester to CHW!

From the HealthSystemCoord perspective, CHW sends CoordinationLevel to MedicalNSocialServicesAndCHAP!

From the HealthSystemCoord perspective, MedicalAndSocialServices sends LBW to MedicalNSocialServicesAndCHAP!

From the HealthSystemCoord perspective, MedicalAndSocialServices sends NormalBirth to MedicalNSocialServicesAndCHAP!

From the HealthSystemCoord perspective, MedicalAndSocialServices sends NormalBirthGivenCHAP to MedicalNSocialServicesAndCHAP!

CHAPEnrollment can be Present or NotPresent in presence!

CHW can be Present or NotPresent in presence!

Listing A.2   MedicalNSocialServicesAndCHAP.ses

```
to start hold in sendConsumeHealthCare for time 0!
from sendConsumeHealthCare go to sendTrimesterEntry!
after sendConsumeHealthCare output ConsumeHealthCare!

hold in  sendTrimesterEntry for time 1!
from sendTrimesterEntry go to sendConsumeHealthCare!
after sendTrimesterEntry output EnterCHAPTrimester!

use ProbOfEntryInFirstTrimester with type double and default ".56"!
use ProbOfEntryInSecondTrimester with type double and default ".20"!
use ProbOfEPatientIDntryInThirdTrimester with type double and default ".24"!
use Rand with type Random and default  "new Random()"!
use PatientID with type String and default  "'ID'"!
use Trimester with type int and default "1"!
use EnterCHAP with type int and default "1"!
use seed with type long and default "28974098"!


generates output on ConsumeHealthCare with type String!
generates output on EnterCHAPTrimester with type String!

initialize variables
<%
        Rand.setSeed(seed);

%>!
output event for sendConsumeHealthCare
<%
 output.add(outConsumeHealthCare, PatientID);
%>!

internal event for sendConsumeHealthCare
<%
Trimester = 1;
determineCHAPEntry();
//EnterCHAP =3;
%>!


output event for sendTrimesterEntry
<%
            if (Trimester == 1 && EnterCHAP == 1) {
                output.add(outEnterCHAPTrimester, PatientID + "
Trimester1");
            } else if (Trimester == 2 && EnterCHAP == 2) {
                output.add(outEnterCHAPTrimester, PatientID + "
Trimester2");
            } else if (Trimester == 3 && EnterCHAP == 3) {
                output.add(outEnterCHAPTrimester, PatientID + "
Trimester3");
            } else {
                System.out.println("Patient output CHAP Error");
            //        System.exit(3);
            }
%>!
internal event for sendTrimesterEntry
<%
                //      PatientID = PatientID+"1";
            if (Trimester == 1) {
                if (EnterCHAP == 1) {
                    holdIn("sendConsumeHealthCare", 1000.);
                } else {
                    holdIn("sendTrimesterEntry", 90.);
                    Trimester = 2;
                }
            } else if (Trimester == 2) {
                if (EnterCHAP == 2) {
                    holdIn("sendConsumeHealthCare", 1000.-90);
```

Listing A.3   PatientGenerator.dnl

```
               }else {
                   holdIn("sendTrimesterEntry", 90.);
                   Trimester = 3;
               }
          } else if (Trimester == 3){
               if (EnterCHAP == 3) {
                   holdIn("sendConsumeHealthCare", 1000.-180);
               }
               else {
//don't output - keep changing trimester to match chap entry
               }
             }

%>!


add additional code
<%
  public void determineCHAPEntry() {
        EnterCHAP = ProbabilityChoice.makeSelectionFrom3(Rand,
               ProbOfEntryInFirstTrimester,
           ProbOfEntryInSecondTrimester);
    //Functions.determineCHAPEntry(Rand,ProbOfEntryInFirstTrimester,
    }
%>!
add library
<%
import java.util.Random;
%>!
```

Listing A.3   (Continued)

```
accepts input on LBW !
generates output on CostOfLBW !
accepts input on NormalBirth !
generates output on CostOfNormal !
accepts input on NormalBirthGivenCHAP !
generates output on CostOfNormalGivenCHAP !

use CostOfNormalBirth with type double and default "0"!//1."!
use CostOfLBWBirth with type double and default "29000."!
use CostOfNormalGivenCHAP with type double and default ".01"!

to start,passivate in waitforBirth !
when in waitforBirth and receive NormalBirth go to sendCostOfNormal!
hold in sendCostOfNormal for time 1!
after sendCostOfNormal output CostOfNormal!
from sendCostOfNormal go to waitforBirth!

when in waitforBirth and receive LBW go to sendCostOfLBW!
hold in sendCostOfLBW for time 1!
after sendCostOfLBW output CostOfLBW!
from sendCostOfLBW go to waitforBirth!

when in waitforBirth and receive NormalBirthGivenCHAP go to
sendCostOfNormalBirthGivenCHAP!
hold in sendCostOfNormalBirthGivenCHAP for time 1!
after sendCostOfNormalBirthGivenCHAP output CostOfNormalGivenCHAP!
from sendCostOfNormalBirthGivenCHAP go to waitforBirth!

output event for sendCostOfNormal
   <%
    output.add(outCostOfNormal, CostOfNormalBirth);
   %>!
  output event for sendCostOfLBW
   <%
    output.add(outCostOfLBW, CostOfLBWBirth);
   %>!
  output event for sendCostOfNormalBirthGivenCHAP
   <%
    output.add(outCostOfNormalGivenCHAP, CostOfNormalGivenCHAP);
   %>!
```

Listing A.4   EFOutcome.dnl

```
accepts input on CostOfNormal with type Double!
accepts input on CostOfLBW  with type Double!
accepts input on CostOfNormalGivenCHAP  with type Double!
generates output on SavingsDueToCoordination  with type Double!

use TotalCostOfLBW with type double and default ".0"!
use TotalCostOfNormal with type double and default ".0"!
use SavingsDueToCoordination with type double and default "0."!

to start,passivate in waitforCost !
when in waitforCost and receive CostOfNormal go to
sendSavingsDueToCoordination!
hold in sendSavingsDueToCoordination for time 0!
after sendSavingsDueToCoordination output SavingsDueToCoordination!
from sendSavingsDueToCoordination go to waitforCost!

external event for waitforCost with CostOfNormal
<%
  double CostOfNormal = messageList.get(0).getData();
SavingsDueToCoordination = CostOfNormal;
TotalCostOfNormal += CostOfNormal;
%>!

when in waitforCost and receive CostOfLBW go to sendSavingsDueToCoordination!
external event for waitforCost with CostOfLBW
<%
double   CostOfLBW = messageList.get(0).getData();
SavingsDueToCoordination = 0;//=29000 - CostOfLBW;
TotalCostOfLBW += CostOfLBW;
%>!

when in waitforCost and receive CostOfNormalGivenCHAP go to
sendSavingsDueToCoordination!
external event for waitforCost with CostOfNormalGivenCHAP
<%
 double CostOfNormalGivenCHAP = messageList.get(0).getData();
SavingsDueToCoordination = (29000-0);
TotalCostOfNormal += CostOfNormalGivenCHAP;
%>!

output event for sendSavingsDueToCoordination
<%
  output.add(outSavingsDueToCoordination,SavingsDueToCoordination );
  %>!
```

Listing A.5   EFOutcomeCost.dnl

```
accepts input on CoordinationLevel with type Integer!
generates output on CoordinationCost with type Double!


use CostForEntryInFirstTrimester with type double and default "751."!
use CostForEntryInSecondTrimester with type double and default "751."!
use CostForEntryInThirdTrimester with type double and default "751."!
use Cost with type double and default "0."!

to start,passivate in waitforCoordinationLevel !
when in waitforCoordinationLevel and receive CoordinationLevel go to
sendCoordinationCost!
hold in sendCoordinationCost for time 1!
after sendCoordinationCost output CoordinationCost!
from sendCoordinationCost go to waitforCoordinationLevel!


external event for waitforCoordinationLevel with CoordinationLevel
<%
 int CoordinationLevel  =    messageList.get(0).getData();
 if (CoordinationLevel ==1){
      Cost = CostForEntryInFirstTrimester;
 }
else if (CoordinationLevel ==2){
      Cost = CostForEntryInSecondTrimester;
 }
else {// if (CoordinationLevel ==3){
      Cost = CostForEntryInThirdTrimester;
 }
 %>!
  output event for sendCoordinationCost
  <%
   output.add(outCoordinationCost, Cost);
  %>!
```

Listing A.6   EFCoordinationCost.dnl

```
accepts input on SavingsDueToCoordination with type Double!
accepts input on CoordinationCost with type Double!
generates output on Ratio  with type Double!

use TotalCoordinationCost with type double and default "0."!
use SavingsDueToCoordination with type double and default "0."!
use SavingsOverCHAPCost with type double and default "0."!
use ExpectedSavingsOverCHAPCost with type double and default "1."!
use reportTime with type double and default "1000.0"!
use NumberOfPatients  with type int and default "0"!

to start,hold in waitforInput for time "reportTime"!
when in waitforInput and receive CoordinationCost go to waitforInput
eventually!
after waitforInput output SavingsOverCHAPCost!
when in waitforInput and receive SavingsDueToCoordination go to waitforInput
eventually!
from waitforInput go to waitforInput!

initialize variables
<%
ExpectedSavingsOverCHAPCost= 0.47*0.13*29000/751;

//N*ProbOfNormalGivenCHAP*ProbOfLBW*CostOfLBWBirth/N*CoordinationCost;
 %>!
```

Listing A.7   EFROI.dnl

```
external event for waitforInput with CoordinationCost
<%
NumberOfPatients++;
double cost =  messageList.get(0).getData();
  TotalCoordinationCost +=    cost;
    SavingsOverCHAPCost=SavingsDueToCoordination/TotalCoordinationCost;
  %>!
 external event for waitforInput with SavingsDueToCoordination
<%
double cost =  messageList.get(0).getData();
  SavingsDueToCoordination +=   cost;
SavingsOverCHAPCost=SavingsDueToCoordination/TotalCoordinationCost;
  %>!

output event for waitforInput
<%
output.add(outRatio, SavingsOverCHAPCost);
  %>!

graph TotalCoordinationCost with label "TotalCoordinationCost"!
graph SavingsDueToCoordination with label "SavingsDueToCoordination"!

graph SavingsOverCHAPCost with label "SavingsOverCHAPCost" on graph
"Convergence to Expected Savings/Cost Ratio"!    graph
ExpectedSavingsOverCHAPCost with label "ExpectedSavingsOverCHAPCost" on graph
"Convergence to Expected Savings/Cost Ratio"!
```

Listing A.7   (Continued)

```
From the HealthSystemCoord perspective, MedicalNSocialServicesAndCHAP is made
of  CHAPEnrollment, CHW, and MedicalAndSocialServices!
From the HealthSystemCoord perspective, MedicalNSocialServicesAndCHAP sends
ConsumeHealthCare to MedicalAndSocialServices!
From the HealthSystemCoord perspective, MedicalNSocialServicesAndCHAP sends
EnterCHAPTrimester to CHAPEnrollment!
From the HealthSystemCoord perspective, CHW sends ConnectToServices to
MedicalAndSocialServices!
From the HealthSystemCoord perspective, CHAPEnrollment sends
EnterCHAPTrimester to CHW!
From the HealthSystemCoord perspective, CHW sends CoordinationLevel to
MedicalNSocialServicesAndCHAP!
From the HealthSystemCoord perspective, MedicalAndSocialServices sends LBW to
MedicalNSocialServicesAndCHAP!
From the HealthSystemCoord perspective, MedicalAndSocialServices sends
NormalBirth to MedicalNSocialServicesAndCHAP!
From the HealthSystemCoord perspective, MedicalAndSocialServices sends
NormalBirthGivenCHAP to MedicalNSocialServicesAndCHAP!
CHAPEnrollment can be Present or NotPresent in presence!
CHW can be Present or NotPresent in presence!
```

Listing A.8   MedicalNSocialServicesAndCHAP.ses

```
use idNtrimester with type String!

accepts input on EnterCHAPTrimester with type String !
generates output on EnterCHAPTrimester with type String!

to start,passivate in waitforEnterCHAPTrimester !
when in waitforEnterCHAPTrimester and receive EnterCHAPTrimester go to
sendEnterCHAPTrimester!
hold in sendEnterCHAPTrimester for time 1!
after sendEnterCHAPTrimester output EnterCHAPTrimester!
from sendEnterCHAPTrimester go to waitforEnterCHAPTrimester!


external event for waitforEnterCHAPTrimester with EnterCHAPTrimester
<%
  idNtrimester =    messageList.get(0).getData();

 %>!
  output event for sendEnterCHAPTrimester
  <%
   output.add(outEnterCHAPTrimester, idNtrimester);
  %>!
```

Listing A.9   ChapEnrollment.dnl

```
accepts input on EnterCHAPTrimester with type String!
generates output on CoordinationLevel with type Integer!
generates output on ConnectToServices  with type String!

use idNtrimester with type String!
use CoordinationLevel with type int!


to start passivate in waitforEnterCHAPTrimester !
when in waitforEnterCHAPTrimester and receive EnterCHAPTrimester go to
sendCoordinationLevel!
hold in sendCoordinationLevel for time 1!
after sendCoordinationLevel output CoordinationLevel!
from sendCoordinationLevel go to sendConnectToServices!
 hold in sendConnectToServices for time 1!
after sendConnectToServices output ConnectToServices!
from sendConnectToServices go to waitforEnterCHAPTrimester!


external event for waitforEnterCHAPTrimester with EnterCHAPTrimester
<%
                idNtrimester = messageList.get(0).getData();
             String Level = idNtrimester.substring(idNtrimester.length()-1,
                    idNtrimester.length());
             CoordinationLevel = Integer.parseInt(Level);
 %>!
  output event for sendCoordinationLevel
  <%
   output.add(outCoordinationLevel, CoordinationLevel);
  %>!
    output event for sendConnectToServices
  <%
   output.add(outConnectToServices, idNtrimester);
  %>!
```

Listing A.10   CHW.dnl

```
accepts input on ConsumeHealthCare with type String !
accepts input on ConnectToServices with type String !
generates output on NormalBirth with type String !
generates output on NormalBirthGivenCHAP with type String !
generates output on LBW with type String !

use PatientID with type String!
use Trimester with type int and default "0"!
use ProbabilityOfLBW with type double and default ".13"!
use ProbabilityOfNormalBirthGivenCHAP with type double and default ".47"!
use Rand with type Random and default "new Random()"!
//use seed with type long and default "28974098"!
use NumberOfPatients with type int and default "0"!
use NumberOfNormal with type int and default "0"!
use NumberOfLBW with type int and default "0"!
use NumberOfNormalGivenCHAP with type int and default "0"!

to start,hold in setSeed for time 0!
from setSeed go to waitforConsumeHealthCare !

internal event for setSeed
<%
nextSeed();
Rand.setSeed(seed);

  %>!
passivate in waitforConsumeHealthCare!

when in waitforConsumeHealthCare and receive ConsumeHealthCare go to
waitforConnectToServices!
hold in waitforConnectToServices for time 270!
from waitforConnectToServices go to sendLBW!

external event for waitforConsumeHealthCare with ConsumeHealthCare
<%
 PatientID =     messageList.get(0).getData();
  %>!
when in waitforConnectToServices and receive ConnectToServices go to sendLBW
eventually!
external event for waitforConnectToServices with ConnectToServices
<%
                String idNtrimester = messageList.get(0).getData();
                String number = idNtrimester.substring(idNtrimester.length()-
1,
                        idNtrimester.length());
                Trimester = Integer.parseInt(number);
                // holdIn("sendLBW",.0);
  %>!

hold in sendLBW for time 0!
after sendLBW output LBW!

  output event for sendLBW
  <%
              NumberOfPatients++;
              double choice = Rand.nextDouble();
              if (!(
              ProbabilityChoice.lessThan(choice, ProbabilityOfLBW)
          //  Functions.lessThan(choice, ProbabilityOfLBW)
              )) {
```

Listing A.11   MedicalAndSocialServices.dnl

```
                    output.add(outNormalBirth, PatientID);
                    NumberOfNormal++;
            } else if (Trimester == 0) {
                    output.add(outLBW, PatientID);
                    NumberOfLBW++;
            } else { //CHAP
                    choice = Rand.nextDouble();
                    if (
                    ProbabilityChoice.lessThan(choice,
ProbabilityOfNormalBirthGivenCHAP)) {
                            //Functions.lessThan(choice,
ProbabilityOfNormalBirthGivenCHAP)) {
                            output.add(outNormalBirthGivenCHAP, PatientID);
                            NumberOfNormalGivenCHAP++;
                    } else {
                            output.add(outLBW, PatientID);
                            NumberOfLBW++;
                    }
            }
    %>!


from sendLBW go to waitforConsumeHealthCare!

add additional code
<%
static long seed = 222998722;
public static void  nextSeed(){
      seed = seed+11111;

      System.out.println(seed);
}
%>!
add library
<%
import java.util.Random;
%>!
```

Listing A.11   (Continued)

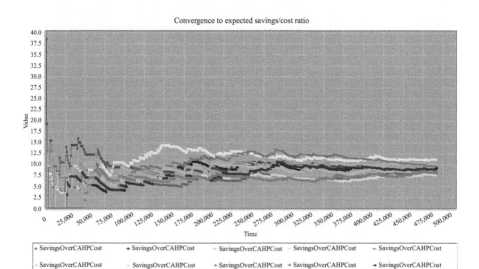

Figure A.3   Plot of expected savings/cost ratio over time

Chapter 8

Health information technology support for pathways-based care

8.1 Introduction

As indicated in Chapter 7, provider companies that are trying to follow new models of healthcare delivery and payment are finding that their electronic systems are inadequate to this task. Also as indicated, even an information technology (IT) infrastructure that improves health record keeping will not provide significantly greater care coordination since it does not provide transparency into the threads of transactions that represent patient treatments, their outcomes, and total costs. In this chapter, we examine the role of pathways in support of care coordination in the HDSoS context that was introduced in Part I. This role is examined through the lens of the health IT (HIT) that addresses the current shortcomings of electronic systems just mentioned. For concreteness, our primary focus is at the community level in relation to the Pathways Community HUB model [1]. However, the implications derived apply more generally to municipal, regional, and national contexts as well.

Our discussion of HIT support for pathways-based care coordination begins with a brief account of the data model underlying the Pathways Community HUB. The client referral process, client–CHW interaction, and billing of payers are integral to the pathways model. These activities have their corresponding data storage and retrieval representations in a database. In the referral process, upon enrollment, a client is interviewed with a checklist of questions that enable assessing the risk level of the client and determining the set of pathways to be initiated. Subsequently, CHWs interact with clients on a one-to-one basis to encourage, monitor, and track them in order to achieve the assigned subgoals of the pathways, thereby improving the prospects of attaining their end goals.

Such interaction is documented in auxiliary data fields associated with pathways steps in the database. We refer to all data related to a client as the *pathways client record (PCR)* and later discuss the relation of such data to more well-known EHRs. Further, billing can involve multiple entities, especially if behavioral health services are involved. These data can be challenging to integrate retrospectively as needed to associate payments with services provided in a pay-for-performance model. Pathways are associated with payment for specific benchmarks along the pathway with the highest payment provided for successful outcomes at completion.

Accordingly, payers, such as Medicaid managed care plans, are associated with clients and pathways and represented in the billing data, thereby enabling payments to be linked to accomplishments.

To date, the Pathways Community HUB model has been implemented in some 16 communities and standards for certifying Community HUB programs have been developed [2]. We turn to discussion of requirements for Data Architecture to Support the Pathways Community HUB model [3].

8.2   Requirements for pathways-based web portal

Figure 8.1 sketches a view of the data architecture to support record keeping and decision-making for care coordinators, CHWs, administrators, and other users implied by the Pathways Community HUB model and the data architecture requirements to be presented. Before proceeding with a discussion of such requirements, we briefly summarize the main points of the architecture in Figure 8.1. This distinguishing characteristic of this architecture is that, except for pathway initialization, user interactions with the system are driven by the steps of pathways currently in play. One or more pathways are initialized when a client is enrolled on the basis of an enrollment questionnaire that seeks to establish the particular needs or risks of the client and prescribes appropriate pathways to address such issues.

Subsequently, at each step in a pathway, the system requests data from the CHW specifically required for completion of that step. Besides, keeping such data immediately available to help with progress to the next step, the web services portal transmits appropriate patient health information for long-term storage in electronic

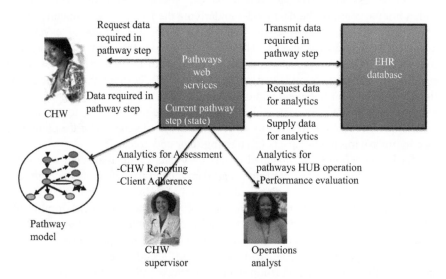

Figure 8.1   Data infrastructure for pathways-based web portal

form in the PCR. Both data stores are then available via the portal for analytics by other users such as supervisors, system operators, operation analysts, and researchers. The PCR database for care coordination is located in the pathways HUB and accessible to CHWs and other HUB personnel directly via the pathways portal. The central role of the pathways concept and its organization of the PCR database illustrates a general approach to address and improve interoperability between the data and systems being used by diverse users.

Having summarized the main points of the architecture in Figure 8.1, we proceed to outline the requirements for data architecture to support the Pathways Community HUB model. These requirements are not derived from existing literature studies because such literature has yet to develop. Rather the requirements are grounded in real world experience in attempting to extract, analyze, and apply pathways data.

Our intent is to stimulate and guide the development and study of pathways, and more broadly care coordination models, with particular attention to their data architecture requirements:

- The pathways web services portal should mediate between the database holding pathway client records and users such as care coordinators (CHWs, supervisors), care providers, managers, and quality improvement analysts. The portal should support establishing a PCR on first registration of the client.
- Care coordinators should be able to enter timely patient information, while other users have access to the information in real time.
- The portal should guide entry of data according to the rules of the instance of the pathway currently controlling user input. The portal can do this, knowing the state of the pathway model and, therefore, the current step of the pathway.
- The portal should ideally eliminate incomplete and inconsistent data entry of pathway steps improving data quality while simultaneously reducing user effort in data entry. Of course, users must still understand the semantics of a pathway (e.g., what the steps and dates represent) to make meaningful entries.
- The portal should enable the care coordinator to access her client caseload with associated pathways for each specific client allowing the coordinator to interview and record the information gained during their client meetings.
- The information collected should be used in the pathways payment process for reporting and invoicing based on pathway billing codes reflecting completed outcomes (in contrast to payment based on activities i.e., pay-for-services).
- The data system should also be the primary repository for the collection of pathways data and interconnected with EHRs in health information exchanges (HIEs) in a secure manner.

8.3 Temporal requirements

In addition to data architecture, support is needed for scheduling, cancellation, reminder, and alert generation to more fully capture Pathway temporal behaviors. The DEVS formalization of the pathways model (Chapters 5 and 6) affords a solid, implementation-neutral basis for enhanced computerized support for coordination of care based on the pathways concept. In this light, we present some additional requirements based on this formalization:

- The software should support the discrete event dynamics required to handle the time management, event scheduling, state transitions, and input/output of the pathways model. Multiple model instances may be active at any time to represent several concurrent pathways of a single client as well multiple such instances of the current set of clients with records resident in the database.
- The software should also provide tools that provide more in-depth analysis based on temporal and dynamic behaviors such as analytics for client assessment (i.e., adherence or compliance) and for organizational operation (activity-based time-driven accounting, reporting quality, outcome evaluation). These functions can be based on the pathways formalization as discussed earlier.

8.4   Implementation approaches

To implement such features, an organization can consider either tailoring off-the-shelf generic systems to meet the requirements or offerings that are specialized for community-based care coordination. To support the growth of the Pathways Community HUB model, care coordination systems (CCS) [4] has developed a database (PCR) that implements the above features. CCS introduced additional modules, the Pathways HUB Connect system and pathways mobile, to provide the Pathways Community HUB model with numerous additional features aiming for streamlined interfaces and better user experience. Care coordinators access the system through the pathways mobile tablet applications, mobile tablets accessing the HUB portal through secure web browsers, and directly via the user-enabled HUB portal used by the HUB administration staff as well as care agencies and their desk-based supervisors/coordinators.

In this way, the HUB administrative staff and the care coordinators are able to accomplish the information tasks outlined in Section 8.3 in a timely manner. A risk module scores the validated pathways client data to rank clients and assist care coordinators in prioritizing services for clients (also available for use by researchers).

More general HIT systems may have generic features that can be specialized for community-based coordination of care. Systems such as Covisint [5] and JProg's CAREWare [6] offer a variety of data organization and aggregation functions and also provide support for HIE initiatives that need to access applications and share information in agnostic manner with existing national, state, or regional EHR systems.

As mentioned above, the data quality issues raised here can be seen within a broader context of problems that have arisen with electronic health record systems as they become more widespread and required for use under federal healthcare policies. Perhaps the most common situation encountered in data quality is that a data element is either overlooked entirely or is entered inconsistently in multiple locations or in different formats within, or across, EHR systems.

This makes it difficult to ascertain the intended relevant data value or even to obtain a close approximation to it. In this context, the remedies suggested above can be viewed as providing approaches to particular problems within larger problem sets. For example, experience in the Beacon Communities suggests that data quality can be improved by providing charts showing missing data as feedback to collectors and customizing their workflow to support standard data collection [7].

The web-based pathways simulator (WPS) mediates between the EHR database holding client data and users of data. Such users include CHWs, care providers, managers, quality improvement analysts, etc. At the first level, the WPS guides CHWs in entry of data according to the dictates of the DEVS Pathway Model currently in focus [8]. The WPS does this, knowing the state of the Pathway Model and therefore, the current step of the pathway. The WPS is built on top of a DEVS simulator operating in real-time mode so that scheduled events occur anchored by a calendar class that properly manages time in terms of current wall-clock seconds, hours, days, and years.

The DEVS simulator handles the time management, event scheduling, state transitions, and input/output of the DEVS pathway models. Multiple model instances may be active at any time to represent several concurrent pathways of a single client as well multiple such instances of the current set of clients with records resident in the database. Other functions that are based on the formalization include analytics for client assessment (i.e., adherence//compliance, see section 6.4.2) and for HUB operation (reporting quality, outcome evaluation). These are of interest to care coordinators (CHWs, supervisors) and operations analysts, respectively.

8.5 Ontology to include care coordination in UMLS

The requirements and features for adequate metadata descriptions and standardized terminology stated above should ideally be based on ontologies that are universally adopted for coordinated care and healthcare more generally. However, the ontology for healthcare systems was developed in Chapter 1; Figure 1.2 recognizes the structure of healthcare systems but does not go into depth in further delineating coordination of care. Moreover, we have to recognize that current ontologies for healthcare are focused on medical terminology and do not address coordination of care.

More broadly, Koppel [9] recently called the universe of electronic health record systems a "Tower of Babel" situation partly due to the lack of adoption of a single standard for EHRs. A quick review shows that continuity of care record is a health record standard intended to provide summary records of a patient's health information that uses eXtensible Markup Language (XML) to provide flexibility that will allow users to formulate, transfer, and view such records in a number of ways. Such means include transport by Health Level 7 (HL7) messages, the most widely adopted HIT message standard [10,11]. Such summaries are much less granular than patient data obtained from an individual visit or encounter as required by the pathways portal. Indeed, a key design feature of the pathways portal is the ability to track a client at each step of the pathway and provide decision-support to CHWs and reporting capabilities to all involved staff.

EHR vendors in the United States are now required by Meaningful Use Stage 2 to enable electronic systems to *understand* each other using the common language Systematized Nomenclature of Medicine-Clinical Terminology (SNOMED-CT) included within the UMLS of the National Library of Medicine [12]. To comply with this requirement, vendors are utilizing "maps" between existing EHR systems based on the International Classification of Diseases, 10th Revision, Clinical Modification (ICD-10-CM) [13] and SNOMED-CT. The nursing services orientation of the Clinical Care Classification (CCC) [14], which is integrated in the UMLS Metathesaurus and SNOMED-CT, renders it a candidate for the consideration in the coordination of care context discussed here.

Figure 8.2 displays a potential approach to developing ontology for Pathways Community HUB model that can support the enhanced PCR data infrastructure discussed above.

To integrate with the UMLS, existing terminologies can serve as inputs for the terms that constitute the pathway descriptions. In particular, the clinical terminology of SNOMED-CT and the nursing terminology of CCC are considered as an input to a potential pathway ontology. Although the CCC considers coordinated care, the type of care it concerns differs substantially from that under consideration here. Unlike nurses, care coordinators (e.g., community health workers) in the Pathways Community HUB model coordinate, but do not provide, care directly to

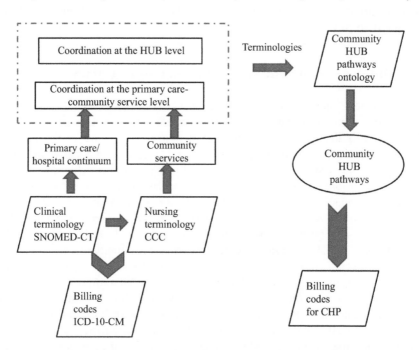

Figure 8.2   Approach to developing pathways ontology

the patient. Indeed, in this model, the care recipients are typically referred to as clients, rather than patients.

One of the roles of the care coordinators is to facilitate connections to evidence-based care that is provided by clinicians (e.g., prenatal care services). In addition, they facilitate connections to services provided by community-based agencies (e.g., transportation and social services). While care coordinators are not direct service providers of either the clinical or community-based services, they are involved in finding those at risk, ensuring that they are treated with evidence-based medical and social interventions, and measuring the health outcomes and costs of these efforts. The pathways in this model concern the representation of some of the activities in this kind of coordination. Therefore, we recommend that the pathways ontology employ UMLS terminologies rather than attempt to subsume it within the CCC.

Figure 8.2 recognizes the levels of clinician-provided care and community service agencies coordinated by the HUB. This is consistent with the multi-level framework for coordination within and across organizations, which identifies key dimensions that can be altered within and across organizations to enable or improve communication and coordination [15]. Such dimensions include structure of the organization, knowledge and technology employed, and administrative operational processes (e.g., by introducing cross-organization care pathways) and were found to be effective in enhancing information exchange, alignment of goals and roles, and improved quality of relationships in service of better care outcome quality and cost.

The HUB pathways are distinct from the pathways developed across the *primary-hospital care continuum* shown to enhance the components of care coordination of the multi-level framework [16]. Existing billing codes are shown in Figure 8.2 as being correlated with UMLS terminologies by vendor mappings (as mentioned earlier). Likewise, billing codes are being developed for the HUB model to be consistent with pathways steps which will allow each such step to be linked to payment based on outcomes rather than activities.

8.6   Implications for EHR systems and care coordination in general

The limitations encountered with the original HIT support for the Pathways Community HUB model are germane to the problems encountered more generally in trying to implement new healthcare models and payment methods. One source of problems is electronic health record systems as they become more widespread and required for use under federal healthcare policies. Many physicians were doubtful about the effects of EHR use on the quality, costs, or efficiency of healthcare in the United States. The Tower of Babel situation [9] is partly due to the lack of adoption of a single standard for EHRs [11]. In this context, the remedies suggested above can be viewed as providing approaches to particular problems within larger problem sets.

The need for standardized data schema in HIT can be viewed within the general limitations of HL7 which does not include elements that relate to coordinating care services under consideration here. The suggestions above to base standardization on ontologies have to recognize that current ontologies for healthcare, notably NIH's Unified Medical Language System [6], are focused on medical terminology and do not address coordination of care. Currently, there are efforts to develop standards for exchange of EHR data and for improving the interfaces that enable healthcare users to better visualize the data and to explore and query data sets of interest to them [17–19].

Most relevant to this chapter is the development of support for data entry that encourages accurate and complete entry of pathway events by CHWs as they occur in interacting with the client. The formalization and associated proposed implementation of the pathways model address these limitations with the intention of providing a design for an improved implementation. The lack of ontologies for coordination of care also motivate criteria for the proposed formalization to lay the basis for standards for pathways semantics and pragmatics that eventually can be incorporated into EHRs [20].

8.7   Conclusions and recommendations

We have presented an approach to ensuring pathways client data validity and improving the design of associated health information infrastructure for community-based coordination of care. Beyond making explicit the aspects in which data quality can fall short, the approach has been shown to inform the design of decision support systems for community health workers and other participants in the Pathways Community HUB. We exploited the existence of pathways structures to organize the data using formal systems concepts to develop an approach that is both well-defined and applicable to support general standards for certification of such organizations. The presentation showed that pathways structures are an important principle not only for organizing the activities of coordination of care but also for structuring the data stored in electronic health records in the conduct of such care. We also showed how it encourages design of effective decision support systems for coordinated care and suggested how interested organizations can set about acquiring such systems.

A major goal of data architecture design for data validity is to reduce the errors in PCR data. In our previous analysis of the CHAP database, we have developed metrics for measuring the completeness and consistency of data entered by CHWs. As reported in Chapter 6, by computing these metrics, we were able to correlate quality of CHW performance with eventual client pregnancy outcomes. In general, measurement of errors and correlation to types of users and outcomes is an aspect of EHR data that seems to be a topic that is often overlooked in HIT policy and programs and needs to be better understood. We recommend further research on metrics such as percentage of records with errors, different types of errors and how they are distributed among records, and how such errors affect the outcomes and implementations of HIT data systems.

Although we have suggested how to expand standard Unified Medical Language System ontologies to include community-based care coordination, we have not provided a detailed plan for such expansion which remains for future work.

We recommend that it be undertaken by the Pathways Community HUB certification committee along with the associated formalization of pathways.

Although the presentation is focused on the Pathways Community HUB model, the principles for data architecture are stated in generic form and are applicable to any health information system for improving care coordination services and population health.

References

[1]   Zeigler B.P., Carter E.L. Redding S.A., and Leath B.A. Pathways Community HUB: A Model for Coordination of Community Health Care. *Population Health Management*. 2014; 17(4): 199–201.

[2]   Leath B.A., Brennan L., Pope A., Nieva V., Redding. M., and Redding S. *HUB Certification Pre-requisites & Standards*. Rockville Institute, Kresge Grant: 245873, Revised 22 Jul 2013.

[3]   Zeigler B.P., Carter E.L., Redding S.A., Leath B.A., and Russell C. 2015, Guiding Principles for Data Architecture to Support the Pathways Community HUB Model. *eGEMs – The Journal of Electronic Health Data and Methods*. 2016; 4(1): 1182.

[4]   Care Coordination Systems. Available at http://carecoordinationsystems.com [Accessed on Nov 2013].

[5]   Covisint Corp. Available at https://www.covisint.com/stories/healthcare/ [Accessed on April 2014].

[6]   Jprog Corp. Available at http://www.jprog.com/features.aspx/ [Accessed on April 2014].

[7]   Schachter A., Rein A., and Sabharwal R. *Building a Foundation of Electronic Data to Measure and Drive Improvement*. Beacon Policy Brief, Beacon Community Technical Assistance Program, Contract No. HHSP23320095627WC, 2013.

[8]   Seo C., Zeigler B.P., Kim D., and Duncan K. Integrating Web-based Simulation on IT Systems with Finite Probabilistic DEVS. *Proceedings of the Symposium on Theory of Modeling & Simulation – DEVS Integrative M&S Symposium*, Alexandria, VA, USA, 12–15 Apr 2015. pp. 173–180.

[9]   Koppel, R. Demanding Utility from Health Information Technology, *Annals of Internal Medicine*. 2013; 158(11): 845–846.

[10] HL7. Available at http://en.wikipedia.org/wiki/Health_Level_7 [Accessed on Nov 2013].

[11] Health Information Technology Advisory Committee (HITAC). *Health IT Standards Committee*. Available at https://www.healthit.gov/hitac/committees/health-it-standards-committee [Accessed on 18 May 2018].

[12] National Institutes of Health. *Unified Medical Language System (UMLS)*. National Library of Medicine. Available at http://www.nlm.nih.gov/research/umls/ [Accessed on 18 May 2018].

[13] ICD-10 Clinical Modification. Available at http://en.wikipedia.org/wiki/ICD-10_Clinical_Modification [Accessed on 9 Apr 2014].

[14]   Saba V.K. Clinical Care Classification (CCC) System, Version 2.5, 2nd Edition – User's Guide. New York, NY: Springer; 2012.

[15]   Gittell J.H., and Weiss S.J. Coordination Networks Within and Across Organisations: A Multilevel Framework. *Journal of Management Studies.* 2004; 41(1): 127–153.

[16]   Van Houdt S., Heyrman J., Vanhaecht K., Sermeus W., and De Lepeleire J. Care Pathways Across the Primary-Hospital Care Continuum: Using the Multi-Level Framework in Explaining Care Coordination. *BMC Health Services Research.* 2013; 13:296.

[17]   Rind A., Wang T.D., Aigner W., *et al.* Interactive Information Visualization to Explore and Query Electronic Health Records – Foundations and Trends. *Human–Computer Interaction.* 2013; 5(3): 207–298.

[18]   Roth C.P., Lim Y., Pevnick J.M., Asch S.M., and McGlynn E.A. The Challenge of Measuring Quality of Care from the Electronic Health Record. *American Journal of Medical Quality.* 2009; 24(5): 385–394.

[19]   Anderson K.M., Marsh C.A., Flemming A.C., Isenstein H., and Reynolds J. *Quality Measurement Enabled by Health IT: Overview, Possibilities, and Challenges.* AHRQ Publication No. 12–0061-EF (prepared by Booz Allen Hamilton, under Contract No. HHSA290200900024I), Rockville, MD, USA, Jul 2012.

[20]   Zeigler B.P., and Hammonds P.E. *Modeling and Simulation-Based Data Engineering: Introducing Pragmatics into Ontologies for Net-Centric Information Exchange.* New York, NY: Academic Press; 2007.

Part III

Application of the framework

Overview—This part uses the foundation developed in Parts I and II to discuss the design and application of pathways-based care coordination in various contexts. We start with a case study of the application of the framework to modeling and simulation of access to healthcare resources in rural France where such access presents a problem to most residents. This calls for characterization of care delivery with a spatial dimension that can be supported by the proposed methodology. The case study shows how business process modeling helps to develop suitable models that are then expressed in DEVS for execution. The resulting model development sets the stage for design and testing of workable coordination of pathway-based care mechanisms that would introduce the concept to the French context.

With the case study as background, we consider the general need to develop a suite of simulations to study specific approaches to value-based delivery of services. Also based on the earlier discussion of pathways-based coordinated care, we consider how the multi-perspective modeling and DEVS-based simulation methodology can support the design and development of architectures to coordinate systems and services using health information networks and interoperable electronic medical records. Success in this direction will contribute to the major global health care goal of solving the "iron triangle" of reducing cost while improving quality and increasing access. We set the foundation for development of simulation tools to support the design of coordination architectures and predict important quality metrics that are applicable to diverse populations.

The approach to value-based healthcare reform presented in this book affords a framework for modeling and simulation that is intended to provide evidence-based and simulation-verified proposals for healthcare delivery system reengineering and new system design. Given the complexities to be tackled, we consider several prerequisites that must be in place to enhance the probability of success of such reform efforts. In this vein, we discuss some of the requirements for conducting multi-perspective modeling and DEVS-based simulation methodology in the context of complex adaptive systems.

Chapter 9

Pathways-based coordination
of care in rural France

Authors' note: Mariem Sbayou, PhD student at the University of Bordeaux and lab IMS assisted with this simulation study and Professor Bruno Vallespir from the University of Bordeaux supervised the building of the care pathways and the elaboration of the algorithms.

9.1 Introduction

As introduced previously by the ontology for healthcare systems simulation (O4HCS) in Chapter 1, a healthcare system is known as a set of organizations and resources (providers and provision, respectively) whose primary goal is to ensure healthcare quality services to flows of individuals (patients). Four interrelated perspectives have been identified by O4HCS: resource allocation (RA), health diffusion, population dynamics, and individual behavior (IB).

The healthcare environment for patient following a pathway out of the hospitals in France is characterized as a chaotic work environment not much monitored, greatly let free to patient choices. Recent governmental initiatives have caused groups concerned with healthcare to think about how bring about more rationality regarding complexity, performance, and costs of this environment. To formalize and test recommendations emerging from such groups, it seems appropriate to employ process-oriented approaches and enterprise modeling applied to the management of healthcare organizational operations.

The case study presented in this third part of this book is an instantiation of the generic scheme introduced in Chapter 4. Here, the Business Process Modeling and Notation (BPMN) tool is considered at the domain expert's level, and models are built from the RA and IB perspectives. Subsequently, these models are used with Discrete Event System Specification (DEVS) for simulation.

Figure 9.1 shows how this specialization is done. The RA model is an Agent-based Model (ABM)-based provider model, each healthcare practitioner being represented by an agent. The IB model is a BPMN-based flow model representing a patient in the healthcare system. These models are glued together by encapsulating the flow model in an agent-based entity model. That way, a multilayer network structure appears due to the relationships between agents, which derive from the BPMN specifications. The resulting DEVS model is coupled with an Experimental Framework (EF) that focuses on performance criteria such as the time for a patient

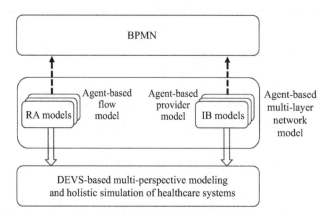

Figure 9.1   Example of multi-formalism approach to holistic M&S of healthcare systems

to complete his/her care pathway, and the distance covered by a patient during this process.

Recently, business process management (BPM) has come to be considered a key valuable asset in the healthcare domain [1]. It is increasingly adopted by healthcare organizations because it helps improving healthcare processes by taking into account the increasing complexity in patient treatment and the continuous reduction of available resources [2]. Various modeling languages have been developed to cover different aspects of business processes and organizations. In this context, the BPMN can be a good tool for improving the healthcare service quality by providing models in an explicit and understandable way which furnish a solution to coordination issue between participants. On the other hand, the use of DEVS formalism may be useful to specify interaction between these participants, as it gives a full specification of behavior of agents and environment as discrete event system models. Furthermore, the use of modeling and simulation approach can be considered relevant for handling these distributed healthcare processes. It might help observing impact of governmental decision and care pathway processes over the population, reducing cost and as final objective to keep providing a good quality of services.

In this chapter, we show how the modeling and simulation framework for value-based healthcare introduced earlier can support modeling the process that describes the patients' courses of treatment and interactions with such systems. Such modeling allows agents involved in the healthcare system to better situate roles and point out the need to rebalance the coordination between actors who treat the patient. The particular issue that we would like to take into consideration in our study is the maldistribution of human resources for health systems which is a worldwide phenomenon that may appear in different dimensions and affect any country. In this chapter, we will take the example of France. Thus, we propose a new M&S architecture, based on BPMN and DEVS methodologies for care coordination modeling (as described in Part 2) that overcomes the coordination issue between healthcare stakeholders, in the case of inequitable distribution of physicians. The architecture is populated with data from National Institute of Statistics and Economic studies (INSEE), community health workers (CHWs), and some perspectives are given to validate by using data gathered from a rural area in France named Dordogne.

9.2 Problem statement

As stated earlier in Parts 1 and 2, one of the major problems of healthcare systems in the world is the lack of coordination of its participants. In healthcare organization, it can be easily admitted that medical professionals have to coordinate care services efficiently. But in the case of extra-clinical patient care, many participants can be involved simultaneously and with no clear priority defined. The question then is how to observe what is going on, when no global view is available.

The coordination issue can be illuminated by modeling and execution of the process that describes the patient pathway; it allows CHWs actors to better know their place in the pathway and facilitates the coordination between actors who intervene on the same patient. Thus, the coordination of these different stakeholders with heterogeneous profiles and occupations is a perspective to ensure the quality and the continuity of care [3].

Another issue faced by the studied healthcare system is the geographic distribution of healthcare professionals and in particular physicians' specialist. In this context, Starkiene [4] describes the maldistribution of human resources for healthcare as a worldwide phenomenon that may appear in different dimensions. Shortages and imbalances in the distribution of the health workforce are social and political problems that reduce the access of the population to the health services [5]. This issue is more significant in areas with higher proportions of low-income and minority residents, such as rural areas, which suffer the most from lower physician supply. Accordingly, Starkiene [4] also states that approximately one-half of the global population lives in rural areas, but these are served by less than quarter of total physician workforce.

Therefore, access to healthcare varies across space because access to healthcare is affected by where health professionals locate (supply) and where people reside (demand) and neither health professionals nor population are uniformly distributed [6]; in general, the supply of physicians in different localities is most commonly measured in terms of the number of physicians per population (physician density, or physician-to-population ratio). Access to healthcare is also influenced by the population's health status and the socioeconomic and financial resources available to the population and health services [7].

Although the density of physicians is increasing in most of the Organization for Economic Cooperation and Development (OECD) countries, the problem of their disparities still exists in most of these countries [8]. One of the OECD countries where the issue of physician disparities has become critical nowadays is France, where healthcare system covers both public and private hospitals, physicians and other medical specialists who provide care services to French residents. It is accessible for all residents, independently of their age, income, or status. In France, the question of vacant post is regularly considered in studies on physicians, particularly for specialties that are less attractive, but today, there is more a problem of geographic maldistribution rather than a lack of physicians [9].

A high level of health workers availability is often observed in capital regions; thus, the desertification issue in rural areas can be divided into the following two types [10]:

- Geographical: As French regions suffer from medical deserts due to the important number of physicians that decreases in these areas because of the non-replacement of retired physicians, and also many hospitals are closed because they are judged not to propose insufficient care quality.
- Sectoral: Some specialties are stricken; there is a real shortage of gynecologists, pediatricians, anesthetists, and psychiatrists add to this the incredible delays to have a physician consultation.

Consequently, the establishment of Regional Health Agencies (Fr. *Agence Régionale de Santé* (ARS)), created by the "Hospital, Patients, Health and Territories" (*Hopital, Patients, Santé et Territories*) law, reinforced the need of having knowledge about the accessibility of the various health services. Indeed, accessibility to health services covers several concepts: geographical accessibility, affordability, and timing accessibility when queues are taken into account [11]. In France, the inability to find a physician or the high wait times in some regions may be attributed to the lack of coordination and disparities in physician density. In order to study these issues, we propose to use simulation for comparing average distance traveled by patients and average waiting time to get appointment in rural and urban areas located in a France region.

9.3   Case model

After giving an overview about the problem statement and before tackling a use case, it is required to introduce the structure of the French healthcare system. In France, each resident has to register a general practitioner (GP) "Fr. *Medecin Traitant,*" in order to ensure full eligibility to reimbursement of health costs. In addition, for visiting a physician (GP or specialist), the patient has to request for an appointment before. In this section, we will provide the case model of our study which focuses on Dordogne region located in France.

9.3.1   Physicians' distribution

As mentioned previously, one of the major problems in healthcare sector is the maldistribution of physicians. This issue will likely continue to be at the forefront of health workforce policy concerns as the population ages. For example, 6% of people over the age of 65 who reported unmet health needs in 2013 indicated that the travel distance and transportation limitations were the main reasons for these unmet care needs in the 28 EU countries. This is expected to be a particular challenge in countries where elderly populations are more concentrated in underserved regions [8].

9.3.1.1 The French case

In France, the medical density is high but they are poorly distributed; this poor distribution is now a major issue in the regulation of medical demography [12]. According to INSEE, there are 334 physicians per 100,000 inhabitants in 2016. This is 60 more than in 1985 (275 physicians per 100,000 inhabitants). The problem is not their number but how they are distributed in the country; the unequal distribution of physicians makes accessibility to care difficult for the population [13].

Figure 9.2 shows the unbalanced distribution of GPs compared with the population [14]; we can observe that there are many regions that face the lack of physicians' issue, in this respect; we will focus on Dordogne county which contains an elderly population compared to other counties in Nouvelle Aquitaine region.

9.3.1.2 Focus on the French Dordogne county

The Dordogne county (Fr. Department) located in southwest of France and considered as mostly representative of "medical desert" zone among the counties in Nouvelle Aquitaine region with the low population, as with the most difficulties in

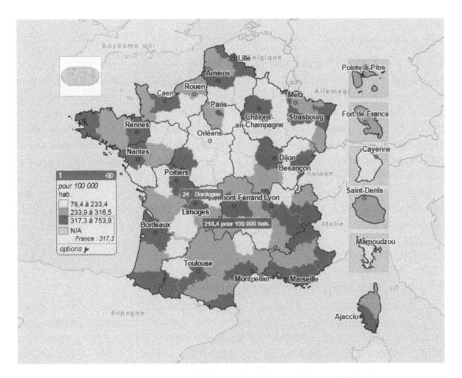

Figure 9.2 Medical densities in France [14]

accessing to health services. It is affected by 63% of the accessibility difficulties, which are due to the low urbanization in this region, according to a survey provided by the Consumer Science and Analytics Institute to the Regional Agency of Health (ARS), also 24% of the health problems in the Dordogne county concern the remoteness of health, the rest concerns problems of long delays and the complexity of the administrative procedures [13].

As shown previously, the medical density in Dordogne department is low compared with other areas in France. In our case study, we will use a very simplified pathway that involves only generalists, radiologists, and otolaryngologists. Figure 9.3 presents the map of Dordogne populated with the chosen French healthcare professionals. Their locations have been extracted from the French yellow pages that contain healthcare stakeholders' information, and it is used to instantiate simulation physicians' agents.

For the needs of the study, only geographic information about GPs, otolaryngologists (ENT physicians), and radiologists (R) have been extracted and represented in Figure 9.3, where we position these physicians in a Google Map, and we can observe that the center part of the Dordogne county mostly gathers the actors in and around the most populated cities, Périgueux and Bergerac.

In Figure 9.3, bubbles illustrate, respectively, GPs, R, and ENT. At the opposite, a great part of this territory is with a low level of practitioners and almost no healthcare resources.

9.3.2   Population

9.3.2.1   Population pyramid in Dordogne

In addition to medical resources shortage, economic and social factors can impact the accessibility to care, namely, the characteristics of the population. Thus, Figure 9.4 shows the population pyramid of Dordogne county generated according to INSEE data [15]. Persons over 60 years old represent one-third of the population. This county cumulates this problem of senior population with disadvantaged socio-professional categories: Dordogne is ranked 89 over 100 counties (Fr. *Département*) in France in terms of average monthly income reported by tax household in 2013. These factors generate limitations to access to unsupported care costs and transportation service to reach the health services. This socio-geographical configuration with low number of health services, social aspects, and age structure predetermine together potential health access issues. This explains the choice of the Dordogne for our study.

9.3.2.2   Generating Dordogne population

The Dordogne population was generated according to the previous statistics. At the level of each municipality, inhabitants of the population have been randomly generated respecting the pyramid constraints recalled in Figure 9.4. Also as we want to study the shortage issue of physicians, we need population's location, for this we give to each individual a random position included in the boundaries of the municipality where he lives; as a result, each inhabitant has been located in the

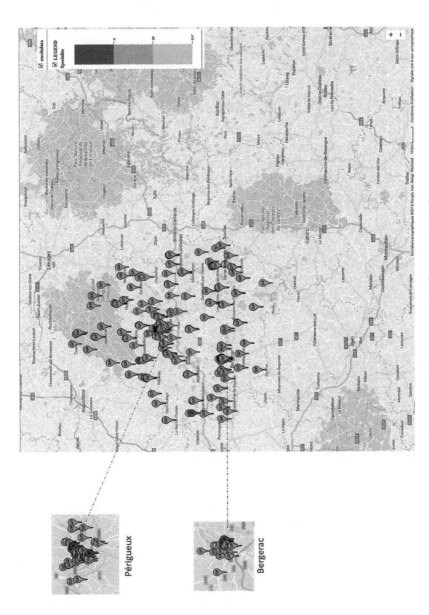

Périgueux

Bergerac

Figure 9.3 Physicians' distribution in Dordogne

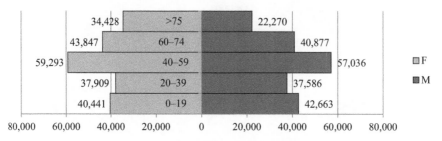

Figure 9.4   Dordogne population pyramid (age and gender)

municipality geographic zone with longitude and latitude position. This information is used to study the accessibility to health services for population in both rural
and urban areas. After giving an overview about the different resources used in our
model, we will explain the recommended approach for our study.

9.3.3   Care pathways

Healthcare environment is characterized as a very dynamic work environment, in
which clinicians rapidly switch between work activities and tasks. The process is
partially planned but at the same time driven by events and availability [16]. The
dynamism and complexity of this environment led to the adoption of process-
oriented approaches and enterprise modeling to the management of organizational
operations.

One of the most known healthcare processes is patient pathway or patient care.
Therefore, a lot of researchers, professionals, and health administrators have given
various definitions of patient pathways. Benabdejlil [17] defines a patient pathway
as a description of process elements of the disease management by the actors of the
care network. It corresponds to a trajectory composed of steps that vary not only
according to the actors of the network who can interact but also according to the
place of residence and the medical history of the patient. We note that this definition omits critical aspects of the pathway concept, such as verification of steps and
accountability of performances that were introduced in Chapter 6. However, in this
chapter, we work within the bounds of this simplified definition and return later to
discuss how coordination can be introduced to improve delivery of healthcare in
distributed settings.

Various modeling languages have been developed to cover different aspects of
business processes and organizations. In this context, even if healthcare CHWs are
not used to work with graphical models [17], we assume that there is a need of
more accessible specification languages. We believe that graphical expression
makes easier the handling of modeling language. The BPMN standardized by the
Object Group Management can be considered as one of them. The BPMN is
emerging as a graphical standard language for modeling processes, especially at the
level of domain analysis and high-level systems design. In consequence, it can be
an appropriate way to specify the patent pathways.

9.3.3.1   Patient pathway modeled with BPMN

In this section, we present an example of the BPMN workflow of a patient pathway in order to approve the feasibility of our approach. The BPMN model illustrates a general case that does not belong to any specific epidemic case in reality, but it has been elaborated in order to require the set of physicians introduced previously (GP, R, and ENT).

Figure 9.5 shows detailed steps followed by the patient to achieve the pathway, when a patient faces health issue he first contacts a GP who is his referred physician, and according to the physician's availability, he gets an appointment, then, according to the patient's severity conditions, the GP suggests a radiology exam, and depending on the results, an intervention of specialist physician may be recommended (ENT in this case).

The drawn BPMN in Figure 9.5 is intended to promote understanding of how our system works and to use some of its elements as inputs of the simulated model; for instance, the message flows "Appointment Request" are represented as input condition in the simulated model. In the simulation, the XML of this model is read to trigger a specific steps the search for appropriate resources.

9.3.3.2   Simplified model of the patient pathway

In order to easily express the simulated process of the pathway, we propose a simplified version (Figure 9.6). Moreover, we split the population to be cared into three sets assuming that one-third of this population needs to meet a GP only (basic consultation), one-third needs a radiological examination too, and the last third must see an ENT (extensive consultation).

9.3.4   Management rules

We propose three management rules for choosing physicians. We also suppose that the patient is the coordinator of his pathway, i.e., he applies the rules by himself.

Figures 9.7–9.9 present the algorithms. The three rules are applied only for choosing ENTs and Rs because, as already mentioned, the GP is chosen once for all in France and then it has not to be changed anymore.

In order to simplify the case, we assume that all persons have chosen the closest GP. From a practical point of view, that means that only the Rule 1 is used for the GP.

9.3.5   Patient's satisfaction

The concept of patient's satisfaction has many figures in the literature. Pascoe [18] proposes a comprehensive examination of the findings of patient satisfaction research, by exploring the conceptualization and measurement of patient satisfaction. Furthermore, Janicijevic *et al.* [19] define parameters that have the greatest impact on healthcare worker job and also on patient satisfaction with services.

Another perception of patient satisfaction was given by [20], where authors define patient satisfaction as a complex and a multidimensional concept, that defines the degree to which the individual regards a healthcare service as useful,

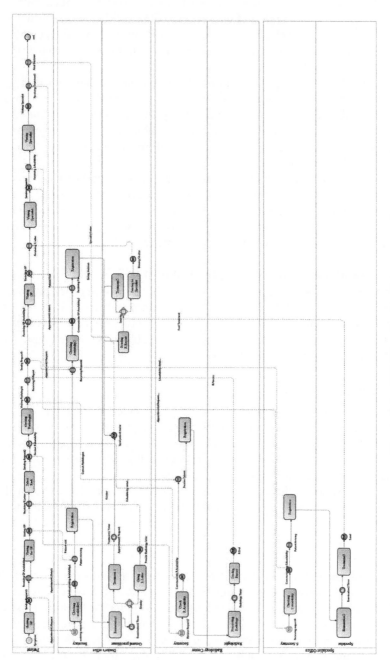

Figure 9.5 BPMN patient pathway example

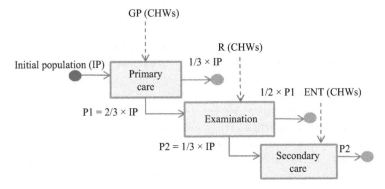

Figure 9.6   Simplified pathway and population splitting

Rule 1: The patient chooses the nearest physician, whatever the proposed appointment date

Algorithm of Rule 1: distance criteria
For all patients and physicians do
Calculate distance between each patient's situation and physicians' situation;
Grant appointment according to the most available;
Return WaitingTime and Distance;

Figure 9.7   Algorithm of Rule 1

Rule 2: The patient chooses the soonest possible appointment, whatever the distance between the patient and the corresponding physician

Algorithm of Rule 2: time criteria:
For all patients and physicians **do**
Calculate WaitingTime between each patient **and** the soonest availability of each physicians
Grant appointment according to the most available;
Return WaitingTime and Distance;

Figure 9.8   Algorithm of Rule 2

Rule 3: It is a hybrid rule where the patient chooses according to a distance radius and a waiting time threshold

Algorithm of Rule 3: hybrid rule
For all patients and physicians **do**
If distance is equal to Radius **and** WaitingTime < Threshold1
Then grant appointment according to the condition values
Else execute Algorithm 2

Figure 9.9   Algorithm of Rule 3

effective, or beneficial [21]. Overall, patient satisfaction has been widely used as a measure of health system performance [22].

9.3.5.1   Measurement of the patient's satisfaction

Surveys of patient satisfaction have been commonly used in most studies to determine if patients are happy with the quality of care provided or not. In this context, Calitri *et al.* [23] used a single questionnaire from the English GP patient survey [24] which classifies patient satisfaction within 5-point levels: "very satisfied," "fairly satisfied," "neither satisfied or dissatisfied," "fairly dissatisfied," or "very dissatisfied," to define how distance from practice may affect the patient satisfaction with care. Janicijevic *et al.* [19] used method based on data collected via questionnaire-based surveys, from 18,642 healthcare workers and 9,283 patients across 50 secondary healthcare institutions in Serbia to calculate patient and healthcare staff satisfaction. The results of this study show that the satisfaction of healthcare workers with the time needed to perform assigned tasks has a significant effect on patient satisfaction. A brief instrument to measure patients' overall satisfaction with primary care physicians was introduced by Hojat *et al.* [25], based on a mailed survey that included ten items, and also a name of physicians.

Although these surveys gave an overview about patient satisfaction, they still have limitations such as a suboptimal participation rate of patients, who may be unable to participate, because of language barriers, physical limitations, or mental problems [26].

9.3.5.2   Our definition of the patient's satisfaction

In our case study, we link patient's satisfaction with the distance traveled to get a physician and also with the average waiting time. We propose that patient's satisfaction is affected by distance and time thresholds. Consequently, in our case study, the patient's satisfaction takes into consideration geographic distance from practitioners and the overall waiting time to perform the patient pathway, as these two parameters are measurable by our approach.

9.4   General overview of the proposed architecture

BPM is undoubtedly beneficial to all types of organizations. Healthcare is no exception; on the contrary, it is perhaps one of the areas that could benefit most from process management. It heavily relies on process models to identify, review, validate, represent, and communicate process knowledge [27]. The modern healthcare system is a complex system that combines different entities such as institutions, healthcare professionals, and patient information, where types of shared information depend on the physicians' profession, patient's pathology, so neglecting such multilayer structure, or in other words working with the projected network, may alter our perception of the topology and dynamics, leading to a wrong understanding of the properties of the system [28]. Hence, a multilayer representation may be helpful for having a meaningful representation of the system under study. Also, for meeting the individual needs of different roles of the patient pathway, BPMN can be considered the most efficient one, since it provides a good understanding of healthcare process [29]. Nonetheless, the utilization of BPMN as modeling language is still facing some limitations such as limited flexibility during process enactment [30], inability to cope with dynamic changes in resource levels and task availability, limited ability to predict changes, due to external events, in both the volume and composition of work, lack of performance, scalability, and reliability as well [31].

In order to overcome some of the BPMN's drawbacks mentioned previously, we propose a new architecture that draws an overview about the network of healthcare professionals by using BPMN processes with an agent-based model architecture. The proposed architecture uses the BPMN to describe different roles, and an open database where information about different participants who perform the process is collected from the INSEE. The collected data are then used for creating the DEVS network. Connection between agents is obtained according to the convenient decisional task in the BPMN process and time is conditioned by treatment tasks.

We call each decisional task BPMN task that provides information from a participant to another one, and which is responsible for creating connections between agents. On the other hand, we mean by treatment tasks each task that needs time to be performed and affects the availability of physicians and patients' waiting time as well.

To emphasize the structure of our architecture, we draw an overview about it in Figure 9.10. The first step is to establish the BPMN process of the current health issue and then generate the XML file that contains information about the decisional and treatment tasks; besides, input information about population and healthcare stakeholders are collected and used to create the DEVS network of the healthcare system.

Agents in the DEVS network are described by a set of attributes distinguished into the following two categories:

- Static attributes that time has no effect on, i.e., identifier, gender, age, number of patients/physicians, latitude, longitude, etc. and
- Dynamic attributes (variables that evolve over time or in reaction to an event), i.e., transaction time, average waiting time, etc.

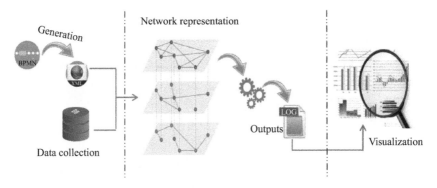

Figure 9.10   General overview of the healthcare global architecture

Our global DEVS model is divided into two parts [32]: coupled model of agents that perform tasks and XML model to orchestrate and schedule the network.

9.4.1   Patient behavior and connection to multilayer network

In order to provide a comprehensive way of the used approach, we propose a multilayer representation of the system. Furthermore, multilayer network has become a crucial tool for understanding different relationships and their interactions in a complex system; it is often composed of many single-layer networks with nodes linked within edges. Each layer is sharing the same set of nodes, and edges are divided into two categories: intra-layer networks within each layer and inter-layer edges between vertices of different layers [33].

In our case and since connections between healthcare stakeholders and patients are different, it is better to give a simplified illustration of the global system as a multilayer representation. We model the global health studied system according to three layers of Figure 9.11, the first one represents the possible connection between the patient and his GP (G), and pharmacies (PH and PH1), and in general cases, patients choose the nearest GP to their residence or working place. The second

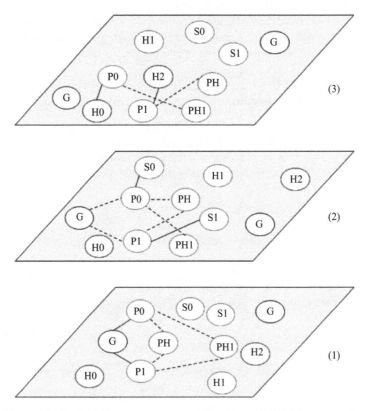

Figure 9.11   Multilayer representation of French healthcare system

layer represents the possible relation between patients and a specialist (S0, S1); in the cases when the pathology needs deeper examination, the third layer is created according to the relation that can be established when hospital intervention is required (H, H1).

9.4.2 Class diagram of the architecture

The proposed DEVS models are implemented using VLE (virtual laboratory environment). VLE software [34] implements DEVS M&S and supports multi-modeling, simulation, and analysis. It is based on an extension of DEVS, the Dynamic Structure Discrete Event formalism—DSDEVS [35]. The implementation of the DSDEVS abstract simulators gives VLE the ability to simulate distributed models and to load and/or delete atomic and coupled models at runtime.

The class diagram in Figure 9.12 shows different classes with the major used functions to create the architecture; our BPMN is used to ensure coordination between resources at first, and then an .xml file is generated where we will extract tasks that would be able to decide which resources to choose. For example, the task choosing GP which we call decisional task will be a condition for calculating distance between each patient and physicians who have free appointment on their calendars where they can be free or busy at a specific time; on the other hand, the

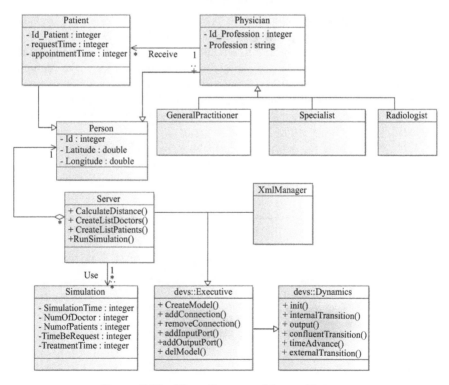

Figure 9.12 Class diagram of the architecture

task visiting GP which we call scheduled task will be a condition to set physicians to busy so that they will not accept another appointment request at this time. In order to handle these tasks, we propose an xml manager atomic model which will provide information to other DEVS models through output ports.

The implementation of DEVS models under VLE is done by inheriting from the "Devs::Executive" class that contains methods for controlling dynamics the models. The behavior of the XML file generated from the BPMN diagram is described by the class "XmlManager"; on the other hand, the "Server" class describes the behavior of the system and allows the development of case studies scenarios.

9.4.3   XmlManager behavior

As previously elaborated, an xml file is generated from the BPMN diagram and then read for extracting useful information to launch and couple the global DEVS model. The operation model of the XmlManager is described in Figure 9.13 and it is initialized in "Init" state where it reads the Xml file that contains tags with ids and checks for decisional xml tags according to their ids. Then the model state will move to the next state "Phase_0" and according to the convenient information contained in the Xml tag, it will provide information to other models.

9.4.4   Scenarios and simulation results

To implement the proposed approach, we have collected statistics about population from the literature. More precisely, we extracted data from INSEE [36], then we generated the population according to a random approach bound by first age and gender pyramid and second the geography for each municipalities of the county of Dordogne. In this chapter, we have chosen to focus on four municipalities. The four municipalities have been chosen according to different population's densities levels and location. They represent relatively low, average, and high population densities.

Table 9.1 presents the selected municipalities and their respective number of inhabitants used for this study in Dordogne. Also, the locations of these municipalities

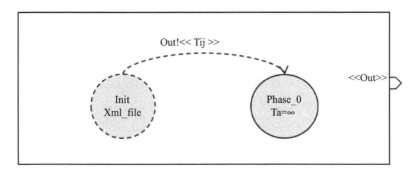

Figure 9.13   Operation of the XmlManager

Table 9.1   Municipalities and number of inhabitants

City's name	Saint-Priest-les-Fougères ●	Saint-Front-de-Pradoux ▢	Boulazac-Isle Manoire ◉	Périgueux ○
Number of inhabitants	383	1,170	10,104	31,540

are represented in Figure 9.14. It shows that we consider remote cities, well-served middle size cities, suburban middle cities, and bigger urban cities.

Therefore on the one hand, we have generated an artificial population for the Dordogne county municipality by municipality; on the other, Health Care Worker (HCW) physicians (GP, R, and ENT) have been populated according to yellow pages information in the model. They are described by their ids, profession, latitude, longitude, and also appointment calendar.

In order to run the simulation, each physician has his own appointment calendar where he can be busy or free. We assume that the physicians' calendars are not completely empty at the simulation launch but instead they are already filled initially with ongoing appointment values described in Table 9.2 [37,38].

Moreover, the examination time varies from one physician to another, for example, GPs perform on average 22 consultations per day with an average duration of 17 min [39]. Also, an exam with a radiologist lasts about 10–15 min [40]. For ENT physicians, the duration of the consultations varies little from one patient to another: half of the sessions last between 15 and 20 min [41]. For this reason, we have particularized the agenda by healthcare resource specialty.

The previously presented statistics are used to set the global simulated model. In addition, authors in [42] show that the weekly incidence rate of illness like influenza per 100,000 inhabitants is 1,000, this value is not for a specific influenza epidemic; so for our case study and as we will run simulation for a duration of 4 weeks of appointment requests, we will take into consideration the rate of 4,000 per 100,000 inhabitants for the initial population. The simulation of this geographically distributed population will permit to observe the required time to reach the healthcare resources and the covered distance at the level of municipalities. Then those results will be compared globally at the level of the county of Dordogne regarding the evolution of two criteria: the average time to complete the pathway per patient and the average distance covered to reach each physician participating in the process.

To run the simulation, we applied the three algorithms described in Figures 9.7–9.9. In the following, we present three figures. They display the simulation results according to three dimensions. The horizontal axis is presenting the four studied municipalities of Figure 9.14 and Table 9.1. The right vertical axis is presenting the average traveled distance (km) with blue color and the average waiting time (days) with green color. The left vertical axis is presenting with pink bars percentage of patient that manage to achieve the full pathway workflow within the

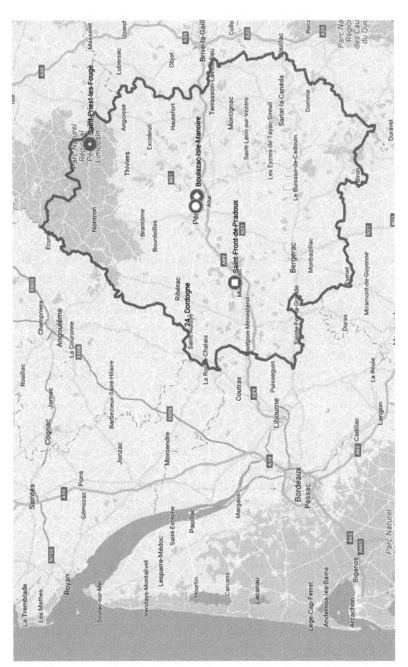

Figure 9.14   Locations of the chosen municipalities

Table 9.2   Time for getting an appointment

	GP (days)	R (days)	ENT (days)
Urban area	2	6	31
Rural area	1	6	21

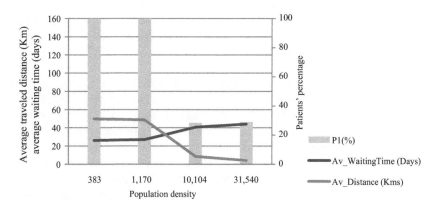

Figure 9.15   Simulation results of Rule 1

maximum duration allowed. Figure 9.15 shows simulation results of Rule 1, where patients request for appointment according to the nearest physician to their positions.

We note that in municipalities with a high density, patients will travel less than 10 km to meet the three physicians; on the other hand, it exceeds 40 km for the other two municipalities with a low density, where the average waiting time to perform the whole process is less than 30 days; in this case, 100% of sick patients will be able to complete the process in a duration equal or less to 40 days. It is not the case for the patients of the other two urban areas, where the average waiting time exceeds 38 days. Also, less than 40% of people who requested for an appointment during the simulation period will be able to complete the process in a duration equal to or inferior to 40 days.

As for the second rule for choosing physicians, described in Figure 9.16, the results denoted that in the case where the choice of the physician is made following its availability, an important distance needs to be traveled by patients of the two zones with low density. In the two rural areas, the covered distance is more than 80 km, and in the other municipalities, it can be less than 40 km. However, the average waiting time is lower than the Rule 1 for the four municipalities. Also, as a positive result, more than 85% of sick patients in the dense municipalities will be able to complete the process within a maximum of 40 days. In the third rule presented in Figure 9.16, we run simulation according to three radius values. Waiting time thresholds were set to 40 days for ENT in urban areas, 31 days for rural areas, and 10 days for both areas for radiologists. These data were taken from the barometer of access to care [37,38]; then, we implement the hybrid rule previously introduced in Figure 9.9.

Figure 9.17(a) presents the results with a first radius value equal to 30 km. In this case, none of the patients of both rural areas could have an appointment taking into account the conditions applied to each physician; however, in the other municipalities, a bit less than 10% of patients could validate the condition with an average global waiting time of almost 45 days and a distance less than 10 km. It is due to maximum radius of coverage allowed that does not permit the inhabitants to reach all the services.

As for the second chosen value of the radius (40 km) described in Figure 9.17(b), patients in the region with a density of 1,170 could all have an appointment and the average distance to travel for the three physicians is about 66.39 km; however, the average waiting time is about 32 days. These results demonstrate the interest of being a medium size municipality close to bigger urban areas. Here, the patient can first access to primary care with a reasonable amount of time and distance and then go to bigger center in case of need of more specialized healthcare.

Finally, in Figure 9.17(c), the limit has been pushed to a radius equal to 50 km; patients in the least populated areas could quite easily have an appointment with physicians. Nevertheless, they will have to travel about 95.78 km to visit the three physicians in order to perform the whole process in a maximum of 40 days.

One conclusion is no approach can fit for all; the strategy depends highly on the locations of the population. Also, the conclusion depends greatly on parameters that have been set at the beginning of the simulation regarding the waiting times of the different category of healthcare resources. Those results cannot be considered

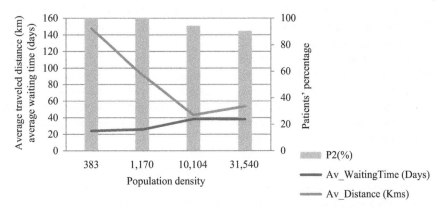

Figure 9.16   Simulation results of Rule 2

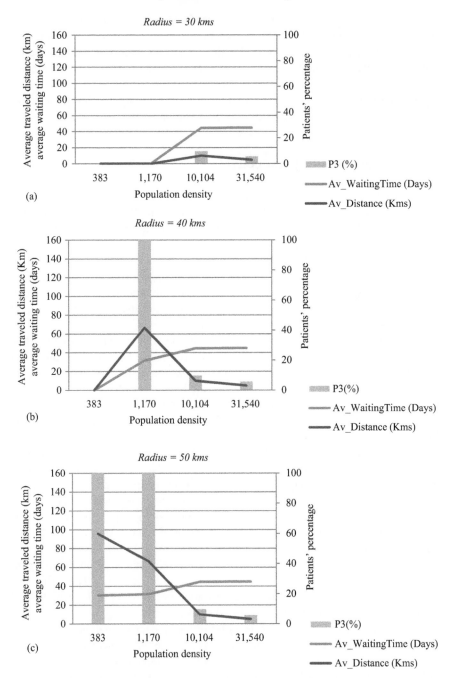

Figure 9.17   Simulation results of Rule 3

as individually quantitatively pertinent and verified but they tend to give an idea of the value of one solution relatively to others. They can be considered as the first step to build the patient satisfaction function. We recall that here patients are the only coordinators of their pathways. In France, regarding external healthcare pathway, only the patient drives the process.

In the future, some orchestration might prevent some agenda bottlenecks and identify closer or more available resources for patients.

9.5   Conclusion and perspectives

In this chapter, we proposed a new architecture that covers two different aspects: technical aspect related to the interoperability between BPMN and DEVS and a health aspect related to the inequality of the geographical physicians' distribution, and to the long waiting times.

As a case study, we have chosen a county in the southwest of France which contains resources that are poorly distributed geographically, and which leads to considerable distances to be traveled by population of rural areas. In urban areas, the problem is more closely linked to waiting times and the impossibility to take care of all sick patients in reasonable periods.

The simulation results observe and confirm the well-known information that people are not equal in their access to healthcare resources. Nevertheless, this study tends to quantify this phenomenon. It appears that the better access according to this simulation is given to people living in satellites cities of bigger urban zones. They benefit first of close service of primary care and then can access quite rapidly to specialist regarding more remote cities. The patient satisfaction can be greater in this area.

We are truly conscious of the limits of this approach. Our goal is solely to demonstrate the feasibility of the method. The population generated is still very artificial. The rules to manage the healthcare resources are greatly simplified and the distances covered are also computed as not including the category of roads and traffic. The objective was primarily to give a methodology and observe how the pathways can be faced to population and geographical issues. Also, this model and simulation represent the current situation in France. The orchestration of the external pathway is only driven by the patient himself; this is called the independent care in the other chapters. In the future, we can add in CHWs that can potentially support patients for better orchestration of pathways on the way toward the accountable care models discussed in the book.

As a perspective, we propose to observe the impact of assigning new geographical positions to physicians, as well as to propose the implementation of new healthcare homes. Then, simulation results will be affected and the performance will be deduced. And finally, we will propose a satisfaction index that can balance the performance index; it can be essentially based on the two indicators: *accessibility*, which is represented by the distance traveled, and the *availability* indicator that is expressed by the average waiting time for performing the patient pathway.

References

[1] Borycki E.M., Kushniruk A.W., Kuwata S., and Kannry J. Use of Simulation Approaches in the Study of Clinician Workflow. *Proceedings of the AMIA Annual Symposium*, Washington, DC, USA, 11 Nov 2006. pp. 61–65.

[2] Antonacci G., Calabrese A., D'Ambrogio A., Giglio A., Intrigila B., and Levialdi Ghiron N. A BPMN-Based Automated Approach for the Analysis of Healthcare Processes. *Proceedings of the IEEE 25th International Conference on Enabling Technologies: Infrastructure for Collaborative Enterprises*, Paris, France, 13–15 Jun 2016. pp. 124–129.

[3] Lamine E., Zefouni S., Bastide R., and Pingaud H. A System Architecture Supporting the Agile Coordination of Homecare Services. *Proceedings of the 11th IFIP Working Conference on Virtual Enterprises*, Saint-Etienne, France, 11–13 Oct 2010. pp. 227–234.

[4] Starkiene L. Inequitable Geographic Distribution of Physicians: Systematic Review of International Experience. *Health Policy and Management*. 2013; 5(1): 101–117.

[5] de Oliveira A.P.C., Gabriel M., Dal Poz M.R., and Dussault G. Challenges for Ensuring Availability and Accessibility to Health Care Services under Brazil's Unified Health System (SUS). *Ciência & Saúde Coletiva*. 2017; 22(4): 1165–1180.

[6] Wei L., and Wang F. Measures of Spatial Accessibility to Health Care in a GIS Environment: Synthesis and a Case Study in the Chicago Region. *Environment and Planning B: Planning and Design*. 2003; 30(6): 865–684.

[7] Aday L.A., and Andersen R. A Framework for the Study of Access to Medical Care. *Health Services Research*. 1974; 9(3): 208–220.

[8] Ono T., Schoenstein M., and Buchan J. Geographic Imbalances in the Distribution of Doctors and Health Care Services in OECD Countries. *Health Workforce Policies in OECD Countries: Right Jobs, Right Skills, Right Places. OECD Health Policy Studies*. Paris: OECD Publishing; 2016, pp. 129–163.

[9] World Health Organization. *World Health Organization Assesses the World's Health Systems*. Geneva: Press Release WHO/44, 2000.

[10] Etienne L. La Désertification Médicale. Le blog Du Docteur Loic Etienne. Available at http://www.zeblogsante.com/la-desertification-medicale/#comment-572 [Accessed on 18 May 2018].

[11] Evain F. A Quelle Distance de Chez Soi Se Fait-on Hospitaliser o Available at http://fulltext.bdsp.ehesp.fr/Ministere/Drees/Publications/2011/etabsante2010-distance.pdf [Accessed on 18 May 2018].

[12] Delattre E., and Samson A.-L. Stratégies de Localisation Des Médecins Généralistes Français: Mécanismes Économiques Ou Hédonistes? *Economie et Statistique*. 2012; 455(1): 115–142.

[13] Dejean J.-P. En Aquitaine, l'Accès Aux Soins Est Meilleur En Montagne. Available at https://objectifaquitaine.latribune.fr/politique/2015-07-02/en-aquitaine-l-acces-aux-soinsest-meilleur-en-montagne.html [Accessed on 18 May 2018].

[14] Conseil National de l'Ordre des Médecins. *Cartographie Interactive de La Démographie Médicale*. Available at https://demographie.medecin.fr/ [Accessed on 18 May 2018].

[15] Ferret J.-P. 2017. *5 879 144 Habitants En Nouvelle-Aquitaine Au 1er Janvier 2014*. Available at https://www.insee.fr/fr/statistiques/2530016 [Accessed on 18 May 2018].

[16] Dahl Y., Sørby I.D., and Nytrø Ø. 2004. Context in Care – Requirements for Mobile Context-Aware Patient Charts. *Proceedings of the 11th World Congress on Medical Informatics*, San Francisco, CA, USA, 7–11 Sep 2004. pp. 597–601.

[17] Benabdejlil H. *Modélisation Des Processus de Soins: Vers Une Implantation de Nouveaux Services à Valeur Ajoutée*. PhD Thesis, Bordeaux, 2016.

[18] Pascoe G.C. Patient Satisfaction in Primary Health Care: A Literature Review and Analysis. *Evaluation and Program Planning*. 1983; 6(3–4):185–210.

[19] Janicijevic I., Seke K., Djokovic A., and Filipovic T. Healthcare Workers Satisfaction and Patient Satisfaction – Where Is the Linkage? *Hippokratia*. 2013; 17(2): 157–162.

[20] Fitzpatrick R., and Hopkins A. Problems in the Conceptual Framework of Patient Satisfaction Research: An Empirical Exploration. *Sociology of Health & Illness*. 1983; 5(3): 297–311.

[21] Coulter A. *The Autonomous Patient: Ending Paternalism in Medical Care*. London: Stationery Office (for the Nuffield Trust), 2002.

[22] Kumari R., Idris M.Z., Bhushan V., Khanna A., Agarwal M., and Singh S.K. Study on Patient Satisfaction in the Government Allopathic Health Facilities of Lucknow District, India. *Indian Journal of Community Medicine*. 2009; 34(1): 35–42.

[23] Calitri R., Warren F.C., Wheeler B., *et al*. Distance from Practice Moderates the Relationship between Patient Management Involving Nurse Telephone Triage Consulting and Patient Satisfaction with Care. *Health & Place*. 2015; 34: 92–96.

[24] England N.H.S. *GP Patient Survey 2014*. Available at: www.england.nhs.uk/statistics/2015/01/08/gp-patient-survey-2014/ [Accessed on 2 March 2015].

[25] Hojat M., Louis D.Z., Maxwell K., Markham F.W., Wender R.C., and Gonnella J.S. A Brief Instrument to Measure Patients' Overall Satisfaction with Primary Care Physicians. *Family Medicine*. 2011; 43(6): 412–417.

[26] Gayet-Ageron A., Agoritsas T., Schiesari L., Kolly V., and Perneger T.V. Barriers to Participation in a Patient Satisfaction Survey: Who Are We Missing? *PLoS One*. 2011; 6(10):e26852.

[27] Van Der Aalst W.M.P., Ter Hofstede A.H.M., and Weske M. Business Process Management: A Survey. *Proceedings of the International Conference on Business Process Management*, Eindhoven, Netherlands, 26–27 Jun 2003. pp. 1–12.

[28] Kivelä M., Arenas A., Barthelemy M., Gleeson J.P., Moreno Y., and Porter M.A. Multilayer Networks. *Journal of Complex Networks*. 2014; 2 (3):203–271.

[29]   Sang K.S., and Zhou B. BPMN Security Extensions for Healthcare Process. *Proceedings of the IEEE International Conference on Computer and Information Technology – Ubiquitous Computing and Communications – Dependable, Autonomic and Secure Computing – Pervasive Intelligence and Computing*, Liverpool, UK, 26–28 Oct 2015. pp. 2340–2345.

[30]   Bolcer G.A., and Taylor R.N. Advanced Workflow Management Technologies. *Software Process: Improvement and Practice*. 1998; 4(3): 125–171.

[31]   Pang G. Implementation of an Agent-Based Business Process. *Technical Report*, University of Zurich, 2000.

[32]   Sbayou Y.B., Zacharewicz G., Ribault J., and François J. DEVS Modelling and Simulation for Healthcare Process Application for Hospital Emergency Department. *Proceedings of the 50th Annual Simulation Symposium*, Virginia Beach, VA, USA, 23–26 Apr 2017. Article 4.

[33]   Murata, T. Comparison of Inter-Layer Couplings of Multilayer Networks. *Proceedings of the 11th International Conference on Signal-Image Technology Internet-Based Systems*, Bangkok, Thailand, 23–27 Nov 2015. pp. 448–452.

[34]   Quesnel G., Duboz R., and Ramat E. The Virtual Laboratory Environment – An Operational Framework for Multi-Modelling, Simulation and Analysis of Complex Dynamical Systems. *Simulation Modelling Practice and Theory*. 2009; 17(4): 641–653.

[35]   Barros F.J. Modeling Formalisms for Dynamic Structure Systems. *ACM Transactions on Modeling and Computer Simulation*. 1997; 7(4): 501–515.

[36]   Institut National de la Statistique et des Etudes Economiques (INSEE). Available at https://www.insee.fr/ [Accessed on 18 May 2018].

[37]   Baromètre De L'accès aux Soins. Available at http://apps.smartsante.com/barometre-acces-soins/ [Accessed on 18 May 2018].

[38]   Ophtalmos, Cardiologues, Radiologues... Des Délais d'attente Très Disparates. Available at http://www.sudouest.fr/2015/10/10/rendez-vous-medicaux-des-delais-dattente-disparates-selon-les-specialites-2150712-4696.php [Accessed on 18 May 2018].

[39]   22 Consultations par Jour de 17 Minutes en Moyenne: Comment Travaillent les Généralistes. Available at https://www.lequotidiendumedecin.fr/actualites/article/2017/05/02/22-consultations-parjour-de-17-minutes-en-moyenne-comment-travaillent-les-generalistes_847151 [Accessed on 18 May 2018].

[40]   Assurance Maladie en Ligne. Comment se Déroule un Scanner? Avalaible at https://www.ameli.fr/assure/sante/examen/imagerie-medicale/deroulement-scanner [Accessed on 18 May 2018].

[41]   Gouyon M. *Consulter Un Spécialiste Libéral à Son Cabinet: Premiers Résultats d'une Enquête Nationale*. Direction de la Recherche, des Etudes, de l'Evaluation et des Statistiques (DREES), Études et Résultats 704, 2009.

[42]   Carron M. Surveillance de La Grippe et Des Infections Respiratoires Aigües En Collectivités de Personnes âgées – Saison 2014–2015. https://www.cpias-nouvelleaquitaine.fr/wp-content/uploads/2015/06/AQ_04112015_Bilan_IRA_2015.pdf [Accessed on 18 May 2018].

Appendix B

B.1 Architecture modeling and simulation of propagation phenomena

Here, we describe the conceptual architecture and the modeling and simulation platform proposed in [1] and used in the chapter. We describe the three modules for integrating the models. The simulation part is developed under the virtual laboratory environment (VLE) platform [2] where each model is a component of the VLE framework. We expose the bases of the development of the simulation models, with a definition of the generation cycle of agents and their network. Social multilayer and its translation under VLE and part of visualization and exploitation of simulation results have been developed under R.

B.2 Background and general architecture

B.2.1 Context

As part of the work, Bouanan *et al.* [1] have developed an M&S framework to model social connections and propagation phenomena within a multiplexed social network. The main topic of this appendix is the description of the modeling structure adopted in this approach. The work to represent individuals and social networks comes from a multidisciplinary development between computer science and sociology proposed in the frame of the French DGA SICOMORES project. These works are exploiting the VLE platform [2] based on the work of Zeigler [3]. In this appendix, we outline the basics of the proxy-server architecture and services of the M&S model, with a definition of the network generation cycle, the discrete event simulation services, and the analysis and visualization of simulation results.

B.2.2 General architecture

We distinguish three main phases in the study process of information dissemination phenomena:

- Pre-simulation: the generation of the static model of the population;
- Simulation: the instantiation of the models of the agents and their networks (to represent the behavior of the individuals and their interactions) as well as the simulation of the propagation of information in the network; and
- Post-simulation: the analysis of results (accompanied by actions).

With the increasing complexity of systems, modeling and simulation require more and more powerful computer tools. In particular, research fields such as sociology use modeling and simulation software to better understand the dynamics of the studied systems. The main challenge is to respect the modeling and simulation cycle presented by moving in a model-based approach from different steps: social theory conceptual → formal models → simulation → and then tools to visualize for analysis.

Indeed, the modeler must be able to define his models via the use of several formalisms. He must be able to define experimental frameworks by manipulating the inputs and outputs of his simulations. Then he needs to refine the parameters of his models according to his observations (interpretation of the simulation results) and potentially to change the behavior of his models if they do not provide the expected values.

The steps have the following translations in our model:

- Conceptual models: corresponds to the social data of the population and to the topology of the social network. These data are transformed into formal models.
- Formal models: atomic models, or model compositions provided by the platform or developed by the modeler.
- Simulations: VLE is the simulation engine of the platform; it provides the translation of the formal part into an operational version based on the abstract DEVS simulator. The outputs of the simulation are of two types: during simulation or a posteriori. They are obtained through observation components and data streams.
- Analyses: a set of tools for real-time observations of simulation, by the platform of statistical tools R [4].

B.3   Population generation for network simulation

B.3.1   General presentation

Most of the models implemented in the field of simulation of complex systems require a large number of agents to initialize, calibrate, and run a studied system. Accurate reproduction of the initial states is extremely important to obtain reliable predictions from the model. In this module, we propose static graph generation algorithms by combining different data sources to obtain an individual representation of the agents, approaching as closely as possible the correlation structure of a real population. Another important element that most models also need is the location of the agents in a geographic study environment.

B.3.2   Dimensions and algorithms

Here, we present the three dimensions (family, friendship, and neighborhood) constituting the primary group of a society [5]. The author differentiates between the primary group that groups individuals who spend more time together (e.g., family, friendship, or neighborhood dimension) and the larger and more specialized secondary group in which members engage in impersonal, goal-oriented relationships (e.g., formal organizations, associations, etc.).

B.3.3   Nodes

The general algorithm for generation of the population starts by creating the number of nodes desired by the user for his grouping (region, city, and village within the framework of a geographically identified company or department, service within the framework of an organization). Each node represents an individual. In the rest of this section, we illustrate our approach with examples of the civilian population type.

Nodes are defined by a number of characteristics; some allow to create relationships, others will only be used for the dissemination of information.

The characteristics to describe an individual are as follows:

- Vital data: sex, age class, social level, intellectual level, language, etc.;
- The roles of everyone in the family: head of family, husband, wife, child, related individuals;
- Situation data: need for safety, food, housing, health, social relations;
- Accessibility data: television, radio, illiteracy, mobile phone, and internet;
- The opinion concerning an event; and
- Cultural characteristics.

These data make it possible to characterize the individuals according to the information given by the user through the configuration file (input data).

B.3.4   The family dimension

During the family generation phase, the algorithm will assign the nodes' different attributes such as sex, age class, etc. These works have been proposed in the frame of the SICOMORES project [1]. The results concerning the family generation have been published in [6]. The structure that connects the family will be used to recommend a medical resource within the family or to find in a professional network a close specialist.

In this dimension, we define three types of family: the nuclear family, the extended family, and the expanded family. Each family will be created from a function. The algorithm for generating the family dimension will call each function to build the family dimension. The generation of these three family groups must be taken into account in our algorithms to create the desired final population. For the phenomenon of information propagation, it is important to define the role of individuals in the family.

The algorithm asks for the number of single persons in the group, the proportions of each type of family in the group. Then it chooses the attributes that will be generated. Finally, the algorithm generates single people whose number may be more or less important depending on the type of grouping.

The nuclear family: also known as the conjugal family, is composed of a head of household (H), his or her spouse (W), and their children (C). In a first step, the function that generates a family of nuclear type begins by assigning the roles of husband and wife with their respective characteristics. Finally, the algorithm randomly chooses a number between one and five representing the number of children.

The extended family: is a broad residence group consisting of several married men (H and W) and their offspring (C). The "GenerationFamilyExtended" function proposed by [6] generates an extended family whose number of members varies

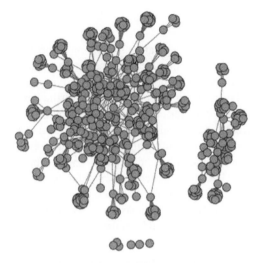

Figure B.1   Family size of a population of 550 individuals with 10% of individuals alone

between 9 and 20. The algorithm first creates the couple or the person (in the case of matriarchal families) including the head of the family. Then, it generates the other couples being part of the family and finally the children.

The expanded family: consists of a nuclear family and often related individuals (I). The generation algorithm calls the function that makes it possible to generate the nuclear family and then creates the individuals related to the family. The general algorithm will call each of the functions according to the proportion of each family type (Figure B.1).

The friendship dimension: The social bond of friendship is one of the three relationships of primary sociability [5]. The relationship of friendship is based on homophilia: people tend to bond with people who look like them: same age, same social class, etc. Thus, to transcribe the homophilic nature of friendly relations in our algorithms, we have created in [6] an equation taking into account the characteristics of individuals. Each parameter of the equation can be weighted to increase or decrease its importance. Some societies may favor friendship relationships between people of the same sex, ethnicity, etc. The equation is used in the algorithm for generating friendship links.

The neighborhood dimension: uses the geographic locations of individuals to create the links. It should also be noted that this geographical location may be used later to define reception areas of a message according to the means of transmission (broadcasting a message by radio or distributing leaflets over a specific geographical area). It should also be noted that since each family is considered to be a household in the family dimension, the geographical position is identical for each member of the family. The neighborhood link generation algorithm requires as input the maximum distance between two families to be considered as potential neighbor. The geographical visualization of the individuals is done from the software R, the example detailed after is mostly based on this dimension.

B.4   Conclusion

We have presented some of the population generation algorithms that are used in the chapter, focusing on the primary group containing the dimensions: family, friendship, and neighborhood. The social network is of multilayer type, i.e., it consists of several differentiated social layers. Each layer represents a distinct and specific sociability space; communications will differ in family, among friends, or with neighbors. The types of relationship between individuals vary from one layer to another and also from one society to another. Outside the primary group, we can generate other interconnection links between individuals depending on the application, for example, political, religious, etc. The global structure for population generation is composed of two elements: the input data file that allows the parameter values to be configured for the desired population and the database that will make it possible to link the module of the population generation and simulation of the propagation of information.

B.5   Simulation of information propagation module: simulation

B.5.1   *Operational modeling agent*

Operational modeling leads to the development of an operational model involving choices of modeling and simulation architectures. This step proposes a solution of implementation of the conceptual level of the studied phenomenon.

We propose to integrate the individual characteristics of the agents during the development of the operational model. For the representation of agents and their behavior at the operational level, we propose a modeling approach to differentiate business processes and decision-making processes. Each agent is composed of two types of models: the server model for the cognitive process and the proxies models for the operational processes. In the agent decision module, we implement the cognitive processing of messages and events. This part allows agents to act autonomously to achieve their goals by adopting a complex behavior through the implementation of influence processes. Proxies adopt simple behaviors to carry out the filtering activities of external events. We propose an architecture describing the links between the different models.

It is from this observation, namely, taking into account and exploitation of events, actions, and interactions between the entities of the real system, that our thinking was built. Complex dynamic systems like multilayered social networks are composed of distinct entities and several means of interaction. An actor represents a decision-making entity, for decision-making, or an operational entity, for carrying out actions and relations for the transmission of information. In order to understand the dependency relationships between these types of activities, namely, the diffusion and influence model, we propose a modeling and simulation architecture based on the DEVS formalism. Figure B.2 illustrates the approach chosen for modeling actors in a multilayered social network.

Agent server model behaviors are defined as states and are related to the behaviors of the associated proxies. The states specify the actions performed following message reception from the proxies. The interactions between the servers and

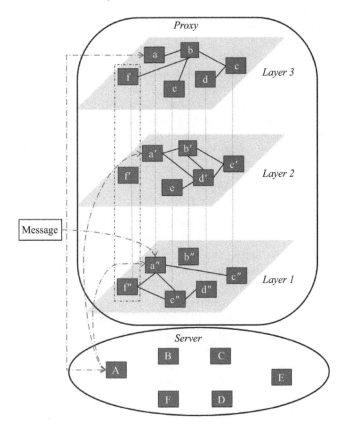

Figure B.2   Agent model architecture in a multiplexed social network

proxies modules are formalized by sending and receiving messages. These interactions can be broken down into three possibilities: (1) the interactions between proxies agents of the same layer, (2) the interactions between proxy and server, and (3) the interactions between proxies of an agent but between different layers.

B.5.2   Representation of agent behavior

Agents can return intelligent behaviors, results of autonomous actions that they realize within their environment. Agent behaviors are specified as state graphs using the DEVS formalism. From the information available, the agent processes the information, updates its variables (opinion, needs, etc.), and decides whether to propagate the message to the other agents or block the broadcast at this level.

Two types of events can appear to dynamically animate the graph: internal events and external events. External events are related to the receipt of messages and correspond to the communication aspect of the multi-agent system. Internal events are the internal activities of the agent. Transitions allow the transition from one state to another. These states can be active (the execution of an action) or passive (waiting states). Passive states are modified when an external event arrives (e.g., arrival of a message).

The proposed example shows the behavior of an individual when receiving a message and checks whether the message is interesting to accept and process it. In the event that the message is not interesting, the agent will not propagate it to other individuals. The description of the dynamic behavior of the system requires the implementation of the IB of the agents as well as the implementation of the inter-actions between the agents. An example of cognitive behavior will be presented in detail in the first part of Chapter 5.

B.5.3   Simulation environment

B.5.3.1   Presentation of the VLE

VLE [2] is a multi-modeling and dynamic systems simulation platform based on DEVS discrete event formalism [3]. VLE allows you to specify complex systems in terms of reactive objects and agents, to simulate the dynamics of the system, and to analyze the results of the simulations. The libraries provided also allow the development of custom programs. We can mention the RECORD project that uses VLE to model and simulate agroecosystems [7].

Modeling of a system in the VLE platform is related to the GVLE program. GVLE is a graphical application for specifying the behavior of atomic models, composing new models via coupled models, and defining model hierarchies (Figure B.3). In addition, it has the ability to set up the experiment plans with the initializations and the outputs of the simulations.

The behavioral development of atomic models uses the concept of object-oriented programming. Indeed, VLE proposes the class DEVS::Dynamics as functional interface for the development of behavior of atomic models [3]. VLE uses object-oriented programming concepts such as inheritance and polymorphism to simplify

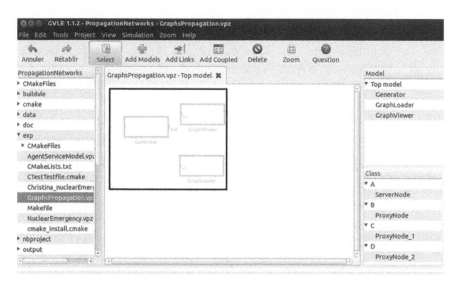

Figure B.3   Graphical interface of the GVLE modeling environment

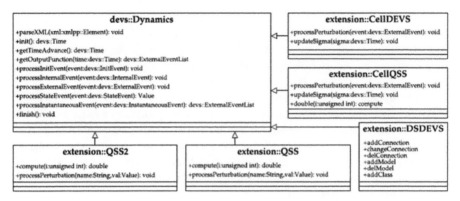

Figure B.4   Simplified UML class diagram of the inheritance tree for developing the behavior of atomic models

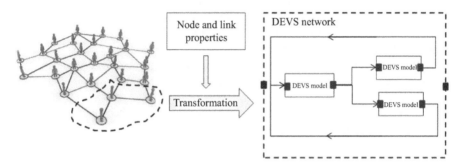

Figure B.5   Transforming a graph into DEVS

development. The platform provides a set of classes inheriting from DEVS::Dynamics with a predefined behavior. For example, the extension::QSS class is suitable for solving differential equations [8] and the extension ::DSDEVS class is adapted to the change of structure in the model hierarchy in DEVS [9]. These classes overload and protect the virtual methods of DEVS::Dynamics and provide a different functional interface but adapted to the model to be developed. Figure B.4 illustrates the simplified UML class diagram of the inheritance tree implemented in VLE.

B.5.3.2   Implementation of propagation phenomenon models in a multilayer social network

The implementation of a DEVS model under VLE is achieved by a legacy of the DEVS atomic model class, the DS-DEVS model class, or another DEVS extension of the VLE framework. We present here the auxiliary models (GraphLoader, Generator, and Observer) introduced in our model to model and simulate the propagation phenomena within a multilayered social network.

Figure B.5 shows the classes illustrating the dynamics and behavior of the auxiliary models introduced in the simulation model.

Generator: This model is connected to the agents according to the means of communication chosen to transmit the information. In this class, we initialize and adapt the message to be propagated within the network according to the configurations set by the user of the model. The behavior class of this model inherits from the DEVS::Dynamic class to exploit DEVS formality specifics.

GraphLoader: The behavior class of this model inherits the extension:: DSDEVS class and allows the manipulation of graphs: transform a static graph into a DEVS model network automatically (either in the initial state or even in execution). We use GraphLoader to load the various information stored in the database, create the different DEVS models (server models and proxies' models), and connect them according to the network topology presented in the database (relationship table) as shown in Figure 9.16.

The dynamics of the "GraphLoader" model controls the network of agents and their connections. When he receives a change order, it builds the new connections and destroys the old ones. This template inherits the DEVS::Executive class in order to be able to manipulate the connection graph in runtime. As the class "GraphLoader" inherits from the class DEVS::Executive (which is based on the DS-DEVS specification), it can use a specific Application Programing Interface (API) that allows you to build, delete, or modify connections.

Observer: The "Observer" model (Figure B.6) is connected to all agents. It saves in the output file the current state of the model at a given time and for a specific port.

Figure B.7 summarizes the mechanism to transform a static graph into a dynamic DEVS network. This diagram also illustrates the proxy-server approach developed in this work.

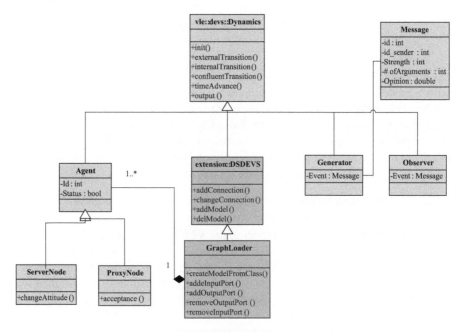

Figure B.6   Simplified UML simulation module diagram

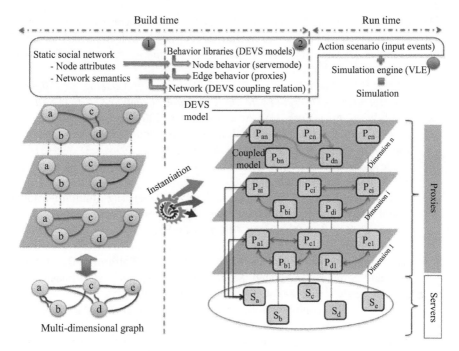

Figure B.7   Conceptual overview

B.6   Conclusion

In this section, we presented the concepts of development of the architecture and the platform for modeling and simulation of multiplexed social networks which integrates the works presented originally in [1]. The model allows in a first step to generate an agent society characterized by a set of attributes and variables based on the demographic and sociocultural characteristics of a designated area. Then, within the VLE platform, we have developed an architecture to dynamically transform a network of actors into instances of DEVS network models. We have also proposed auxiliary models to initiate the simulation of the propagation of a message within the network. This conceptual methodology has been developed in [1] as summarized in Figure B.7.

References

[1]   Bouanan Y., Zacharewicz G., Ribault J., and Vallespir B. Discrete Event System Specification-Based Framework for Modeling and Simulation of Propagation Phenomena in Social Networks: Application to the Information Spreading in a Multi-layer Social Network. *Simulation*. First Published: 1 June 2018, https://doi.org/10.1177/0037549718776368.

[2]   Quesnel G., Duboz R., and Ramat E. The Virtual Laboratory Environment –
An Operational Framework for Multi-Modelling, Simulation and Analysis of
Complex Dynamical Systems. *Simulation Modelling Practice and Theory.*
2009; 17(4): 641–653.

[3]   Zeigler B.P., Praehofer H., and Kim T.G. *Theory of Modeling and Simulation: Integrating Discrete Event and Continuous Complex Dynamic Systems.*
2nd Edition. New York, NY: Academic Press; 2000.

[4]   Ihaka R., and Gentleman R.R: A Language for Data Analysis and Graphics.
Journal of Computational and Graphical Statistics. 1996; 5(3): 299–314.

[5]   Cooley C.H. *Social Organization: A Study of the Larger Mind.* New York,
NY: Charles Scribner's Sons. 1909.

[6]   Forestier M., Bergier J.-Y., Bouanan Y., *et al.* Generating Multidimensional
Social Network to Simulate the Propagation of Information. *Proceedings of
the IEEE/ACM Conference in Advances in Social Networks Analysis and
Mining*, Paris, France, 25 Aug 2015. pp. 1324–1331.

[7]   Bergez J.-E., Chabrier P., Gary C., *et al.* An Open Platform to Build, Evaluate and Simulate Integrated Models of Farming and Agro-Ecosystems.
Environmental Modelling & Software. 2013; 39: 39–49.

[8]   Kofman E. A Second-Order Approximation for DEVS Simulation of Continuous Systems. *Simulation.* 2002; 78(2): 76–89.

[9]   Barros F.J. Modeling Formalisms for Dynamic Structure Systems. *ACM
Transactions on Modeling and Computer Simulation.* 1997; 7(4): 501–515.

Chapter 10

Surveillance of avian influenza in Vietnam

10.1 Context

The emergence of One Health and EcoHealth approaches over the past decade, recently complemented by the Planetary Health movement [1], illustrates the consolidation of a consensus in the veterinary and public-health domains that there is a need for integrated, interdisciplinary, and inter-sectoral approaches to better understand health issues and to improve the sustainability of interventions targeting individual and population health. The recent publication of the One Health theme issues in the *Philosophical Transactions of the Royal Society B* [2] and *Frontiers in Veterinary Science* [3] journals attest of the importance of developing integrated approaches to health (IAH).

IAH call for transdisciplinary efforts integrating scientists, citizens, government, and private sectors to collaborate in designing and implementing actions to enhance sustainable health management across human, animal, and ecosystem interfaces. Regarding the complexity of the system to be addressed, such approaches remain sparsely used when it comes to public health related interventions. To deal with the complexity of IAH, Duboz *et al.* [4] propose to root IAH development of methods and tools in systems thinking. Ross and Wade [5] present systems thinking as a set of skills used to improve the capability of identifying and understanding systems, predicting their behaviors, and planning change to produce desired effects.

Systems thinking is a practice based on systems theory. It addresses concrete problems where the complexity of the system makes it hard to grasp. When adopting system thinking, one recognizes that participation is key in gathering the relevant knowledge to describe the system and to obtain stakeholders adherence to the decisions of change (ownership). Therefore, to root IAH development in systems thinking, and its formalization in systems theory, we propose to use participatory modeling combined with the Discrete Event Systems Specification (DEVS) framework, as explained in Chapter 4.

Numerous participatory modeling methods exist, all derived from the field of collaborative learning which appeared in the late 1960s [6]. We do not present in detail one particular participatory modeling technique in this book. Instead, we illustrate it through an example. The interested reader can refer to a recent and comprehensive review done by van Bruggen [7]. To summarize, participatory modeling mobilizes the implicit and explicit knowledge of different actors to build a shared representation of reality [8–11]. It facilitates knowledge sharing and the generation of new knowledge to support negotiation and planning. As such, participatory modeling supports decision-making and adaptive management [12]. In participatory modeling, *the model is not the final product but an intermediary object used to foster dialog.*

Although participatory and integrative modeling has been highlighted recently for its potential to deal with IAH complexity [13,14], its use in the context of IAH research and intervention remains minimal [4]. When considering zoonosis, i.e., contagious diseases that are transmitted from animals to human (60% of the contagious diseases), an IAH is needed as both the human and animal sectors have to be mobilized to control the spread.

In this chapter, we illustrate how an IAH applied to zoonosis is achieved by using participatory modeling and a DEVS approach. We focus on particular components of healthcare systems, namely the surveillance systems. The World Health Organization (WHO) defines public health surveillance as "the continuous and systematic collection, analysis and interpretation of health-related data needed for the planning, implementation, and evaluation of public health practice" [15].

Surveillance has the following three main objectives:

• To serve as an early warning system for impending public health emergencies;
• To document the impact of an intervention, or track progress toward specified goals; and
• To monitor and clarify the epidemiology of health problems, to allow priorities to be set, and to inform public health policy and strategies.

Surveillance systems are associated with control strategies. If we are able to rapidly detect an outbreak, intervention measures can be triggered quickly, thus reducing the disease burden (mortality, morbidity, and associated costs). Surveillance and control together form a mitigation system, which is a combination of financial, material, and human organizations and resources [16]. Controlling epidemics implies the isolation of the contagious fraction of the population from the one which is susceptible. It can be done using vaccines, drugs, quarantine measures, or culling in case of animals.

These control measures take time, involve manpower and equipment, and therefore have associated costs. Surveillance activities such as surveys, sampling, analysis in laboratories, reporting, etc. have associated costs as well. The capacities of the sanitary authorities or the private sector to implement mitigation are limited. This is particularly true in developing and emerging countries, where funding and competencies are scarce resources.

In this chapter, we present the surveillance of a zoonosis in Vietnam, the highly pathogenic avian influenza (HPAI). Birds are the natural hosts for avian influenza (AI) viruses and some subtypes, such as H5N1 and H7N9, can be transmitted to human with severe consequences in term of morbidity and mortality. HPAI outbreaks have direct and immediate impacts through morbidity, mortality, and private and public prevention, surveillance, and control costs [17]. An outbreak of HPAI (subtype H5N1) virus occurred in 1997 in poultry center of Hong Kong. Since 2003, this AI virus has spread from Asia to Europe and Africa. In 2013, human infections with the AI subtype H7N9 virus were being reported in China and as recently as 2017, human cases were still reported by Chinese authorities.

Dr. Margaret Chan, Director-General of the WHO, raised an alert on AI risk of pandemic [18]. In November 2016, nearly 40 countries have reported outbreaks of HPAI in poultry or wild birds. Therefore, concerned countries should react by investing in HPAI mitigation systems. Vietnam had an active control program against H5N1 AI since the disease was first detected in 2003. This control program has been successful and cases of H5N1 in poultry and people have declined progressively and significantly. Nevertheless, HPAI is endemic in Vietnam and viruses are naturally mutating and reasserting. New outbreaks of a new subtype occurred, as the one started in February 19, 2018, in Hai Phong province in the north. Therefore, surveillance of HPAI is still a major concern.

In 2014, two workshops were organized by the "Centre International de Recherche Agronomique en partenariat pour le Dévelopement" (CIRAD) in Hanoi, Vietnam, in the frame of two research and development projects: REVASIA (research on innovative tools for the evaluation of zoonotic disease surveillance systems in South East Asia) and READI, the Regional European Union—ASEAN Dialog Instrument. The first workshop took place the 20th of March, and the second, the 28th and 29th of October, both hosted by the National Institute of Health and Epidemiology (NIHE) in Hanoi. The general objectives were to discuss on the links between animal and public health surveillance systems regarding AI.

In the following, we describe the participatory modeling activities that have been done and the associate results.

10.2 Method

Surveillance systems are composed by autonomous and loosely connected subsystems and therefore can be seen as systems of systems. Their composition varies regarding the objective of the surveillance. Generally, the main actors/components of a surveillance system are physicians, national and private hospitals, national and private laboratories, and dedicated national agencies.

The preparation phase of any participatory method includes the selection of the set of participants. This set should be composed by representatives of the main components of the whole system of interest. In our case, we invited participants from departments and organizations under the Ministry of Health (MoH) or the Ministry of Agriculture and Rural Development (MARD) of Vietnam: the NIHE, the National Institute of Veterinary Research, the Department of Animal Health and the National Institute of Animal Husbandry. We also invited national research organizations working on AI in Vietnam: The Hanoi University of Agriculture and the Nong Lam University. One representative of the Emergency Center for Transboundary Animal Diseases of the United Nations Food and Agriculture

Organization was invited as an observer. Experts in participatory modeling from the Agricultural Faculty of Chiang Mai University, Thailand, and from CIRAD, France, were in charge to facilitate the workshop.

Participatory modeling is an iterative process. It can be described as a cycle with five consecutive steps, with the possibility to reiterate the steps as much as desired. The process ends when participants agree they have reached the objective, or when the elapsed time-frame allocated for it is over. Figure 10.1 presents the participatory modeling cycle used in this work. We coupled the PARDI method that has been developed in the frame of the ComMod participatory modeling approach applied natural resources management [11], with the implementation of the model and the simulation of scenarios. PARDI facilitates exploring of complex issues with heterogeneous stakeholders managing and using common resources. "P" stands for the problem, a common/specific issue of interest. "A" represents the actors involved in using and/or managing the common resources "R." "D" stands for the dynamics that drive the system to change over time. "I" is the set of interactions among the actors as well as between the actors and the resources, the infrastructures or the environment. PARDI employs facilitation techniques to animate the meetings in such a way every participant can express her/himself. Classically, a participatory meeting is animated by one or two facilitators, and the number of participants should not exceed 20. If more participants are required, it is preferable to split the group into subgroups corresponding to subsystems.

Table 10.1 gives a summary of the steps in Figure 10.1 and their relation to the methodology of Chapter 4. The reader can refer to the table for an overall perspective as each step is outlined below.

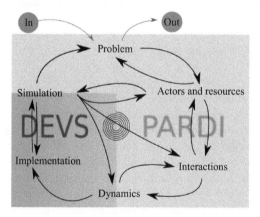

Figure 10.1   The participatory modeling cycle. The PARDI method is associated with implementation and simulations steps. The arrows indicate the possible paths between the different steps in the method. The cycle starts at the "In" circle, defining a problem or an issue. It should end when a solution has been found to the problem or a decision made regarding it. The light grey part of the figure involves all the stakeholders, including the modelers, the dark grey section involves mainly the modelers

Table 10.1   Participatory modeling steps and actions

Steps	Participant actions	Relation to methodology of Chapter 4
Step 1	Participants agree on a common problem/issue	The system of interest is identified
Step 2	Identify relevant actors/resources/ variables and choose the ones to observe/measure/control	• The components of the system are identified • The experimental frame is characterized
Step 3	Concept maps for interactions are drawn	The outlines of a base model for the system are defined
Step 4	Activities/dynamics are identified	The state transition function is defined
Step 5	Specification as a simulation model Also, the activities are pruned to those most relevant to the problem solution	The model is implemented in a DEVS-simulation environment. Also, for selected EF, base model is reduced to lumped model

10.2.1   Problem

The first step is for the participants to agree on a common problem/issue to address. This step can be very time consuming. It is important to inform participants they have to reach a consensus in a restricted time. It is important to recall that this step, as the others, can be repeated as many times as possible. Every participant proposes one sentence, preferably a question, to summarize the issue to address. The facilitator collects them and engages the group in a discussion to agree upon one definitive sentence. This sentence can be revised later.

10.2.2   Actors and resources

The problem stated in the previous step frames the set of actors and resources directly concerned by it. Actors are autonomous entities; they can behave and make decisions on their own. They can be individuals (human and animals) or groups, institutions, companies, etc. They are motivated by goals, intentions, which are explicit or implicit. Resources can be natural (water, gas, forests, etc.) or manufactured (electricity, information, money, tools, etc.). They may have an inherent renewing dynamic but cannot make decisions or be proactive toward an actor. The actors use or manage the resources. At this stage, participants can start identifying actors and resources variables to act on, and/or to observe and measure, to be able to answer the question or to solve the problem. This step is the starting point for the specification of the experimental frame described in Chapter 4.

10.2.3   Interactions

The PARDI method uses Actors diagrams to draw the set of actors and their relations with each other and the different resources. Actors and resources are labeled boxes. Relations are lines connecting the boxes. The relations symbolize actions and should be labeled with a verb or active sentence. The relations are drawn by participants during a session where the facilitators pay attention to the discussions between participants. Indeed, at this stage, they will start to define the dynamic of the whole system by rendering explicit when, how, and why these interactions between actors and resources exist. The diagram can be simply drawn on a board, or a flip chart, or using a graphical modeling language or a mind mining tool. In this chapter, we used UML and the Cmap tool developed at the Florida Institute for Human and Machine Cognition. Cmap enables to easily draw diagram to figure out the knowledge regarding a particular issue. It empowers participants to build, share, and criticize knowledge models represented as concept maps [19].

10.2.4   Dynamics

At this step, the participants discuss how the system changes over time. The discussion is organized in two stages. The first stage consists in clarifying the interactions described in the previous step. The participants explain what the interactions are made of (energy, information, and materials) and when and how long do they take place. UML activity diagrams can be used to describe the interactions between actors and resources. They provide a very convenient way to record the elapsed time spent by actors and resources to perform activities, before, during, and after the interactions. As such, by codesigning activity diagrams, the participants identify the set of activity belonging to each actors and resources. These activities are specified in the second stage of this step, where participants provide their knowledge regarding the internal changes occurring in each actor and resource. They list the set of internal configurations and explain how, when, and why the actor and resource transit from one configuration to the next one. These dynamic changes can be captured using states diagram or state charts for discrete changes, or with Forrester diagrams for continuous changes for instance. This step is very technical. The participation of an expert in dynamical systems modeling is required. By going through steps 1–4, we move up in the specification hierarchy described in Chapter 4.

10.2.5   Implementation and simulation of the model

The implementation step is not participatory. It is realized by one or several experts in modeling and simulation, which should be involved in the previous steps either as observers or as facilitators. For the formalization and the implementation of the model, we choose the holistic multi-perspective modeling and simulation framework described in Part 1 of the book. We indeed need to model two different perspectives, the avian flu surveillance system and the spread of the disease in animal population. Several types of this last model can be found in the scientific literature and can be reused and coupled in a multi-perspective framework. The implementation of the model has been done using the virtual laboratory environment (VLE) [20]. VLE is a modeling and simulation software, which implements a parallel and dynamic structures DEVS simulator. It enables developing plugins for particular formalisms implemented in DEVS simulators, such as the Forrester's diagrams mentioned above.

The simulation step should be participatory. The stakeholders propose scenarios in the form of set of parameters values and variable to observe. By confronting the simulation results, the engage in discussion regarding the whole system and its dynamic. At this step, the may decide to change the EF or to review the model, and reengage in an additional turn in the participatory modeling cycle.

10.2.6   Relation to the methodology developed Chapter 4

A perspective corresponds to a particular point of view on the system. It corresponds to a particular set of stakeholders. For any perspective, participatory modeling can be engaged, and, as stated in Chapter 4, the merging of the different perspectives should be participatory too. After one participatory cycle is achieved, one specific model is produced. Depending on the problem posed, it can be desirable to produce different versions of models, representing alternatives to test. If the problem changes, then the EF can be questioned and altered. It can eventually lead to a new perspective. Moreover, it may become necessary to limit the complexity of the model developed, so that a validity-preserving simplification process is undertaken, as will be seen in the next section.

Participatory modeling navigates in the specification system levels described Chapter 4, starting at level 0 with the definition of the system and the experimental frame, it jumps to the coupled network level, focusing on the interactions between the different elements of the focal level without detailing internal dynamics of the elements. Then, by discussing the dynamic of the interactions, participants provide the knowledge at the I/O function relation observation level. Thereafter, they detailed the internal dynamics of the components considered independently as open systems. Table 10.1 summarizes participants' actions through the modeling steps and shows their relation to the methodology presented in Chapter 4.

10.3   Results

In this section, we present the results of the two participatory meetings mentioned above. During the two meetings, participants went from step 1 to step 4 in the modeling process, continuously revisiting the problem, the list of actors, the interactions, and the dynamics. Hereafter, we present the final product of the meetings. A total of 13 participants were present [8 from public health and 5 from animal health (AH)].

10.3.1   Problem

Participants agreed on the following common issue: "How to improve AI surveillance and control in Vietnam?" This question encompassed the aspects of "developing mechanism of information sharing between AH and public/human health (PH) sectors," as well as "communication with policy maker" and "promoting inter-sectoral collaboration." Using PARDI framework, the participants started working on "zoonotic diseases surveillance issues," collectively narrowed down, and defined the common issue of interest that is relevant to zoonotic surveillance system in Vietnam.

10.3.2   Actors and interactions

During the first workshop, participants started by drawing two conceptual models, one for AH and one for PH. They finally joined the two models in a unique one presented in Figure 10.2. This model is very complete and necessitated to narrow down the system by focusing on the issue.

In the second workshop, participants agreed on a second version presented in Figure 10.3. Then they decided to join the two models to describe the interactions between AH and PH surveillance systems.

Finally, a simplified version was drawn to focus on the main actors and interactions describing the interactions between the two systems when an outbreak of avian flu occurred. To do that, the participants were asked to address the following question: "What are the most important actors for the collaboration between AH and PH regarding the surveillance of avian influenza in Vietnam?" The participants had to list the actors they considered important on a color card and to provide justification. These actor cards were reviewed during a group discussion:

- Actors mentioned by all the participants were kept and considered as "critical."
- The actors which were not mentioned by any of the participants were removed from the diagram from a collective consensus.
- The relevance of actors mentioned by few participants was discussed until reaching a general consensus to keep or remove them.

Figure 10.4 presents the conceptual model resulting from the second workshop.

During the second workshop, participants, assisted with a modeler, drew a UML class diagram reflecting the conceptual model previously designed. They have to label the different actors (e.g., farmers, patients, MoH, etc.) and interactions, and choose the variables and functions which define them (e.g., "report," "consult," and "send samples").

Figure 10.5 presents the UML class diagram elaborated based on the relational diagram validated during the workshop and on the information generated on each type of actors by the workshop participants.

Finally, participants selected indicators that needed to be considered, such as the following:

- Timing in reporting AI cases and delays between actions
  - Delay between a case report and its confirmation,
  - Delay to inform the other authority,
  - Time intervals between all the actors actions;
- Early announcement/delay
  - Media to public; and
- System activity
  - Duration of internal tasks for each actors.

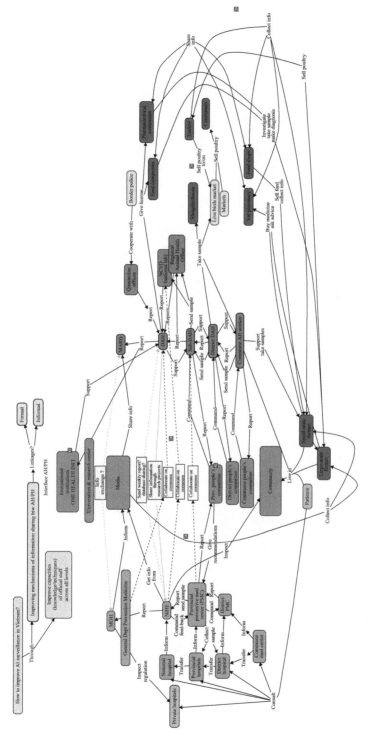

Figure 10.2   Full conceptual model of the animal and public health surveillance systems of avian influenza in Vietnam generated during the first workshop

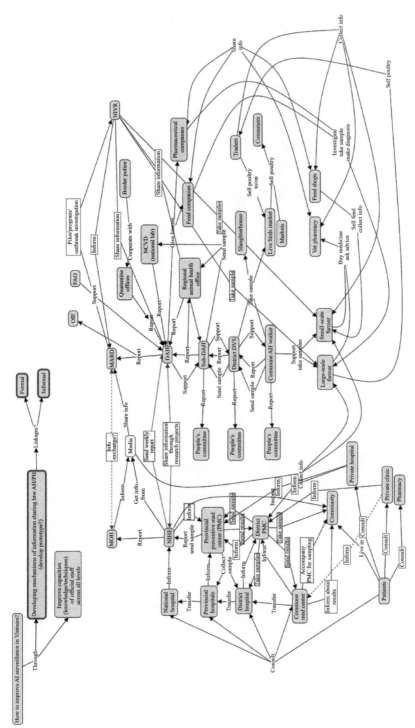

Figure 10.3   Animal and public health surveillance network of avian influenza in Vietnam. Conceptual model revisited and validated during the second workshop

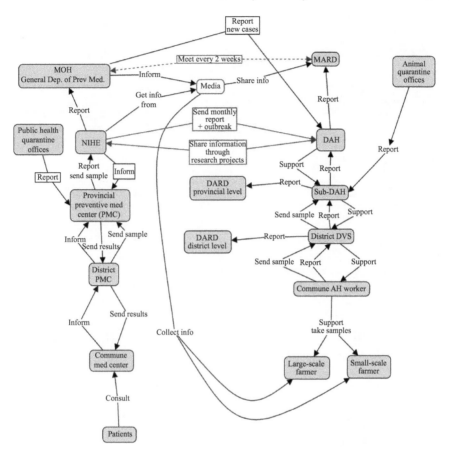

Figure 10.4   Simplified actors and interactions diagram produced using the Cmap tool

10.3.3   Dynamics

With the help of a modeler, the participants draw two UML sequence diagrams. They described the set of interactions between the actors and identified the tasks these interactions trigger in the actors. To make it, they discussed the sequence of actions taken upon a suspicion of a case of AI in human and in animals. Figure 10.6 (a) and (b) presents the resulting sequence diagrams.

The sequence diagrams provide a view of the system dynamic. In a DEVS context, they figure out the necessary coupling between models (here the actors), the type of events that will be exchanged between them, the scheduling of events, and partially design the external and internal transitions and output functions of the DEVS models.

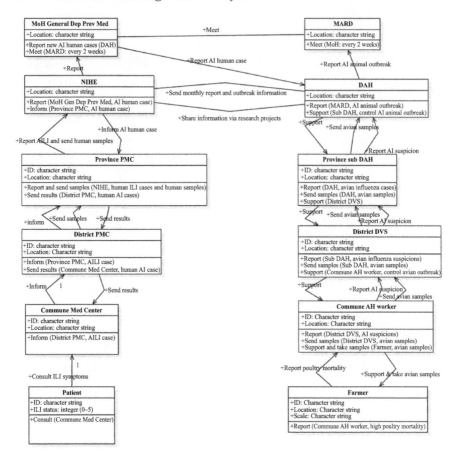

*Figure 10.5   Class diagram of the animal and public health avian influenza
surveillance networks in Vietnam*

10.3.4   Implementation

We followed the methodology described in Chapter 4. Figure 10.7 presents the
pruning of O4HCS (ontology for healthcare systems simulation) for the surveillance of AI in Vietnam, while Figure 10.8 presents the graphical interface
of VLE with the implementation of the corresponding multi-perspective
model.

It is noteworthy that surveillance and control in epidemiology are part of
community care. It appears in O4HCS under the umbrella of home care. A further
distinction could be made between home care and community care, the former
being restricted to care given at home and the latter involving public health concerns such as surveillance, information/vaccination campaign, etc.

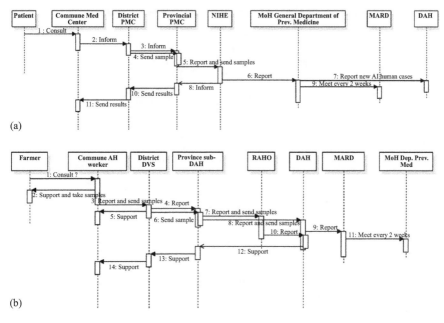

(a)

(b)

Figure 10.6 Sequence diagrams of actions taken upon suspicion of a case of avian influenza either in human (a) or in poultry (b). The labeled boxes represent the actors. The dashed lines are the time lines spanning from the top to the bottom and the rectangles on them figure activities with a duration proportional to their length

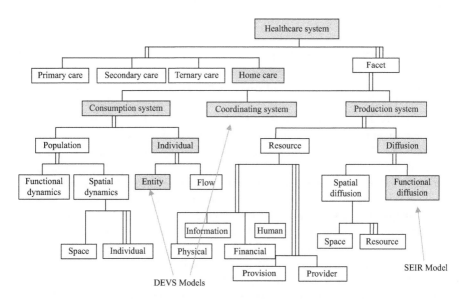

Figure 10.7 Pruning of O4HCS for the surveillance of avian influenza in Vietnam

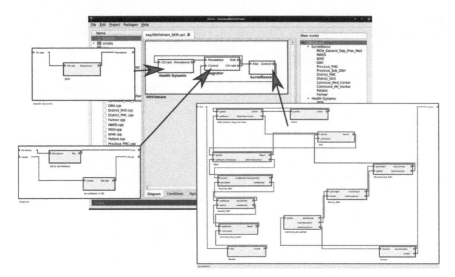

*Figure 10.8   Implementation of the multi-perspective model in the virtual
laboratory environment (VLE)*

10.4   Conclusion

In this chapter, we illustrated how participatory modeling involved the stakeholders in the design of a multi-perspective model and how the participatory cycle is related with the methodology developed in this book. The ambition was not to provide the full formalization of the dynamics and simulation results but to show with a concrete example how the inclusion of the necessary knowledge owned by non-modelers to describe a system can be harvested and formalized to produce a simulation model that is useful for them.

As the stakeholders are involved in the model design, the model is not black box for them. They have a sense of ownership making the model useful for them. They can discuss simulation results and propose model structure alternatives. In the case presented here, a pilot simulation model has been built based on the class and sequence diagrams to assess the timing of information exchanges and actions between the AH and human health surveillance systems, regarding different detection thresholds in human and animals. Participants can learn from their respective actions and collectively innovate to improve the system considering its timeliness to detect an outbreak and control an epidemic.

Participatory modeling also highlights the gap in knowledge that the participants have on the system, and the necessity to involve new participants, to conduct new investigations and researches. For instance, during the second workshop, many discussions were ongoing on how the information is being exchanged at all levels. The means of exchange of information between MoH and MARD was not clear ("they have official collaboration and exchange information"). This point highlighted the need to involve a representative from all the actors mentioned in the diagram to clarify the different links and actions. In the first workshop, it was clear from the development phase of the conceptual diagram of the surveillance processes that essential actors were missing from the group of participants. These actors were from the implementation of the surveillance systems at local level. This limitation was previously identified by the experts during the preparatory phase but could not be addressed at this stage because of language issue.

The training of Vietnamese researchers in participatory modeling would have been the best approach to involve all the actors mentioned in the diagram (local level and ministerial level); an alternative would have been to ensure simultaneous translation during the workshop. The Vietnamese researchers involved in the work took responsibility for running some of the discussion and had many inputs on how the collaboration between public health and AH sectors should/could work.

A multi-perspective approach to modeling and simulation in the context of IAH is becoming of primary importance since, as for any other holistic effort to understand or manage complex adaptive systems; IAH is multi-sectorial and interdisciplinary. It implies the participation of the relevant stakeholders and the use of advanced methodologies and tools to merge and articulate heterogeneous and distributed knowledge. The methodology proposed in this book fits these requirements and is a clear support for participatory modeling in the context of IAH.

References

[1] Horton R., Beaglehole R., Bonita R., Raeburn J., McKee M., and Wall S. From Public to Planetary Health: A Manifesto. *The Lancet*. 2014. 383:847.

[2] Cunningham A., Scoones I., and Woodthe J. (eds.). One Health for a Changing World: Zoonoses, Ecosystems and Human Well-Being. *Philosophical Transactions of the Royal Society B*. 2017: 372–1725.

[3] Lerner H., and Berg C. A Comparison of Three Holistic Approaches to Health: One Health, EcoHealth, and Planetary Health. *Frontiers in Veterinary Science*. Sep 2017, doi:10.3389/fvets.2017.00163.

[4] Duboz R., Echaubard P., Promburom P., *et al*. Systems Thinking as a Foundation for the Design of Integrated Approaches to Health. *Submitted to Frontiers in Veterinary Science*. 2018.

[5] Ross D.A., and Wade J.P. A Definition of Systems Thinking: A Systems Approach. *Procedia Computer Science*. 2015; 44: 669–678.

[6] Voinov A., and Bousquet F. Modelling with Stakeholders. *Environmental Modelling & Software*. 2010; 25: 1268–1281.

[7] van Bruggen A. *Transformative Modeling: Building Capacity for Transformation in Large Scale Socio-Technical Systems Using Computer Modeling with Stakeholders*. Master Thesis, TU Delft Technology, Policy and Management, Delft, Netherlands, Dec 2017.

[8] Bousquet F., Barreteau O., Le Page C., Mullon C., and Weber J. An Environmental Modelling Approach. In Blasco F., Weill A. (eds.). *Advances*

in Environmental and Ecological Modelling. The Use of Multi-Agents Simulations. Paris: Elsevier; 1999. pp. 113–122.

[9]   van Eeten M.J.G., Loucks D.P., and Roe E. Bringing Actors Together Around Large-Scale Water Systems: Participatory Modeling and Other Innovations. *Knowledge, Technology & Policy*. 2002; 14(4): 94–108.

[10]   Becu N., Bousquet F., Barreteau O., Perez P., and Walker A. A Methodology for Eliciting and Modelling Stakeholders' Representations with Agent Based Modelling. *Lecture Notes in Artificial Intelligence*. 2003; 2927: 131–149.

[11]   Etienne M. *Companion Modelling – A Participatory Approach Supporting Sustainable Development*. Versailles: Quae (Collection Update Sciences & Technologies); 2011.

[12]   Jacobson C., Hughey K., Allen W., Rixecker S., and Carter R. Toward More Reflexive Use of Adaptive Management. *Society and Natural Resources*. 2009; 22: 484–495.

[13]   Binot A., Duboz R., Promburom P., *et al*. A Framework to Promote Collective Action within the One Health Community of Practice: Using Participatory Modelling to Enable Interdisciplinary, Cross-sectoral and Multi-level Integration. *One Health*. 2015; 1:44–48, doi: 10.1016/j. onehlt.2015.09.001.

[14]   Scoones I., Jones K., Lo Iacono G., Redding DW., Wilkinson A., and Wood JLN. Integrative Modelling for One Health: Pattern, Process and Participation. *Philosophical Transactions of the Royal Society B*. 2017, doi: 10.1098/rstb.2016.0164.

[15]   World Health Organization. *Public Health Surveillance*. Available at http://www.who.int/topics/public_health_surveillance/en/ [Accessed on 1 Jun 2018].

[16]   Collineau L., Duboz R., Paul M., *et al*. Application of Loop Analysis for the Qualitative Assessment of Surveillance and Control in Veterinary Epidemiology. *Emerging Themes in Epidemiology*. 2013; 10(7): 11 pages.

[17]   Otte J., Hinrichs J., Rushton J., Roland-Holst D., and Zilberman D. Impacts of Avian Influenza Virus on Animal Production in Developing Countries. *CAB Reviews: Perspectives in Agriculture, Veterinary Science, Nutrition and Natural Resources*. 2008; 3(080): 18 pages.

[18]   Chan M. *Address to the Executive Board at World Health Organization. 140th Session*; 23 Jan 2017. Available at http://www.who.int/dg/speeches/2017/140-executive-board/en/ [Accessed on 8 Jun 2018].

[19]   Florida Institute for Human & Machine Cognition (IHMC). *Cmap*. Available at http://cmap.ihmc.us/ [Acceded on 12 Jun 2018].

[20]   Quesnel G., Duboz R., and Ramat E. The Virtual Laboratory Environment. An Operational Framework for Multi-Modelling, Simulation and Analysis of Complex Dynamical Systems. *Simulation Modelling Practice and Theory*. 2009; 17(4): 641–653.

Chapter 11

Multi-perspective modeling in relation to complex adaptive systems

Advocating that healthcare should be considered as a complex adaptive system (CAS), Kuziemsky [1] indicates, "An acknowledged shortcoming in much of the existing research is that it is descriptive in nature without guidance on how to study healthcare delivery as a CAS."

Kuziemsky offers an informal system model that is intended to establish understanding of how the system works with respect to the relevant concepts and the relationships. He illustrates how various and costly unintended consequences have emerged from steps taken to reform healthcare systems due to the complexity of the pertinent healthcare processes [1]. Before proceeding, we briefly review some relevant background on CAS. We then consider fundamental requirements for M&S of CAS in the context of health care.

11.1 Background on complex adaptive systems and systems of systems

Healthcare systems involve a plethora of systems, from traditional legacy IT systems where the human element maintained information on paper to next generation automation systems that involve the human element as an active participant.

Boardman and Sauser [2] discuss the following five characteristics of systems of systems (SoS):

- Autonomy—Component system's ability to decide and act independently; ranges from conformance to independence; autonomy is ceded by parts in order to grant autonomy to the system;
- Belonging—Component's functional ability to qualify for participation in SoS; ranges from centralized to decentralized;
- Connectivity—Component's capability to exchange information with other components; ranges from platform centric to network centric;
- Diversity—Degree of notable processing differences among components managed, i.e., reduced or minimized by encapsulation, modular hierarchy; ranges from homogeneous to heterogeneous; and

- Emergence—The appearance of new properties in the SoS not present in the components, both good and bad behavior, and designed in or tested out as appropriate; ranges from foreseen to indeterminable.

An illustration of the application of these concepts to healthcare is given in Table 11.1, where the SoS is formed by the integrated practice unit (IPU) of Chapter 5.

Subsequent papers have used these characteristics to provide a space of types of SoS, with examples, such as directed, acknowledged, collaborative, and virtual, and how different governance policies apply to different instances in this space. A SoS also becomes CAS when a set of components inside a SoS display intelligent, adaptive, or autonomous or semiautonomous behavior. One way to measure complexity is the number of binary questions that have to be asked to determine the state of the network when one is operating within its boundaries; healthcare systems appear near the top of such as scale of complexity along with other sociotechnical systems such as aerospace, retail, and telecommunication [3]. A complex system is characterized by dynamic processes where the interactions and relationships of different components simultaneously affect and are shaped by the system. Complexity of this kind appears at many levels of healthcare, from the daily interactions between patients and staff to the overarching interaction of healthcare system components such as payers and insurers.

The approach to value-based healthcare reform presented in this book affords a framework for modeling and simulation (M&S) that is intended to provide evidence-based and simulation-verified proposals for healthcare delivery system reengineering and new system design. Given the complexities to be tackled, it is worth acknowledging that several prerequisites must be in place to enhance the probability of success of such reform efforts.

Table 11.1   System of systems characteristics in the context of healthcare

SoS characteristic	Application to physicians forming an IPU
Autonomy	Assuming that "as is" physicians are independent practitioners: how much autonomy do they need to give up to form the IPU team? Can they retain their economic sustainability if the IPU fails?
Belonging	Do physicians join the IPU on a cost/benefit basis? Do their skill sets match those needed by the CDVC? Can they adapt to provide needed services rather than the sophisticated ones they brought to the table?
Connectivity	Will the HIT infrastructure support the increased communication needed to enable the increased coordination?
Diversity	Are the EHR systems employed by the providers either the same or interoperable at the syntactic, semantic, and pragmatic levels? Do they provide the ability to synchronize to achieve common goals among the constituent systems?
Emergence	How strongly can the emerging CDVC be predicted from the formal composition of the individual CDVCs?

11.2 Fundamental requirements for M&S of complex adaptive systems and SoS

Following are some of the requirements for M&S that supports complex healthcare delivery system reengineering, new system design understanding and minimizes the risk of unintended consequences.

11.2.1 Deal with system-of-systems (SoS) M&S nature

The predominant characteristic that healthcare M&S must face is its SoS nature with the central organizational shortcoming—coordination that rises no higher than pairwise interactions. As described earlier, healthcare systems are fragmented, loosely coupled in structure, but tightly cohesive in required functionality. The system is organized around pairwise interactions between physicians and patients with discrete maladies. Interactions between physicians are discretional rather than institutionalized, and clinical care is very weakly influenced by community determinants of health.

Therefore, dealing with the SoS-nature of healthcare where organizational structure does not meet functional objectives constitutes the first requirement for M&S of healthcare.

11.2.2 Develop an organizational ontology for M&S-based applications

We have seen that considerable amount of M&S research work has been devoted to healthcare in recent decades, seeking to address problems related to its management, including clinical and extra-clinical aspects [4–6]. There is no doubt that

1. healthcare M&S has multiple facets; and
2. there is a lack of ontology that structures all the knowledge about existing models in a hierarchy, allowing one to easily derive new models.

To address this issue, an ontology for healthcare systems simulation is required. Such an ontology should not just aim at establishing a unified vocabulary, such as formalizing and reorganizing healthcare terminologies and taxonomies. It is essential that the ontology provide, at some general level, a formal way to capture all the knowledge that might be in the range of healthcare M&S for which it is likely to be used. The healthcare ontology for M&S should capture and share a common understanding of the abstractions necessary/used for the simulation of the entire healthcare domain (beyond unit specific and facility specific modeling).

11.2.3 Enable the ontology to support combinatorial model compositions

The ontology should be targeted to support the *plan–generate–evaluate process* in simulation-based systems design. The plan phase recaptures all the intended objectives of the modeler. The generate phase reproduces a candidate design model that will meet the initial objectives. The evaluate phase assesses the performance of the generated model through simulation. As such, the ontology organizes a family of alternative models from which a candidate model can be generated, selected, and evaluated through system design repeatedly until the model meets an acceptable objective.

While complex systems are composed of large components and their structural knowledge can be broken down and systematically represented, their behaviors can be specified in either atomic or coupled models and saved in a model base (MB, an organized library) for later use. Once the models are saved, they can be retrieved from their repository and reused to design complex systems.

The system entity structure (SES), an instrument for variability management, can also support run-time model recommendation in a real-time plan–generate–evaluate cycle [7]. The ontology should be supported by a standardized knowledge base and an MB for seamless on-demand construction and sharing of models in a healthcare SoS composition.

11.2.4   Include the major facets at the top level to ensure macro behavior

As suggested before, healthcare simulation in the current literature usually focuses on supply and demand as driving forces. Broadening the scope to all of healthcare forces us to both deepen these abstractions and add more top-level ones as needed. Therefore, we need to formalize a healthcare system as a whole made up of production, consumption, and coordinating components.

Demand and supply, in healthcare-related literature, usually refer respectively to patients seeking for care services and providers of these services. Replacing "supply" with the concept of "production," we also include health phenomena that generate or amplify health concerns, such as disease spreading, as well as mechanisms used to supply healthcare, such as vaccination or information diffusion. Similarly, with the concept of "consumption," we extend health demand beyond the individual patients, and we include the population as a dynamic entity.

Pulling the major facets at the top level allows the M&S study to evaluate the holistic behaviors that are the macro behaviors. Such macro behaviors are characteristics of complex systems and must be identified a priori for the overall effectiveness of the healthcare reform.

11.2.5   Include a large spectrum of models for combinatorial composition

The MB for healthcare systems M&S must include a large spectrum of models, such as theoretical models provided in the literature ranging from functional dynamics models to spatial dynamics models, functional diffusion to spatial diffusion, provision, provider, entity, and flow-type models. This spectrum is needed to support the vast number of compositions that are possible with selections from ontologically established alternatives.

Healthcare consumption focuses on consumers seeking for care and the dynamics of their demands. Two facets must be covered: population models and individual models, corresponding respectively to macro and micro approach to healthcare consumption modeling. A population model is related to births, deaths, and demographic flows such as immigration and emigration. An individual can be modeled as an autonomous entity or as a flow that captures scenarios the individual can undergo (e.g., patient flow and care pathways), parameterized by his needs (such as required health services/resources).

Healthcare production deals with the generation and diffusion of health phenomena, whether positive or negative. Positive phenomena (like vaccination campaign) produce ease, while negative ones (like disease spreading) produce disease. Two facets must be covered: resource models and diffusion models. The first facet deals with how resources are transformed into services. Resources (and therefore services) can be physical (beds, rooms, etc.), human (physicians, nurses, etc.), financial, or information. Models of such transformation explicitly describe the dynamics of providers, or provision or both. Diffusion processes are described as either spatial or functional phenomena: Spatial phenomena explicitly describe space, while functional phenomena formulate the dynamics of the diffusion process in the form of mathematical equations, such as compartmental models.

Care coordination can be seen as cross-organization coordination managing the entities and resources of existing ones. It is needed to the extent that existing organization is lacking.

Models at multiple resolutions, from an individual, to an organization including the artifacts shared across these resolutions along with the business processes must be made available for healthcare SoS model composition. This should also incorporate the inherent stochasticity and variability at these resolutions.

11.2.6 *Instrument the system to support continuous on-going high-quality data*

There must be *systems implemented to observe, measure, and record*, in a *continuous* ongoing manner, the entities, processes, and protocols called out in the experimental frames that characterize the different modeling objectives. These measurements should be amenable to aggregation to validation of model hypotheses, calibration of parameters, and computation of metrics defined for system performance at multiple levels of resolution.

At the SoS level, this can involve an additional layer of instrumentation that cannot be synthesized from current infrastructure that supports component system instrumentation in isolation from the rest of the SoS. However, there may be managerial and operational issues in implementing such an instrumentation layer as it will involve coordination at SoS level and an acknowledgment of *the* SoS by the participants, to begin with.

The complex system designed to observe, measure, and record the very system in near real time must be supported with sound data engineering practices for a responsive event-driven dynamical system to enable healthcare coordination across multiple levels of the system. Accordingly, the system under design must reflect the architecture using model-based engineering practices.

11.2.7   *Include pervasive incremental automated learning*

The instrumentation just discussed can produce data-rich environments where sheer amounts of data are created in records of care delivery processes, pathway tracked interventions, and patients at the population and individual levels. Activity-based credit assignment and computer-based machine learning (ML) can present a key enabling factor to exploit such data repositories to provide insights for value-based healthcare [8].

Another key point aimed in the extended framework is the role of incremental learning. The framework considers training and fitting of ML models as an incremental process, rather than one-off. The appendix at the end of this chapter (Appendix C) provides a case study [9] showing how ML can discover models to support the assignment of pathways in care coordination.

11.3   Summary

Table 11.2 summarizes how the multi-perspective modeling and holistic simulation framework presented in this book address these requirements. The discrete event system specification formalism and the SES framework provide the conceptual and theoretical background.

Table 11.2   How the multi-perspective framework addresses the requirements

Requirement	Feature provided by the framework
Deal with SoS nature	The framework provides a multi-perspective methodology for developing coupled models of components from various formalisms capable of expressing the different perspectives needed for healthcare delivery SoS together with holistic abstractions that support integration and coordination
Develop an organizational ontology	The framework provides the ontology for healthcare (O4HCS) based on system entity structure (SES) which includes the macro-level facets that properly organize the healthcare system domain and support refinement into more detailed components at the meso and microlevels
Enable the ontology to support combinatorial model compositions	The SES/MB (model base) supports hierarchical composition of the coupled model resulting from pruning that selects from the combinatorial family of possible compositions described by the SES. The framework is supported by the discrete event system specification (DEVS) formalism, which can encompass models expressed in various formalisms typically found to be useful in simulation studies

Table 11.2   (Continued)

Requirement	Feature provided by the framework
Include the major facets at the top level to ensure macro behavior	Provides Ontology for Healthcare based on SES that ensures suitable macro behavior involving the production, consumption, and coordination facets and their specializations and further decompositions
Include a large spectrum of models for combinatorial composition	A variety of examples have been demonstrated, spanning health diffusion, resource allocation, provider and provision modeling, population diffusion, and spatial models, including agent-based models at individual and higher level abstractions. The focus was on coordination mechanisms such as the pathways model
Instrument the complex system to support ongoing high-quality data	There must be systems implemented to measure, in a continuous ongoing manner, the elements of clinical and extra-clinical interventions that can be aggregated to compute quality of service
	The DEVS formalism includes experimental frames that can specify, collect, and aggregate the information for higher levels in a multi-perspective model. The simulation infrastructure guarantees correct execution of the composed model and the behaviors in a transparent manner
	A properly designed data architecture needs to support record keeping and decision-making that serve the needs of all participants such as care coordinators, community health workers, and administrators. The pathways model data architecture supports data acquisition based on a web portal that guides entry of data based on the state of the currently active pathway and supports entering timely, consistent information, so that other users have access to the information in real time
Include pervasive incremental automated learning	The activity-based credit assignment learning approach described in Chapter 5 is based on correlation between the activity of component systems and the behavior achieved at the SoS composition level. We expand the framework by including a component for unsupervised ML, which can serve the purpose of knowledge elicitation at early stages of problem formulation. Unsupervised techniques (e.g., clustering and rule mining) can discover patterns that represent abstractions of the system that in turn reflect on the simulation model used to design and evaluate the coordinated care architecture. The idea of incremental learning is based on the premise that

(Continues)

Table 11.2   (Continued)

Requirement	Feature provided by the framework
	new system states are being continuously captured in timely snapshots of data and added to an accumulated repository representing the system knowledge. In this manner, ML models can be iteratively trained in order to learn about possible updates in the system behavior

References

[1]   Kuziemsky C. Decision-Making in Healthcare as a Complex Adaptive System. *Healthcare Management Forum.* 2015; 29(1): 4–7.

[2]   Boardman J., Sauser B. System of Systems – The Meaning of "of." *IEEE/ SMC International Conference on System of Systems Engineering*, Los Angeles, CA, USA, 24–26 Apr 2006. pp. 118–123.

[3]   Rouse W. *Health Care as a Complex Adaptive System: Implications for Design and Management.* National Academy of Engineering, Volume 38, Number 1, Spring 2008.

[4]   Almagooshi S. Simulation Modelling in Healthcare: Challenges and Trends. *Procedia Manufacturing.* 2015; 3: 301–307.

[5]   Bountourelis T., Ulukus M.Y., Kharoufeh J.P., and Nabors S.G. The Modeling Analysis and Management of Intensive Care Units. In *Handbook of Healthcare Operations Management.* New York, NY: Springer; 2011, pp. 153–182.

[6]   Powell J.H., and Mustafee N. Widening Requirements Capture with Soft Methods: An Investigation of Hybrid M&S Studies in Healthcare. *Journal of the Operational Research Society.* 2016; 68(10): 1211–1222.

[7]   Pawletta T., Schmidt A, Zeigler B.P., and Durak U. Extended Variability Modeling Using System Entity Structure Ontology Within MATLAB/Simulink. *Proceedings of the 49th Annual Simulation Symposium*, Pasadena, CA, 3–6 Apr 2016. Article 22.

[8]   Aldenderfer M.S., and Blashfield R.K. *Cluster Analysis.* Quantitative Applications in the Social Sciences 07–044. London: Sage Publication; 1984.

[9]   Elbattah M., Molloy O., and Zeigler B.P. Designing Coordinated Care Pathways Using Simulation Modelling and Machine Learning. *Accepted in the Winter Simulation Conference*, Gothenburg, Sweden, 9–12 Dec 2018.

Appendix C

C.1 Case study of machine learning based assignment of pathways in care coordination

Elbattah *et al.* [8] demonstrated how the knowledge learned from machine learning (ML) experiments can be used within the process of healthcare modeling. The data-driven knowledge was used to reflect on the structure and behavior of the patient's care journey through surgery in different respects as follows.

Initially, a system dynamics (SD) model was built, representing the three clusters of patients. In particular, the model was disaggregated into three stocks corresponding to the clusters of patients. Furthermore, the auxiliary variables were decided based on the cluster analysis. For instance, the first and second patient clusters were set to undergo the same TimeToSurgery (TTS) delay (i.e., TTS1), while the third cluster was assigned a different delay (i.e., TTS2).

Similarly, the inflows of elderly patients were structured based on the age variation within clusters. Both the first and third patient clusters were modeled to include more elderly patients (i.e., aged 80–100 years), while the second cluster was associated with less elderly patients (i.e. aged 60–80 years). This reflected the age groups within the patient clusters.

In general, the SD model can be used to provide projections of hip-fracture patients discharged with a focus on different patient characteristics and care-related outcomes. Figure C.1 illustrates the cluster-based SD model.

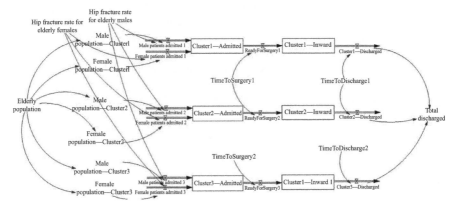

Figure C.1   Cluster-based systems dynamics model

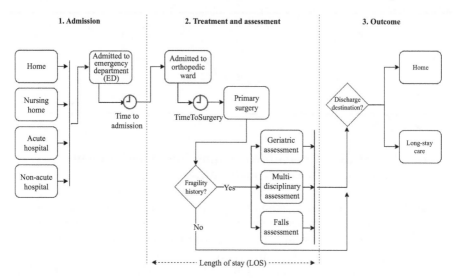

Figure C.2   The patient journey simulated by the discrete event model

Table C.1   Summary of cluster characteristics

	Patient count	Avg. age	Gender distribution		Avg. LOS (days)	Avg. TTS (days)
			Males	Females		
Cluster1	2,735	86.78	640	2,095	8.65	1.62
Cluster2	1,111	83.39	363	748	31.27	2.32
Cluster3	2,164	71.67	720	1,444	8.57	1.62

A fine-grained perspective of the patient's care journey is illustrated in Figure C.2. Such a model helps to deal with patient entities, rather than aggregate populations. Each patient entity could be treated individually in terms of characteristics (e.g., age, gender, type of fracture, etc.) and care-related factors (e.g., time-to-admission, TTS, etc.).

The model can also be utilized to produce a realistic sequence of events corresponding to those within the care journey.

Clusters derived by ML are shown in Table C.1 that can be employed to assign patients to pathways in Figure C.2.

Chapter 12

Extending the framework to value-based coordination architectures

12.1 Introduction

In this chapter, we discuss the need to develop a suite of simulations to study specific approaches to value-based delivery of services. The focus is on coordinated care with the goal of enabling design and development of architectures that treat patients as agents interacting with systems and services that are coordinated using health information networks and interoperable electronic medical records. Success in this direction will contribute to the major global healthcare goal of solving the "iron triangle" of reducing cost while improving quality and increasing access. Our specific goal is to allow organizations, such as public health departments, hospital-based, and emerging ones, Accountable Care Organizations (ACOs), to evaluate specific coordinated care strategies, payment schemes, and trade-offs such as between quality and cost of coordinated care. A range of simulation tools must be developed to support the design of coordination architectures and predict important quality metrics that are applicable to diverse populations.

There is growing recognition that optimization at the system level cannot be based on suboptimization of the component systems, but must be directed at the entire system itself [1]. While current industrial and system engineering models and simulations have targeted the micro (component system) level, it is the macro [system of systems (SoS)] level that such approaches must now address. While trying to model entire healthcare systems is overwhelming, restricted SoS level models can be developed to address specific issues while still enabling a general framework for engineering of the overall SoS to emerge.

It has been found that 15% of the most expensive health conditions account for 44% of the total healthcare costs; patients with multiple chronic conditions cost up to seven times as much as patients with one [2,3]. We have seen that coordinated care presupposes that continuing intervention after patients return home can greatly reduce unnecessary retreatment and declining health with the associated reduction in the extra burdens that such populations place on healthcare resources [4]. Falls and nonadherence to prescribed medical regimens are two areas which have been identified as sources of high cost, which have not been controlled despite introduction of current assistive devices [5,6]. Some relevant questions to be addressed to coordinated care organizations therefore are as follows: (1) how exactly can such care be provided in a coordinated manner, (2) what payment schemes should be instituted, and (3) will the cost of the new care strategies be significantly less than the savings that are achieved by dedicating the resources needed to provide such care?

12.2   Alternative payment models

A variety of payment models aim to improve healthcare value, with varying degrees of accountability and financial risk. Pay-for-performance, often known as value-based purchasing, is the first step from today's predominant form, fee-for-service, on the spectrum toward more comprehensive and higher financial risk programs [7]. The simulation model example in Chapter 7 mentioned some necessary extensions such as inclusion of incentives for Community Healthcare Workers (CHWs) and to alignment of pathways with payments.

Here, we expand the scope of objectives for simulation tool development by exploiting recent elaboration of the kinds of payment schemes that range from fee-for-service to the planned full-blown "population-based" payment. These alternative payment models (APMs) were recently detailed in a report by the Healthcare Payment Learning and Action Network.

They were categorized as follows [8]:

- Category 1 is current fee-for-service. Here, payments are based on volume of services and not linked to quality or efficiency.
- Category 2 is an APM in which at least a portion of payments depends on the quality or efficiency of healthcare delivery.
- Category 3 includes APMs that are built on the fee-for-service architecture in which some payment is linked to the effective management of segments of the population or episodes that involve care requiring coordination of multiple providers. Payments are still triggered by delivery of services, but there are opportunities for shared savings or risk of not payments not covering expenses.
- Category 4 is population-based payment. Here payment is not directly triggered by service delivery so payment is not inked to volume. Clinicians and organizations are paid and responsible for the care of a patient for a long period (e.g., year).

So far experience with Category 2 payment schemes has been mixed [7], while the more advanced ones are in the formative stages. Consequently, it is very timely to develop a suite of simulations to study specific approaches to value-based delivery of services. One type of organization that would benefit from such tools is known as ACOs. According to the US Centers for Medicare and Medicaid (CMS) website (www.cms.gov), "ACOs are groups of doctors, hospitals, and other healthcare providers, who come together voluntarily to give coordinated high quality care to the Medicare patients they serve." CMS is providing incentives to establish such ACOs through its offer to share any savings produced by an ACO in delivering service without sacrificing quality. However, currently, there is no basis for negotiating with potential ACOs on the structuring of such profit sharing arrangements. An adequate basis for such negotiations would be an accepted means of measuring quality likely to be achieved by an ACO proposal versus the cost that would be involved. Such a basis is currently lacking.

A system-level model that could reliably predict the quality versus cost performance of an ACO's proposal would provide a basis not only in negotiations between CMS and potential ACOs but also serve in numerous additional ways such as in developing and testing tools and services for ACOs. As central players in federal attempts to "bend the cost curve," such organizations will become key to cutting costs while maintaining and even improving quality of delivered care. Healthcare spending in the United States is highly disproportionate, with half of healthcare dollars spent on 5% of the population. ACOs that manage this at risk population coordinate the care of patients who need special attention after leaving the doctor's office or the hospital because they cannot carry out the activities required to complete their care. People with multiple health and social needs are high consumers of healthcare services and are thus drivers of high healthcare costs. The ability to provide the right information to the right people in real time requires a system-level model that identifies the various community partners involved and rigorously lays out how their interactions might be effectively coordinated to improve care for the neediest patients that cost the most.

12.3   Model construction methodology

Parts I and II have provided a multifaceted modeling framework and simulation methodology for coordination of care and Part III has illustrated how such a methodology can be applied to constructing models for coordination of care in particular settings. However, there remains much development needed to extend the methodology to support the design of coordinated care architectures as well as to predict important quality versus cost metrics of such architectures. Here, we will outline an approach to develop such a methodology, first outlined by [9], that will follow the approach developed [10] to construct multilevel models to address ecosystem problems. Aumann's main idea is to determine and characterize the primary (*focal* level) of model development and those levels immediately above and below together with their experimental frames. The focal level is where the behaviors of the system of interest reside. The "mechanism" for a model behavior arises at the lower level, while its "purpose" is found at the higher level. Defining the relationships among these levels will determine the means by which data and parameter values and constraints will flow between them. For example, if the focal level is the individual level, as in agent-based models, then models and frames at the immediate levels above (e.g., society) and below (e.g., decision and action components) this level must be considered and related to those at the focal level.

Before proceeding, we mention some issues that specifically challenge the methodology.

12.3.1   The crucial role of system-level integration

The ability to provide the right information to the right people in real time requires a system-level model that identifies the various community partners involved and rigorously lays out how their interactions might be effectively coordinated to improve care for the neediest patients that cost the most. Although frameworks such as that of the Institute for Healthcare Improvement (IHI) [11] provide a starting point, currently a rigorous predictive integration model is lacking. For example, the IHI framework does not take account of emerging health information networks and electronic medical records, nor does it integrate system concepts and agent modeling concepts extended to include human behavioral limitations. Consequently, it does not provide a basis for developing and testing architectures for healthcare services with application to coordinated care systems as examples.

A SoS model for coordinated care would enable systems design, engineering, and evaluation of alternative coordinated care structures. Such a model would focus on information technology support and apply to a wide variety of structures of possible interest to ACOs, accounting for the populations that ACOs serve, and the community resources available in their environments. Application of the SoS concept to healthcare recognizes that it includes myriad stakeholders involving multiple large-scale concurrent and complex systems (that themselves comprise complex systems) to address the key challenges such as cost effectiveness, functionality and secure, and ease of use. Wickragemansighe *et al.* [12] advocate a net-centric approach to linking such loosely coupled systems. This approach involves linking up hospitals and physicians through networked information systems. Progress is being made along such lines. For example, in Maryland, the Chesapeake Regional Information System Project known as the CRISP project (www.crisphealth.org) is the designated statewide Regional Extension Center and Health Information Exchange helping nearly 1,500 providers adopt electronic healthcare records, securely share health data, and achieve meaningful use. However, netcentricity is not sufficient in itself to achieve overall system goals such as cost reduction, while maintaining and improving care. Indeed, as argued in Chapter 5, coordination of component systems is necessary and relies on the ability to interoperate the constituent disparate systems, enabling them to exchange information so that it can be understood and acted upon by each system to achieve mutual objectives.

12.3.2   Personal limitation models

While systems-based modeling and simulation provides a suitable platform for addressing the SoS problems in healthcare, there must also be an ability to include human behavioral modeling in order to deal with the limitations that coordinated healthcare must address. Such agent-based modeling can be viewed as included within agent-directed simulation for systems engineering of large and complex systems. Yilmaz and Ören [13] provide a survey of agent concepts, including agent architectures and taxonomies. From a software development and simulation modeling perspective, agents are advanced objects with intelligent capabilities such as autonomy, and proactive and reactive decision-making. With the exception of a few studies such as [14], agent-based models have focused largely on replicating intelligent features of human behavior rather than those characteristic of human behavior limitations. Therefore, one challenge model construction is to introduce agent-based models of human behavior limitations needed for coordinated care applications into the SoS level of analysis supported by Discrete Event System Specification (DEVS)-based simulation tools.

12.4 Family of models approach

Recall from Part I that the criteria for development of a family of models for healthcare delivery systems should be

- flexible to meet the variety of stakeholders' interests and variety of ACO potential implementations;
- scalable to accommodate increases in scope, resolution and detail;
- able to integrate SoS concepts—system, components, and agent concepts extended to human behavioral limitations;
- suggestive of enhancements to Electronic Health Records as needed to support coordinated care; and
- able to evaluate coordination services and architectures, e.g., patient tracking, medication reconciliation, etc.

Based on Part II, the proposed development of simulation tools for coordinated care takes the following steps:

1. Define the questions to be addressed by the family of models. These questions formulate the objectives of the model development that subsequent work is intended to accomplish.
2. Formalize these questions in terms of a concept of experimental frame which states the conditions under which a model will be experimented with to generate the data intended to address the associated questions. Experimental frames allow precisely stating a set of requirements for model development, validation, and simulation experimentation.
3. Conceptualize a family of models that can collectively meet the requirements of Step 2 and consequently address the questions in Step 1. The concept of model family recognizes that different questions may be best addressed with different models.
4. The family of models that results from the different levels of abstractions and requirements for efficiency of expression may require different types of models or modeling formalisms. Recognizing this heterogeneity may require multiple simulation modeling formalisms to coexist and be interoperated to work together.

Aumann [10] has shown how such concepts as model specification hierarchy, hierarchical coupled models, experimental frames, and morphisms provide a practical methodology for developing credible models of complex systems.

Bergez *et al.* [15] elaborated this methodology emphasizing the roles of experimental frames and model specifications. Recent initiatives such as the

RECORD project illustrate the effectiveness of a DEVS-based approach [16] in the domain of agroecosystem, a SoS domain which shares many of the characteristics of the healthcare application domain such as component model coupling and integration of heterogeneous formalisms.

12.4.1   Model decomposition and hierarchical design

Aumann defines the nature of the hierarchy of levels of scale based on emergent properties. Based on this concept, he describes the process of specifying a model design in terms that involve assigning entities or units used in the synthesis to the levels in the scalar hierarchical framework. The following is quoted from Aumann's paper.

In outline, the process is:

1. **Develop a representation of the focal or primary system level by proposing its spatio-temporal limits, describing its main phenomenological characteristics, and specifying the experimental frames operating at this level.**

In Aumann's example of crab behavior in water low in oxygen and containing clams as well, "the focal level was chosen to be that of the individual. Individual crabs have the ability to move, respond to their environment, attack other crabs, feed on prey, etc. For both crabs and the environmental variables, the focal level is a small region of space over which these variables are assumed homogeneous. This region of space keeps track of the age/size structure of crabs and their abundance, and the dissolved oxygen, temperature, depth and salinity at that location."

2. **Develop a representation of the sub-systems at the immediately lower level (focal − 1) of integration making up the focal level and specifying the relevant experimental frames.**

In the example, "an individual crab consists of sub-models governing movement, reproduction, energy balance, aggression, etc., … while the way in which the environmental variables change over the small region of space is governed by the environmental inputs and equations governing dissolved oxygen."

3. **Develop a representation of the system from the immediately higher level (focal + 1) of integration which the focal level is part of and specifying its relevant experimental frames.**

"In (the) example, we refer to this higher level as the estuary level as it is comprised of all individual crabs, the clams and the environmental variables over the entire estuary. It is at this level that one can talk about population averages for crabs and clams, and also environmental averages like the percent of the estuary low in oxygen over the summer."

---

4. **Defining the processes and relationships linking the focal level to the focal + 1 and focal − 1 levels and determining which processes operating between the focal − 1 and focal levels are most dependent on the processes occurring between the focal + 1 and focal levels. Information needs to be passed down, up and within each of these levels.**

"For example, as individual crabs move throughout the estuary, the environmental conditions encountered need to be passed to the crab's lower-level models. Alternatively, the location of individual crabs needs to be passed up to the estuary level so that local crab–crab interactions can occur."

Aumann simplifies the formulation of a hierarchical design in terms of a linearly ordered set of three levels. However, this approach can be extended to multiple levels and can recognize that the questions and associated experimental frames need not satisfy a simple linear order [17].

Figure 12.1 presents a stratification of healthcare focused on coordination showing four levels. At level 2, we focus on how an individual patient interacts with a specific group of providers for example obstetrics. The immediately lower level, 1, allows us to model the patient to predict his/her response to coordination interventions. The next level up expands the focus to populations of patients and groups of provides as formulated in the ACO concept. Continuing the expansion of scope, we recognize the next level up where the environment in which healthcare operates is not static and needs to be brought in explicitly.

The accompanying table (Table 12.1) outlines the primary stakeholders at each level of Figure 12.1 and the kinds of questions that they would ask. These questions are operationalized as experimental frames to form the basis for a flexible environment to support multiple stakeholders in healthcare.

Figure 12.1 Levels of the modeling framework for coordination of healthcare

Table 12.1　Levels of care coordination with associated questions, stakeholders, and processes

Level	Focal questions	Stakeholders	Processes and variables
1. Patient adherence to provider's care plan	How does patient behave in absence of intervention? How does patient react to care guidance?	Care assistance navigators	Patient behavior characteristics, decision processes
2. Coordination of individual's care in a provider group (physician, pharmacist, hospital, etc.)	Assuming provider's treatments are 100% effective, what is the effectiveness of coordinating patient's interactions with a group of providers?	Strategy designers	Coordination strategies and implementation processes; quality of treatment outcomes and costs/savings
3. Coordinated care architectures	What is the effectiveness of Health IT-assisted coordination of community healthcare providers for a given patient population?	Architecture designers/ Accountable Care Organizations/ Medicare negotiators	Exploitation of HIT infrastructure—networks, medical record standards, information exchange services
4. Healthcare environment	What are the effects of environment change due to government policy? Business, technology, pharmaceutical trends?	Policy decision-makers, insurance risk adjusters	

12.4.2　Top-down design

We now discuss applying the above methodology to develop simulation models at level 3 (coordinated care architectures). Examples of such models were discussed in Chapters 7 and 9. In this framework, each pathway is designed to address a single health or social issue and confirm that the issue has been resolved.

The Community Care Coordination Hub coordinates care for individuals within targeted medical "pathways," such as medication assessment, smoking cessation, and obstetrical and postpartum care. The components of the model include the patient's home, the community care coordination hub, and the community partners such as hospital, physicians, pharmacists, etc. The coordination hub is the key addition to the current situation represented by the patient home and community partners. It is intended to fill the vacuum left after the patient has received the formal, though fragmented, care offered by the community partners. The pathways framework affords the addition of coordination of fragmented services, tracking, monitoring, and assuring that the patient follows physician's directions and medication regimes.

The system-level model starts with a top-down decomposition (Figure 12.2) in which a model of the coordinated care system is coupled with an evaluation frame component that generates external events for the system and evaluates its responses to them.

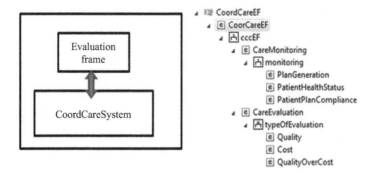

Figure 12.2   Top-down model decomposition and System Entity Structure of evaluation frame

One of the main objectives implemented by the evaluation frame is to assess proposed pay-for-performance schemes that provide financial incentives that are tied to improving outcomes. The model follows a patient starting with his/her physician through interaction with an array of services provided by the coordination hub. Such hub services include risk identification, patient tracking, appointment scheduling, transitional care management, medication therapy management, medical care monitoring, home care, and personal health information management.

The model will characterize these services to a level of abstraction that creates a space of alternative architectures for the coordination hub. Such architecture alternatives will be assessed in the evaluation frame by measuring (1) a patients' health status as they transition through stages of care and (2) accumulating the cost of care (tests, medications, human care managers, and providers) including cost of readmissions to hospitals and emergency departments.

It is important to formulate architectural alternatives to include those that have been identified as best practice representatives [18]. For example, the CareOregon-integrated team includes customer-selected doctor, one or two medical assistants, behaviorist, full-time nurse providing care coordination administrative assistant providing case management support, specialists, and ancillary providers. Such a team is formulated in terms of a relevant subset of services provided by the coordination hub with characterization of their effect on patient health status and level of resources provided. The coordination services should be represented in the model to account for how health information exchange technology (HIT) can provide the basis for tracking patients' use of hospital and other facilities, keeping track of their compliance with medication regimens and general adherence to the plan of treatment developed by the physician. The model includes a component characterizing the population served by coordinated care. Depending on the pertinent geographical region, populations served might include homeless, substance abuse, Medicaid at risk, indigenous, mental health, and patients with multiple problems.

The framework employs the System Entity Structure (SES), introduced in Chapter 1, to enable selection of population served and selection of services provided by the hub to generate appropriate simulation runs to evaluate services in the

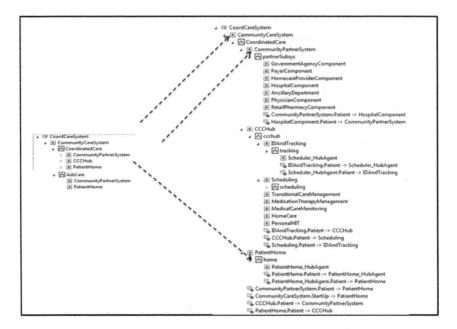

Figure 12.3   System Entity Structure for coordinated care system

context of patient needs. This allows evaluation and discovery of architectures that better match the needs of different populations (Figure 12.3).

The model includes measures of quality of service and cost indicators such as those proposed by Craig *et al.* [11]. For measuring individual experience of care, the model can include survey instruments such as the site-specific experience of care survey, how is your health survey items, and Ambulatory Care-Sensitive Hospitalization survey. For the health of a population, survey instruments are available such as Behavioral Risk Factor Surveillance System and the 12-Item Short Form Health Survey.

The extensive set of metrics offered by the Healthcare Effectiveness Data and Information Set should be examined for selective inclusion in the model. Including such an instrument has a concomitant requirement of including corresponding items in patient health information profiles which will be updated with the patients' interaction with coordinated care and community partner components.

Measures of access to, and utilization of, coordinated care services and community partner resources will arise naturally in the model as it keeps track of patients' consumption of these services and use of these resources. Such utilization variables will then be aggregated into cost outputs. For example, Craig *et al.* [11] suggest measuring per capita cost by aggregating such quantities as number of emergency department visits, readmissions, inpatient days, and behavioral health admissions, as well as hospital-based costs (emergency, inpatient, and detoxification).

12.4.3   Level specification

Given our stated objectives of supporting evaluation of coordinated care of interest to ACOs, it is natural that level 3 (coordinated care architectures) is the focal level.

As in Table 12.1, this allows us to ask questions such as, "What is the effectiveness of Health IT-assisted coordination of community healthcare providers for a given patient population?" For example, the family of models should allow architecture designers to explore how exploitation of Health IT infrastructure can enable different coordination architectures.

With level 3 as focal level, we descend down to level 2 (coordination of individual's care in a provider group) to supply the "mechanics" needed to produce the desired behaviors at level 2. In this case, we need to replicate experiments, each centered on an individual patient interacting with a subset of community healthcare partners. The relationship between level 3 and level 2 is that of sampling of patient encounters generated at level 2 from distributions determined by level 3.

Chapter 7 presented an example of the following process for generating such a sample encounter.

As a general process, it takes the following steps in outline:

1. Select a patient from a specified population using SES pruning,
2. Select a group of providers to interact with the patient (similar pruning),
3. Simulate a trajectory of the patient within a healthcare scenario with the provider group (e.g., a 9-month pregnancy),
4. Gather statistics on quality and cost of care in this encounter, and
5. Repeat these steps over a set of patients and providers sampled from some specific environment.

To supply the "mechanics" of patient behavior in level 2, we descend to level 1 (patient adherence to provider's care plan) to model the patient as an agent with behavior limitations. To supply the context for level 3, we ascend to level 4 (healthcare environment) where the environmental parameters reside and can be specified. These parameters are employed in lower levels.

For example, a poor population in a rural environment determines many of the characteristics (i.e., parameter values) of the patients to be sampled as well as the nature of the providers and services offered.

12.4.3.1 Patient modeling

Our approach applies systems concepts (such as state transitions, inputs/output behavior, and coupling of components) uniformly to all components, regardless of their relation to human or nonhuman elements. Part I presented a start on such integration is to bring discrete event simulation concepts into multi-agent systems. However, little work has been done on individual, as opposed to population, models. An area where the need for models of individual human limitations has been identified is Ambient Assisted Living which encompasses technical systems to support elderly people in their daily routine to allow an independent and safe lifestyle as long as possible [19].

12.4.3.2 Dynamic structure

Besides being needed to effectively implement structural changes reflecting reality (e.g., development and birth of a baby), dynamic structure is used to reduce complexity of model structure at different levels. At level 2, this capability is used to add and remove providers so that only those currently interacting with the patient are included in the active set. Within the patient component, at level 1, discrete behaviors are rolled in and out so that only those that are currently active are present in the model.

For example, behaviors related to interacting with the doctor are present when the patient is coupled to the doctor but not present when interacting with the pharmacy. For large and complex models such as under discussion, such restriction of the field of view helps to increase model comprehension by enabling active components to be the focus of attention, while inactive ones are removed from execution. The beneficial reduction in execution time has been discussed in the literature [20].

12.4.3.3   Model test bed, data calibration and validation

Pathway projects in the United States and France were discussed to develop the models in Chapters 7 and 9. Further development supported by data from real system implementation is needed for model calibration and validation.

12.5   Summary

In this chapter, we discussed the need for extending the modeling and simulation methodology discussed in the book to this point to develop a suite of simulations to study specific approaches to value-based delivery of services. The focus is on coordinated care with the goal of enabling design and development of architectures that treat patients as agents interacting with systems and services that are coordinated using health information networks and interoperable electronic medical records. Success in this direction will contribute to the major global healthcare goal of solving the iron triangle reducing cost while improving quality and increasing access. Our specific goal is to allow organizations such as public health departments, hospitals, and new kinds of organizations (such as ACOs in the United States) to evaluate specific coordinated care strategies, proposed payment models, and potential trade-offs between quality and cost of coordinated care. A range of simulation tools must be developed to support the design of coordination architectures and predict important quality metrics that are applicable to diverse populations.

The multilevel modeling methodology that was discussed shows how to integrate mathematical system and agent modeling concepts to work across multiple levels from the overall healthcare environment to patient behavior. This chapter also illustrated how the DEVS and SES formalisms support the modeling methodology by supporting flexible and accurate representation of families of models. The overall model will significantly extend simulation studies such as described in Chapters 7 and 9 and will need to be validated against data obtained from experiments.

References

[1]   Valdez R.S., Ramly E., and Brennan P.F. *Industrial and Systems Engineering and Health Care: Critical Areas of Research*. Rockville: AHRQ (Agency for Healthcare Research and Quality) Publication No. 10-0079; 2010.

[2]   Conwell L.J., and Cohen J.W. *Characteristics of Persons with High Medical Expenditures in the US Civilian Noninstitutionalized Population, 2002*. Statistical Brief #73. Rockville: AHRQ Publications; Mar 2005.

[3]   Olin G.L., and Rhoades J.A. *The Five Most Costly Medical Conditions, 1997 and 2002: Estimates for the US Civilian Noninstitutionalized Population*. Statistical Brief #80. Rockville: AHRQ Publications; May 2005.

[4]   Gawande A. The Hot Spotters: Can We Lower Medical Costs by Giving the Neediest Patients Better Care? *The New Yorker*. Jan 2011; 24: 40–51.

[5]   McMurdo M.E.T., Millar A.M., and Daly F. A Randomized Controlled Trial of Fall Prevention Strategies in Old Peoples' Homes. *Gerontology*. 2000; 46: 83–87.

[6]   Vik S.A., Hogan D.B., Patten S.B., and Johnson J.A. Medication Non-adherence and Subsequent Risk of Hospitalization and Mortality among Older Adults. *Drugs & Aging*, April 2006, 23(4): 345–356.

[7]   Chee T.T., Ryan A.M., Wasfy J.H., and Borden W.B. Current State of Value-Based Purchasing Programs. *Circulation*. 2016; 133: 2197–2205.

[8]   The Health Care Payment Learning & Action Network (HCPLAN). *Alternative Payment Model APM Framework*. MITRE; 2017.

[9]   Zeigler B.P., Carter E., Seo C., Russell C.K., and Leath B.A. Methodology and Modeling Environment for Simulating National Health Care. *Proceedings of the Autumn Simulation Multi-Conference*, San Diego, CA, USA, 28–31 Oct 2012. pp. 28–31.

[10]   Aumann G.A. A Methodology for Developing Simulation Models of Complex Systems. *Ecological Modelling*. 2007; 202: 385–396.

[11]   Craig C., Eby D., and Whittington J. Care Coordination Model: Better Care at Lower Cost for People with Multiple Health and Social Needs. *IHI Innovation Series White Paper*. Cambridge, MA: IHI; 2011.

[12]   Wickragemansighe N., Chalasani S., Boppana R.V., and Madni A.M. Health Care System of Systems. In Jamshidi M. (ed.). *Systems of Systems – Innovations for the 21st Century*. 1st Edition. Hoboken: Wiley; 2008. pp. 542–550.

[13]   Yilmaz L., and Ören T.I. (eds.). *Agent-Directed Simulation and Systems Engineering*. Weinheim: Wiley & Sons; 2009.

[14]   Gianni D. Bringing Discrete Event Simulation Concepts into Multi-agent Systems. *Proceedings of the 10th International Conference on Modeling & Simulation*, Cambridge, UK, 1–3 Apr 2008. pp. 186–191.

[15]   Bergez J.-E., Chabrier P., Gary C., *et al*. An Open Platform to Build, Evaluate and Simulate Integrated Models of Farming and Agroecosystems. *Environmental Modelling & Software*. 2013; 39: 39–49.

[16]   Duboz R., Bonté B., and Quesnel G. Vers Une Spécification des Modèles de Simulation de Systèmes Complexes. *Studia Informatica Universalis*. 2012; 10: 7–37.

[17]   Zeigler B.P., Praehofer H., and Kim T.G. *Theory of Modeling and Simulation: Integrating Discrete Event and Continuous Complex Dynamic Systems*. 2nd Edition. New York, NY: Academic Press; 2000.

[18]   Institute of Medicine. *Learning Healthcare System Concepts*. Annual Report. Washington: National Academies Press; 2008.

[19]   Kleinberger T., Becker M., Ras E., Holzinger A., and Müller P. Ambient Intelligence in Assisted Living: Enable Elderly People to Handle Future Interfaces. *Universal Access to Ambient Interaction*, Vol. 4555. Heidelberg: Springer; 2007. pp. 103–112.

[20]   Barros F.J. Modeling Formalisms for Dynamic Structure Systems. *ACM Transactions on Modeling and Computer Simulation*. 1997; 7(4): 501–515.

Epilogue

Where do we go from here—Global and National Healthcare Maturity

In this epilogue, we place the technical discussion of the book into global national healthcare perspectives. To set the framework for such a discussion, we introduce the concept of maturity levels. Then, we discuss what it would take for a system to move up from where it is to the top maturity level and how using this book can help.

E.1 Health system capability maturity

Healthcare system structures can be identified at individual hospital, local, regional, national, and global scales. We would like to assign maturity levels to healthcare systems at each scale. Such assignment can serve to characterize the differences in capability at each hierarchical level in part by aggregation of the capability of the organizations below it.

One approach to such stratification is shown in Table E.1 which characterizes stages in transition to Accountable Coordinated Care [1].

This linear progression of discrete stages affords insight into the increasing capabilities that systems would possess as they attained higher levels of maturity. However, its grouping of several distinct factors together may be too coarse to allow more insightful comparison of systems at different scales.

A finer, but still practical, approach can be had by considering the following dimensions of capability:

- **Health information exchange (HIE):** Protocols for sharing of electronic medical records (EMRs) among patient care providers,
- **Coordination** of services for patients with multiple problems and/or special needs, and
- **Value based:** Incentives and methods for performance-based payment (versus fee-for service).

Each of these dimensions might in principle be present or absent independently at an organizational level (Figure E.1). For example, a hospital might have the means to share patient information among its departments but not coordinate operations among them or have in place schemes to impute charges to end-to-end treatments that reflect actual costs of services involved. Furthermore, a capability might be

Table E.1   Stages in transition to Accountable Coordinated Care

Stage	Description
Independent care	Individuals or separate organizations provide much of the patient care with any collaboration falling on the patient
Connected care	Basic care collaboration like sharing patient charts or sending messages about patient care between organizations
Coordinated care	Sufficient patient data to coordinate care by case tracking and health risk management between organizations such as HIPAA business associates
Integrated care	Sharing data between organizations with increased fluidity as if boundaries were removed leads to population management and increased patient engagement
Accountable care	Organizations become accountable for outcomes and costs together as team leading to information therapy and personalized behavior prescriptions

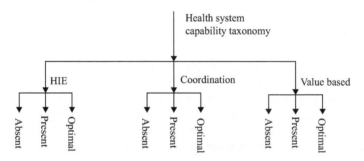

Figure E.1   Health system capability taxonomy

present at one scale but not at another. For example, clinical treatments are well coordinated in a hospital setting but not so well when the patient returns home.

Presence or absence of vector-listed properties offers an easy means of comparing systems, resulting a partial ordering of overall capability. For example, (absent, absent, absent) represents a system with no capabilities, while (present, absent, absent) represents one with ability to share health data. Assigning 0 or 1 for absent or present, respectively, we rank a system higher than another if its assigned values are greater than those of the second in each dimension.

Such an ordering is partial since systems better in one dimension but worse in another are not comparable. A full order is obtained by assigning a score to each vector which is the sum of the numerical values, effectively counting the number of capabilities that are present. To enable compare systems that have capabilities but not equally evolved, we have added an "optimal" label with assigned value of 2 which more refines partial order as well as full ordering by the summation score. In this metric, a system with no capabilities, all capacities present but not optimal, and all optimal, is ranked 0, 3, and 6, respectively.

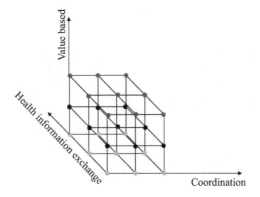

Figure E.2   Health system capability lattice

As depicted by Figure E.2, maturity levels form a 3D lattice, where (a, b, c) denotes a node with a level of HIE ($0 \leq a \leq 1$), b level of coordination ($0 \leq b \leq 1$), and c level of value-based capability ($0 \leq c \leq 1$).

From that perspective, healthcare reform entails the following steps:

1. Identify current (a, b, c) for the healthcare system of interest,
2. Define objectives (a', b', c') such that $c' > c$, and
3. Find a path $(a, b, c) \rightarrow (a', b', c')$.

Possible realistic paths for (a, b, c) to achieve $c' > c$ are shown in Figure E.3:

1. $(a, b, c) \rightarrow (a, b, c')$
2. $((a, b, c) \rightarrow (a, b', c)$ with $b' > b)$ and $((a, b', c) \rightarrow (a, b', c'))$,
3. $((a, b, c) \rightarrow (a', b, c)$ with $a' > a)$ and $((a', b, c) \rightarrow (a', b, c'))$, and
4. $((a, b, c) \rightarrow (a', b, c)$ with $a' > a)$ and $((a', b, c) \rightarrow (a', b', c)$ with $b' > b)$ and $((a', b', c) \rightarrow (a', b', c'))$.

Path [1] signifies value-based capability is brought (if absent) or made optimal (if already present) by additional means than the improvement of HIE and coordination. Path [2] is based on the postulate that more value-based capability requires more coordination. Path [3] means that value-based capability improvement is obtainable through HIE improvement. Path [4] can be stated as "more HIE for more coordination, and more coordination for more value-based." In a practical way, the possibility for following these paths depend on the characteristics of the healthcare system.

Healthcare system reformers and planners should keep Figure E.3 in mind as they contemplate directions in which to purchase Information Communication Technology (ICT) systems and institute care coordination and alternative payment schemes. More than this, the abstractions in the figure can be incarnated with more details specific to the systems under consideration and optimal paths from current state to desired state might be found. We return to consider some ways in which the book's methodology might help in this process in the Epilogue.

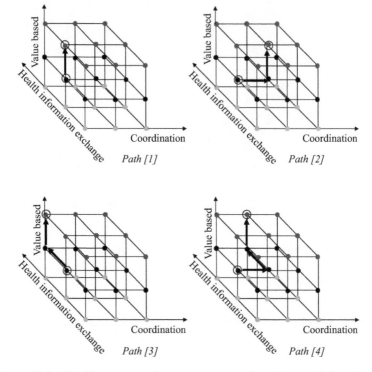

Figure E.3   Health system reform seen as a path in the capability lattice

E.2   Health system performance

The World Health Organization (WHO) developed a framework to measure member countries' attainment of health (how well a country is doing in fostering the health of its citizens) and performance (how well it could be doing with the resources it has) [2]. The employed indicators are as follows:

1. ***Health DALE (Disability adjusted life expectancy)***: A measure of overall population quality of health,
2. ***Health distribution equality***: The extent to which DALE is distributed equally among the citizens,
3. ***Responsiveness-level***: The responsiveness of the system to health concerns including access to services,
4. ***Responsiveness-distribution***: The extent to which responsiveness is equally distributed, and
5. ***Fair-financing***: The fairness of financing in the way in which healthcare services are financed.

The objective of the measurement framework is to determine the impact of a county's attempts at healthcare reform. The above indicators are considered as outputs with inputs of two kinds: controllable and uncontrollable. The primary controllable input is the total healthcare expenditure per capita (both private and public). The primary uncontrollable input is the country's literacy, measured by the average years of schooling of the population over 15. The idea is that health attainment is higher when governments and citizens spend more on healthcare which is something they can choose to do or not. On the other hand, a country with high literacy rate (a proxy for per capita income) naturally does better than one with a lower rate.

Let Composite$_i$ refer to a composite index computed by for country i appropriately normalizing and weighting the basic indicators listed above. This refers to i's level of attainment. However, we can also estimate i's level based only on its estimated level if has the literacy rate it had in 1900. Let this be called its minimum attainment level, Composite$_{i,\text{min}}$. Also, we can estimate how high its level could be if it invested all its gross domestic product on healthcare, called its Composite$_{i,\text{max}}$. Then, its performance is measured by

$$\text{Efficiency}_i = \frac{\text{Composite}_i - \text{Composite}_{i,\text{min}}}{\text{Composite}_{i,\text{max}}\,\text{Composite}_{i,\text{min}}}$$

This estimates the gap between where it is and where it was in 1900 relative to the gap between the best that it can do now and where it was in 1900. A representative ranking of WHO member countries on efficiency is shown in Figure E.4 [2].

It will be interesting to see how the efficiency ranking correlates with Health System Maturity as defined earlier. The belief that the dimensions (of HIE, coordination, and value-based care) are significant for healthcare reform is tantamount to the assumption that the correlation will be high.

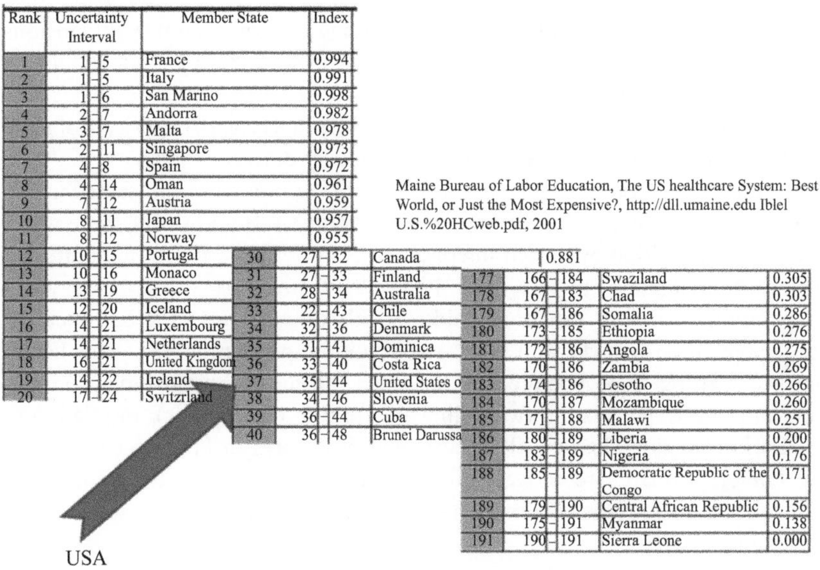

Rank	Uncertainty Interval	Member State	Index
1	1–5	France	0.994
2	1–5	Italy	0.991
3	1–6	San Marino	0.998
4	2–7	Andorra	0.982
5	3–7	Malta	0.978
6	2–11	Singapore	0.973
7	4–8	Spain	0.972
8	4–14	Oman	0.961
9	7–12	Austria	0.959
10	8–11	Japan	0.957
11	8–12	Norway	0.955
12	10–15	Portugal	
13	10–16	Monaco	
14	13–19	Greece	
15	12–20	Iceland	
16	14–21	Luxembourg	
17	14–21	Netherlands	
18	16–21	United Kingdom	
19	14–22	Ireland	
20	17–24	Switzerland	

Rank	Uncertainty Interval	Member State	Index
30	27–32	Canada	0.881
31	27–33	Finland	
32	28–34	Australia	
33	22–43	Chile	
34	32–36	Denmark	
35	31–41	Dominica	
36	33–40	Costa Rica	
37	35–44	United States o	
38	34–46	Slovenia	
39	36–44	Cuba	
40	36–48	Brunei Darussa	

Rank	Uncertainty Interval	Member State	Index
177	166–184	Swaziland	0.305
178	167–183	Chad	0.303
179	167–186	Somalia	0.286
180	173–185	Ethiopia	0.276
181	172–186	Angola	0.275
182	170–186	Zambia	0.269
183	174–186	Lesotho	0.266
184	170–187	Mozambique	0.260
185	171–188	Malawi	0.251
186	180–189	Liberia	0.200
187	183–189	Nigeria	0.176
188	185–189	Democratic Republic of the Congo	0.171
189	179–190	Central African Republic	0.156
190	175–191	Myanmar	0.138
191	190–191	Sierra Leone	0.000

Maine Bureau of Labor Education, The US healthcare System: Best in the World, or Just the Most Expensive?, http://dll.umaine.edu Iblel U.S.%20HCweb.pdf, 2001

USA

Figure E.4   Extract of ranking of WHO member countries on Efficiency

Table E.2   Compared capability maturity

Capability maturity dimension/ country	Health information exchange (HIE)	Coordination of services	Value based: incentives and methods
USA	Mandated motion toward standardized EMR but current systems are not able to support coordination of services	Coordination is increasing in specific aspects of treatment but not yet in end-to-end care value chains (as supportable by pathways)	Motion toward alternative payment models that implement pay-for-performance, accountable care organizations, and risk/reward schemes
France	Full adoption of personal health card that carries patient health and coverage information but no electronic sharing of patient information	Encouraged coordination by general practitioners (GP) of patient's specialized providers and services	Incentives for patients to employ GP selection of patients' follow-on. New propositions to rationalize health public expense by instoring preconized patient pathways
Nigeria	Absent	Absent	National Health Insurance scheme that protects citizens against high costs of treatment and a fair financing of healthcare providers

For example, France as highest in the efficiency ranking should prove to be relatively advanced in capability maturity. USA ranks lower than France and most European countries so it should not be as high along certain capability dimensions as they are. African countries are low ranked in efficiency, and this should be correlated to low healthcare service capability.

Table E.2 summarizes the situation in USA, France, and Nigeria, from the perspective of the Health System Maturity previously defined.

The current position of France in the WHO raking previously presented is due to more than one century of healthcare policies evolution. Before 1900, there was a long standing opposition by vested interests to reform of government to controlled public health in France. In the years 1900–20, it was finally overcome when the recognition took hold that poor public health was acting to the detriment of France's population growth and military recruiting population [3]. The health system was starting to be operated by the state, but the planning dimension still remains an issue when relating extra-clinical care (out of hospitals). In relation to healthcare information sharing, all doctors and drugstores in France can read the *"Carte Vitale,"* a smart card with all information on the patient and the coinsurance company for payment. This information is secured, confidential, and maintained by the government only.

As mentioned in Chapter 9, the coordination is still limited in France for what is called "medicine de ville" healthcare out of hospital. General practitioners act as "gate keepers" who refer patients to a specialist or a hospital when necessary. However, the system offers free choice of the reference doctor, which is not restricted to only general practitioners but may be a specialist or other doctor in a public or private hospital. The goal is to limit the number of consultations for the same illness [4].

In France, almost the entire population must pay compulsory health insurance through taxes on incomes. The insurers are nonprofit public agencies that annually participate in negotiations with the state regarding the overall funding of healthcare in France. A tax is automatically deducted from all employees' pay. At least 75% of a patient's treatment cost is reimbursed and more can be covered also by voluntary health insurance offered by company that subsidizes plans managed by non-for-profit groups. In France, the public incentive to follow healthcare regulation is financial in that expenses are reimbursed at much lower rates for patients who go directly to another doctor than the attributed one (except for dentists, ophthalmologists, gynecologists, and psychiatrists); vital emergencies are still exempt from requiring the advice from the reference doctor, who will be informed later. Because costs are borne by the patient and then reimbursed (most of the time on the spot using the smart card), patients have freedom of choice of where to receive healthcare services. Nevertheless, we have observed in Chapter 9 that free regulation of healthcare resources and patient freedom to choose appropriate resources lead to unbalanced division of resources among territorial units and dissatisfaction of patients in terms of distance and time needed to reach the different resources in a care delivery chain.

The situation in Nigeria contrasts severely with the ones in France and USA. Historically, the policy for Nigeria's Five Year Strategic Plan for healthcare was launched by the government only in 2004. In 2006, a new national health policy was adopted, which includes a redesigned National Health Insurance scheme that protects citizens against high costs of treatment and a fair financing of healthcare providers. Insurance coverage reaches about 5 million people in 2014 (a 3% of the total Nigerian population). Beyond the Nigerian case, a more general observation for almost all of sub-Saharan Africa is the following [5]:

- **Quality-wide:** Health systems in sub-Saharan Africa are organized in a pyramidal fashion and remain weak; consequently, the health situation of the populations remains worrying. Medical research is limited by lack of resources, and brain drain accentuates the staffing deficit. At the same time, the fight against fake medicines remains a challenge. On the other hand, the use of information and communication technology in medicine is very low.
- **Cost-wide:** Health expenditure is reduced, to the detriment of quality and access to care. Free healthcare tends to develop, as well as the promotion of low-cost generic drugs.
- **Access-wide:** Very poor rural people have less access to healthcare; consequently, universal access to care remains a high priority in sub-Saharan Africa. However, paradoxically, the rise of the private sector (private health insurance and private clinics) tends to increase the prospects for improving access to care.

E.3   Model development and application under the multi-perspective modeling framework

The multi-perspective modeling framework expounded in this book (Part 1) underlies the development of Discrete Event System Specification (DEVS)-based pathways-based self-improving and efficient management architecture models (Part 2) and their application (Part 3). Such development proceeds along parallel paths of simulation model development, design of coordination and learning sub-models, testing of the sub-models in the simulation model, and implementation of the sub-models within actual healthcare environments. Progress in the development and implementation can proceed independently in the absence of maturity of the models to support implementation.

Efforts to implement pathway-based care coordination can proceed in parallel with development of supporting system models, which when mature would support more confident implementation. Although ideally the models would proceed the implementations, real world circumstances may dictate that such precedence cannot be respected. In the case of the United States, implementation of pathways-based coordination occurred a decade or so before corresponding modeling was undertaken as described in Part 2. In contrast, as discussed, Europe and Africa are in a position to implement care coordination based on models having been developed and tested.

The situations of model development and implementation in Europe, Africa, and USA can be considered and compared. For Africa, the development of multilayer integrated simulation models is proceeding with focus on West Africa where healthcare is underdeveloped and besieged with epidemic outbreaks. For example, Ebola is a nonstandard infectious disease, due to its very high contamination potential—simple contact is enough, unlike HIV–AIDS which requires fluid transmission inside the body. Such models are not easily analyzed using conventional simulation languages and require DEVS-based methodologies. Hierarchies of scale for both systems (e.g., local units, regional centers, and organizations; cell, individual, and population) and processes (e.g., contamination, disease, and epidemic) are in development. Development of this model family will be employed to test pathways-based care coordination approaches to Ebola outbreak control in West Africa.

In order to improve its efficiency and to guard against "medical deserts," French health working groups have made available some qualitative and quantitative data regarding French "health territorial infrastructure" (including categories of resources and their positioning on the territory). Moreover, in France institutional collections of patients' medical pathways are available. They describe the advocated stages and pathways of patient care in the treatment of a disease. Currently, these processes are semiformal, although some work already aims to formalize them for improved executability. However, French health data that are mostly static and typically do not have a temporal dimension. In Chapter 9, we discussed adding the temporal dimension and showed how to simulate the French health territories with a virtual population but still consistent with demographic and epidemiologist data, for example, to simulate the capacity to contain an epidemic propagation. Clearly, this work's characterization of care value chains in the geographic context lays a foundation for implementation of pathways-based care coordination assure that the chains are implemented as effectively and efficiently as possible.

A hip care and fall prevention program in Ireland serves as an example of modeling prior to implementing coordinated pathways for improving patient health and costs of care. A study similar to that described above is being conducted in relation to designing coordinated pathways within healthcare services in Ireland [6]. Since 2012, the healthcare system in Ireland has been undergoing a radical reform based on a phased strategy. The fundamental goal of the reform is to transition the healthcare system toward an integrated delivery of healthcare. Starting with an outlook on the population, the implications of ageing continue to be a matter of considerable concern for healthcare delivery in Ireland. The Health Service Executive (HSE) of Ireland reported in 2014 that the increase in the number of people over 65 is approaching 20k per year [7]. Population ageing is therefore expected to have profound impacts on a broad range of economic and social areas.

In response to the foreseeable challenges, the healthcare system has been undergoing a substantial reform based on a phased strategy since 2012. The fundamental goal of the reform is to transition the healthcare system toward the integrated delivery of healthcare services. The integrated care is adopted as a means to improve the services regarding accessibility, quality, and user satisfaction of care services. The transitional arrangements included structuring the Irish healthcare system into nine geographic regions named as Community Health Organisations, commonly known by its acronym CHO. The CHOs can be likened to the Accountable Care Organisations (ACO) in the US healthcare system [8]. Similar to the ACOs, the establishment of CHOs focuses on providing coordinated care for patients, such that they get the right care at the right time. To put it in more detailed words, the CHOs are aimed to serve as integrated service areas that can deliver better, more integrated, and responsive services to people in the most appropriate setting.

In this respect, coordinated pathways are embraced as an instrumental artifact in order to realize that goal. The focus of the in-progress study is specifically centered on the incidence of hip fractures among elderly patients. According to the Irish HSE, hip fractures were identified as one of the most serious injuries resulting in lengthy hospital admissions and high costs. A study currently in progress [9] is employing a hybrid approach that incorporates machine learning to extract relevant knowledge from the Irish Hip Fracture Database using data clustering and rule mining to explore possible correlations between patient characteristics, care-related factors, and treatment outcomes. Subsequently, this data-driven knowledge will be used to assist with the development of a simulation model to support the design of pathways-based coordination to improve end-to-end hip fracture treatment while reducing its cost.

Although we have presented a clearly stated DEVS-based multi-perspective modeling framework and supporting simulation computational methodology for addressing healthcare reform, much remains to do before this framework can become widely accepted among all stakeholders. Hopefully, a positive contribution toward such acceptance will be our presentation of metrics for maturity of value-based healthcare systems, putative correlation with measures for national healthcare reform attempts, and the current state of application of the framework presented here.

E.4   Open research

With Figure 12.3 in mind, for healthcare decision-makers and stakeholders to achieve the top level, value-based care, they require coordination, which in its turn requires the capabilities of universal collecting and sharing of health information (HIE). How can they reform their particular healthcare system to advance it in practical steps, starting, with targeted objectives for value-based care, from the current state of the system? This book can help by providing the M&S to plan, develop, and test these transitions. In this vein, we close the presentation with some tentative answers to the question of open research that might be of particular interest to readers of this book: professors and graduate students with background in M&S in healthcare related disciplines, researchers, and developers of health policy alternatives, professionals in information technology, medical technology developers, and other stakeholders.

What research needs to be done to make this M&S possible for people to undertake and apply?

As we have seen, the current research in healthcare is oriented to optimization of patient flow through processing systems rather than toward harnessing data and information technology to improve healthcare quality and patient outcomes and to provide a whole-person view of the patient. More research needs to be done on how to use the methodology to design HIE systems that can best support coordination and value-based care. More research has to be done to determine how to increase the impact of health information technology (HIT) innovation and HIEs on overall costs. One facet is the design of networks at the higher application layers of ICT to enable coordination among patients, providers, and payers for improved care at lower cost with wider access. In particular, we need to have better tools to model and test via simulation the types of alternative payment models for achieving increasing levels of value-based care (Chapter 12).

Based on its application in Chapter 9, more research needs to be done on how to use the methodology to introduce coordination into the French health system (noting its current maturity level according to Table E.2).

More research has to be done on how to use multi-perspective modeling to develop models of shared decision-making that are tailored to the needs of disadvantaged populations, whole-person care to improve health services for persons with multiple chronic conditions.

Not mentioned so far in this book is the use of information systems and data resources to provide meaningful clinical decision support to healthcare professionals and patients and families at the point of care. As discussed in Chapter 5, pathway-based coordination can help capture important actions and outcomes of healthcare to increase evidence on effective practices and support clinical and organizational improvement. More research can be done to support such use based on the book's M&S methodology by extending it toward decision support for providers of primary, secondary, and tertiary (coordinated) care.

Also not heretofore mentioned are preventive services, especially those involving behavioral change (e.g., obesity prevention and substance use prevention). Research can be done on using multi-perspective modeling to develop models of individual behavior (Chapter 2) and pathways-based coordination (Chapter 5) to support design of HIT to increase uptake and implementation of preventive services.

More research needs to be done on how to enhance the ability of healthcare organizations to evolve as learning health systems that effectively apply data and evidence to improve patient outcomes. Porter [10] states,

> Improving performance and accountability depends on having a shared goal that unites the interests and activities of all stakeholders. In health care, however, stakeholders have myriad, often conflicting goals, including access to services, profitability, high quality, cost containment, safety, convenience, patient-centeredness, and satisfaction. Lack of clarity about goals has led to divergent approaches, gaming of the system, and slow progress in performance improvement. Achieving high value for patients must become the overarching goal of health care delivery, with value defined as the health outcomes achieved per dollar spent.
>
> This goal is what matters for patients and unites the interests of all actors in the system.

Tools based on the methodology can be developed to understand the conditions under which a System of Systems (SoS), with a single unified goal, can emerge from a SoS with myriad, conflicting goals [11]. Further, as part of such development, we need to characterize the measurement and evaluation systems of healthcare SoS that support learning and continuous improvement as discussed in Chapter 5 and in Appendix D3.

Not mentioned so far in this book, cost and access to health insurance coverage remains a critical public policy issue. Research can be done on extending the multi-perspective modeling methodology to support obtaining accurate estimates of the size and composition of the insured and uninsured populations, as well as information on how demographic characteristics, economic factors, and health insurance subsidies affect health plan eligibility, decisions to enroll in health insurance plans, and trends in healthcare expenditures and affordability.

How to use the methodology to model emergence and evolution of healthcare as a complex adaptive system of systems (Chapter 11)? Simulations can predict how changes in policy affect the evolution of health insurance markets and the health insurance landscape. Multi-perspective healthcare diffusion models (Chapter 2) can help to understand how changes in the demographics of the population might affect healthcare access, coverage decisions, and healthcare expenditures.

As discussed in Appendix D3, medical, social, and behavioral risk factors help practitioners and policymakers make more informed decisions to improve health and socioeconomic outcomes. The multi-perspective methodology can be marshaled to support development of quantitative information addressing the relative impact of both singular risks and the combined effects of multiple multidimensional risk factors on health and socioeconomic outcomes. To date, pay for value systems most often focus almost completely on medical factors of risk. Simulations based on the methodology (Chapter 7) can be developed to design and test alterative payment models (Chapter 11) that include a broader and more quantitative assessment of the risk factors that are most likely to improve outcomes and value.

As discussed in Appendix D4, possible applications of the multi-perspective methodology encompass a large class of systems whose level of complexity requires a System of Systems approach. However, a series of critical questions immediately arise from such a generalization: *Can the four-layered approach discussed in Appendix D4 that was suggested to drive the Multi-perspective Modeling and Holistic Simulation (MPM&HS) process always be applied, regardless of the domain of interest? When applicable, under which conditions is the holistic integration semantically consistent? How can we validate such an integration?*

The main limitations of the four-layered approach are the a-priori choices made at the facet and the scale levels when building the domain ontology: the assumption that all models from all domains can be treated either as a production system or a consumption system, or a coordinating system, or a combination of the formers has no formal proof. Clearly, the generic MPM&HS cannot apply to domains where systems cannot be seen from these perspectives. However, the principle of building an ontology and applying an ontology-based plan–generate–evaluate process is still possible, provided all facets of the domain systems of interest are identified. The holistic integration principle also holds.

A key issue in developing multi-perspective models is the validity of the bridging components, i.e., the way parameters of a model are modified using outputs of other models. Such an approach promotes the idea that low-resolution processes (captured by parameters) in a given model aggregate high-resolution processes (i.e., the feeding models of the given model). A recurring question is whether it is legitimate (and desirable) to disaggregate or aggregate processes during the course of a given simulation run. Consequently, more research needs to be done to assess the validity of the integrator models. Indeed, integrators are models where issues, such as the how and validity of scale transfer, are addressed. The correlation (whether linear, quadratic, polynomial, or a more complex relationship) between outputs of some models and parameters of others is either an a priori knowledge or need to be established and assessed possibly through several simulation experiments, quantities of data collected, and statistical analysis.

References

[1]   Healthcare IT News. May 2013. Available at http://www.healthcareitnews.com/issue/May%202013 [Accessed on 18 May 2018].
[2]   Tandon A., Murray C.J.L., Lauer J.A., and Evans D.B. *Measuring Overall Health System Performance for 191 Countries*. GPE Discussion Paper Series, No. 30; Jan 2000.
[3]   Wikipedia. Healthcare in France. Available at https://en.wikipedia.org/wiki/Health_care_in_France [Accessed on 16 May 2018].

[4] Assurance Maladie. Choisir et Déclarer Votre Médecin Traitant. Available at http://www.ameli.fr/assures/soins-et-remboursements/comment-etre-rembourse/le-parcours-de-soins-coordonnes/choisir-et-declarer-votre-medecin-traitant.php [Accessed on 16 May 2018].

[5] Performances Management Consulting. *La Santé en Afrique Subsaharienne: Panorama, Problématiques, Enjeux et Perspectives.* Publication PMC; Jan 2010.

[6] Zeigler B.P., Carter E.L., Molloy O., and Elbattah M. Using Simulation Modeling to Design Value-Based Healthcare Systems. *OR58 Annual Conference*, Portsmouth, 6–8 Sep 2016. pp. 33–48.

[7] Health Service Executive. *HSE Annual Report and Financial Statements 2014.* Available at http://www.hse.ie/eng/services/publications/corporate/CHO_Chapter_1.pdf [Accessed on 12 June 2018].

[8] Gold J. *Accountable Care Organizations, Explained.* Kaiser Health News, 14 Sep 2015.

[9] Elbattah M., Molloy O., and Zeigler B.P. Designing Coordinated Care Pathways Using Simulation Modelling and Machine Learning. *Accepted in the Winter Simulation Conference,* Gothenburg, Sweden, 9–12 Dec 2018.

[10] Porter M.E. What is Value in Health Care?. *The New England Journal of Medicine.* 2010; 363:2477–2481. (10.1056/NEJMp1011024) Massachusetts Medical Society. Reprinted with permission from Massachusetts Medical Society.

[11] Zeigler B.P. Contrasting Emergence: In Systems of Systems and in Social Networks. *The Journal of Defense Modeling and Simulation: Applications, Methodology, Technology.* Mar 2016. doi: 10.1177/1548512916636934.

Appendix D1

D1.1 Formalizing Porter's integrated practice unit with system of systems modeling and simulation

https://www.youtube.com/watch?v=6_pmQDc6t2Y&t=1350s

This appendix provides the slides that are discussed in the YouTube video with title and URL given above. Readers may wish to follow the video using the slides.

- US healthcare system, the most expensive in the world,
  - an assemblage of *uncoordinated* subsystems embedded in a market economy
  - promotes fee-for-services without reference to the **end-to-end** quality of care and cost delivered to patients.
- Major aspects of care in need of coordination: *clinical* and *extra-clinical*
- *Clinical care* refers to the medical observation and treatment of patients proffered by physicians, hospitals, labs, etc.
- **Coordination** is required because the various specialists involved in collaborative treatments are too loosely coupled to provide the best care at the lowest cost
- *Extra-clinical care* requires coordination of services to help patients deal with the multiple tasks and decisions *before and after* in-hospital treatment where almost all decisions are made by doctors and nurses

---

Slide 1: Overview

- US healthcare system, the most expensive in the world,
  - an assemblage of *uncoordinated* subsystems embedded in a market economy
  - promotes fee-for-services without reference to the **end-to-end** quality of care and cost delivered to patients.
- Major aspects of care in need of coordination: *clinical* and *extra-clinical*
- *Clinical care* refers to the medical observation and treatment of patients proffered by physicians, hospitals, labs, etc.
- **Coordination** is required because the various specialists involved in collaborative treatments are too loosely coupled to provide the best care at the lowest cost
- *Extra-clinical care* requires coordination of services to help patients deal with the multiple tasks and decisions *before and after* in-hospital treatment where almost all decisions are made by doctors and nurses

---

Slide 2: Healthcare reform—a systems problem

- Porter* advocates radical reform
  - Physicians should re-organize themselves into team-based Integrated Practice Units (IPUs)
  - Provide full cycle of care centered around particular medical condition, e.g., asthma, breast cancer,...
- Porter defines patient value as outcome per unit cost at the output of the end-to-end care delivery value chain (CDVC)
- IPU must be considered as an entire system, currently fragmented into independent systems that include actors and services.
- Our objective: **Formalize Porter's IPU with System-of-Systems Modeling and Simulation**
  - Formulate criteria for creation of IPUs viewed as systems
  - Many questions, e.g., Can a collection of systems with their CDVCs be integrated into a viable system?

Slide 3: Fixing the healthcare system (* M.E. Porter & T.H. Lee, "The Strategy That Will Fix HealthCare," Harvard Business Review October 2013)

Slide 4: Wymore's Mathematical System Framework (T.I. Ören and B.P. Zeigler, "System Theoretic Foundations of Modeling and Simulation: A Historic Perspective and the Legacy of A. Wayne Wymore," SIMULATION September 2012 vol. 88 no. 9 pp. 1033–1046)

Slide 5: Discrete event system specification (DEVS) formalism

- DEVS provides computational basis for modeling and simulation using Wymore's system framework
- DEVS Multi-formalism systems support enables discrete event and continuous models in same simulation environment
- MS4 Systems provides DEVS-based -IDE http://ms4systems.com/
- MS4 IDE supports working directly with systems thinking & systems engineering concepts

Slide 6: MS4 systems DEVS integrated development environment (IDE)

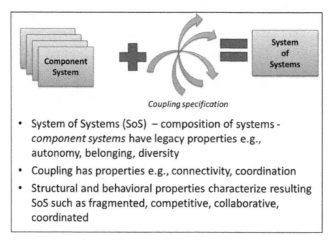

- System of Systems (SoS) – composition of systems - *component systems* have legacy properties e.g., autonomy, belonging, diversity
- Coupling has properties e.g., connectivity, coordination
- Structural and behavioral properties characterize resulting SoS such as fragmented, competitive, collaborative, coordinated

Slide 7: MS4 systems DEVS IDE support of systems of systems (Guide to Modeling and Simulation of Systems of Systems, B.P. Zeigler and H.S. Sarjoughian, Springer; 2013 edition (December 28, 2012))

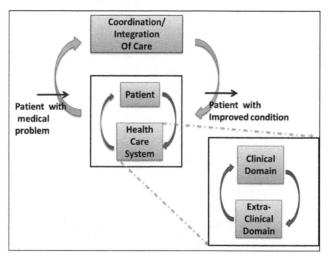

Slide 8: Healthcare system model

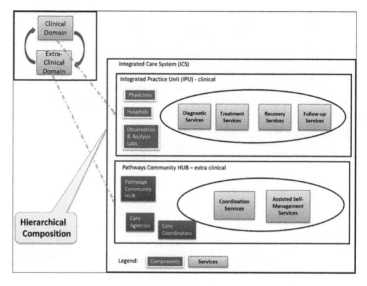

Slide 9: Integrated care system (ICS) model

The CDVC is the architectural blueprint of the IPU – it lays out the organization of its components and their coupling viewed as a SoS

Requirements for a properly constituted CDVC :

- The team includes members with the appropriate skills for the services in the CDVC
- The set and sequence of services is aligned with value
- The services have the right scopes to cover the target medical cluster of conditions and to minimally overlap
- The services are coordinated into a coherent whole with seamless handoffs from one to the other – minimize process delays and "dropping the baton"
- Services in the CDVC are performed in the right locations and facilities
- The right information is collected, utilized, and integrated across the care cycle.

Slide 10: Care delivery value chain for integrated practice unit

- **Autonomy**, Component system's ability to decide and act independently; Ranges from Conformance to Independence; Autonomy is ceded by parts in order to grant autonomy to the system
- **Belonging**, Component's functional ability to qualify for participation in SoS ; Ranges from Centralized to Decentralized
- **Connectivity**, Component's capability to exchange information with other components; Ranges from Platform Centric to Network Centric
- **Diversity**, Degree of notable processing differences among components, managed i.e. reduced or minimized by encapsulation, modular hierarchy; Ranges from Homogeneous to Heterogeneous
- **Emergence**, the appearance of new properties in the SoS not present in the components Foreseen, both good and bad behavior, and designed in or tested out as appropriate; Ranges from Foreseen to Indeterminable

Slide 11: Characteristics of components and couplings in SoS

SoS Characteristic	Application to IPU
Autonomy	Assuming that "as is" physicians are independent practitioners: how much autonomy must they give up to form the IPU team? Can they retain their economic sustainability if the IPU fails?
Belonging	Do physicians join the IPU on a cost/benefit basis? Do their skill sets match those needed by the CDVC? Can they adapt to provide needed services rather than the sophisticated ones they brought to the table?
Connectivity	Will the HIT infrastructure support the increased communication needed to enable the increased coordination?
Diversity	Are the EHR systems employed by the providers either the same or interoperable at the syntactic, semantic, and pragmatic levels? Do they provide the ability to synchronize to achieve common goals among the constituent systems?
Emergence	How strongly can the emerging CDVC be predicted from the formal composition of the individual CDVCs?

Slide 12: Characteristics of a system-of-systems and application to ICS

- Given
  - Medical Condition, e.g., Asthma
  - Patient Population, e.g., Youths
- Select Alternative Architectures to be explored:
  - CDVC and variations
  - Actors and Services Provided
  - Service Assignment to CDVC
- Design Outcome Measurement Component*
- Simulate architectural models to Evaluate Patient Value = Outcome/Cost
- Select Best architecture = model which maximizes outcome per unit cost

Slide 13: M&S-based virtual test of architectures for integrated care systems (Porter, M.E. "Measuring Health Outcomes," New England Journal of Medicine, December 2010)*

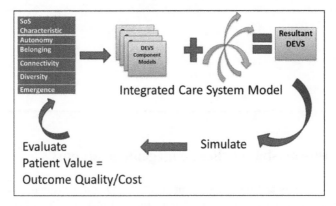

Slide 14: M&S-based virtual test of architectures for integrated care systems

- Health Care Reform is usefully viewed as a Systems Problem
- Porter's Integrated Practice Unit (IPU) within a more inclusive Integrated Care System (ICS) provides needed coordination and integration
- Formalized the ICS using System-of-Systems (SoS) theory expressed in the DEVS modeling and simulation framework
- MS4 Modeling and Simulation Environment based on DEVS supports design and implementation of ICSs in a systems engineering approach

Slide 15: Conclusions

Appendix D2

D2.1   Pathways-based client engagement support

Pathways-Based Client Engagement Support (PCES) is a coordination technology for community health workers (CHW) and care coordinators. CHWs use pathways to guide clients to resources and services and monitor/track their activities. Readers may wish to follow the Slides below while watching the video at the URL:

https://www.youtube.com/watch?v=OCfjasyz46s&t=7s

PCES is a coordination technology for community health workers (CHW) and care coordinators

CHW use Pathways to guide clients to resources and services and monitor/track their activities

A Pathway defines the

- o Problem to be addressed
- o Desired positive outcome
- o key steps required to achieve the outcome

PCES provides

- o Outcome-based accountability
- o Quantitative analysis for quality improvement

PCES is currently being applied to reduce

- o Over utilization of emergency medical services
- o Emergency Room readmissions

Slide 1: Pathways-based client engagement support (PCES) Web/Cloud technology

Client Name	
Date of Birth	
Community Health Worker	
Client's Place of Residence	☐ Capitol Heights ☐ Fairmont Heights ☐ Seat Pleasant ☐ Unincorporated area/town ☐ Other Jurisdiction (specify) _____
Insurance Coverage	
Pathway Start Date	
INITIATION	Identify reliable and affordable means of transportation for client.
Means of current transportation	☐ Family: ☐ Friend: ☐ Charity: ☐ MAT ☐ Metro Access ☐ Call a Cab ☐ Call a Bus ☐ Other Transportation Program. Please specify _____
Education	Date client educated on available transportation resources:

MAT: Free Transportation Service for qualified Medicaid Recipients	**Metro Access**: A shared-ride, door-to-door, paratransit service for people whose disability prevents them from using bus or rail	**Call-A-Cab**: A reduced rate single-ride Cab transportation assistance program for County seniors (age 60+) and/or County persons with disabilities.	**Call-A-Bus**: A demand response curb-to-curb service available to all residents of Prince George's County who are not served by or cannot use existing bus or rail services. However, priority is given to senior and persons with disabilities. Persons with disabilities must provide their own escort, if needed. Service animals are allowed for the visually impaired.
Set up MAT Intake Interview for client. Prep	Print Application from WMATA WebSite	Print application from the https://www.princegeorge	No application required.

Slide 2: PCES automates and replaces paper-based pathways

> • PCES is a coordination platform with 5 types of participants: Clients (patients), CHWs, Nurse Coordinators, Supervisors, and Data Analysts each with their own web browser interface and access rights and restrictions
>
> • Pathways-based analytics helps make decisions to improve community health quality and cost effectiveness

Slide 3: PCES targets community-based managed care organizations with under-served populations

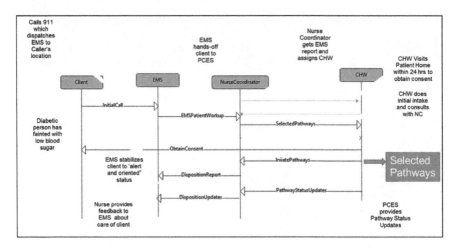

Slide 4: Pathways-based client engagement support (PCES): workflow and pathways

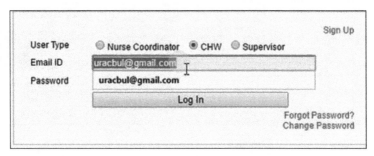

Slide 5: Community healthcare worker login

Slide 6: Nurse coordinator assigns clients to CHWs

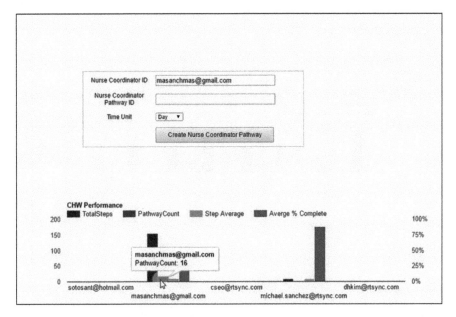

Slide 7: Nurse coordinator gets at-a-glance CWH performance information

Slide 8: CHW generates and assigns a pathway instance to a client

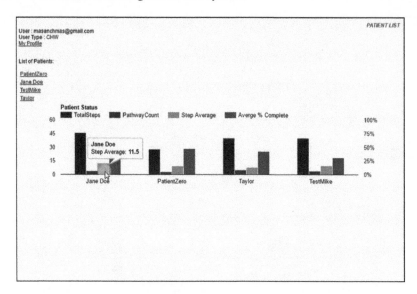

Slide 9: CHW gets at-a-glance client pathway status information

Slide 10: CHW generates and assigns a pathway instance to a client

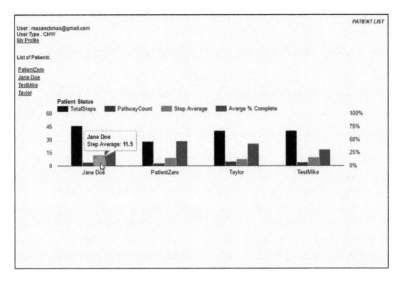

Slide 11: CHW gets at-a-glance client pathway status information

Slide 12: CHW interrogates state of pathway instance for client

Slide 13: CHW (1) inspects current pathway step, (2) confirms its completion, (3) sees next required step

Appendix D3

D3.1   Measuring health outcomes: the outcome hierarchy

In 2010, Porter presented the Outcome Hierarchy introduced in Chapter 5. In 2016, he presented Standardizing Patient Outcomes Measurement (Porter, Larsson, Lee, 2016) to review the concepts and discuss why there had not been as much progress toward adoption as would be expected. He indicated that the lack of outcomes measurement has slowed down adoption of alternative payment models including reimbursement reform and led to hesitancy among healthcare providers to embrace accountability.

The following material is Supplementary Appendix taken from:

Porter M.E. What is Value in Health Care? *The New England Journal of Medicine.* 2010; 363:2477–81 (10.1056/NEJMp1011024) Massachusetts Medical Society. Reprinted with permission from Massachusetts Medical Society.

MEASURING HEALTH OUTCOMES: THE OUTCOME HIERARCHY

Michael E. Porter, Ph.D.

Need for Outcome Measurement

Achieving good patient health outcomes is the fundamental purpose of health care. Measuring, reporting, and comparing outcomes is perhaps the most important step toward unlocking rapid outcome improvement and making good choices about reducing costs. Outcomes are the true measures of quality in health care. Understanding the outcomes achieved is also critical to ensuring that cost reduction is value enhancing (1–3). Thus, outcome measurement is perhaps the single most powerful tool in revamping the health care system. Yet systematic and rigorous outcome measurement remains rare or nonexistent in most settings.

Examples of Outcome Measurement

There are a growing number of examples of comprehensive outcome measurement that provide evidence of its feasibility and impact. At the national level, Sweden and Denmark are the clear leaders in establishing national quality registries covering many conditions (4). In the United States, federal legislation has mandated universal outcome measurement and reporting by all providers in organ transplantation (5), in vitro fertilization (6), and dialysis care. At the provider level, the most advanced largescale efforts are occurring in two German hospital groups and at some U.S. providers (4). Examination of these efforts leads to some clear conclusions:

1. First, in each case, outcome measurement has proven to be practical and economically feasible.
2. Second, accepted risk adjustment has been developed and implemented.
3. Finally, measurement initially revealed major variation in outcomes in each case, but led to striking outcome improvement and narrowing of variation across providers over time.

The feasibility and impact of comprehensive outcome measurement is no longer in doubt. However, the current state of outcome measurement leaves much to be desired.

Limitations of Current Approaches

There is no consensus on what constitutes an outcome, and the distinctions among care processes, biologic indicators, and outcomes remain unclear in practice. Outcome measurement tends to focus on the immediate results of particular procedures or interventions, rather than the overall success of the full care cycle for medical conditions or primary and preventive care. Even the best efforts are often limited to one or a small number of outcomes, frequently those that are most easily tracked. Measured outcomes often fail to capture dimensions that are highly important to patients. Finally, many outcome measurement efforts are ad hoc and not comparable across providers.

Framework for Outcome Measurement

This article offers an overall framework for outcome measurement to guide the development of the full set of outcomes for any medical condition. It introduces the outcome measures hierarchy as a tool for identifying the appropriate set of outcome dimensions, specific metrics, and associated risk factors. It explores the relationships among different outcome dimensions, their weighting by patients, and the relationship of outcomes to the cost of care. I examine the process by which outcomes improve over time as well as the evolution of risk factors. Finally, the article examines the benefits and costs of standardized or monetized outcomes across medical conditions. The detailed steps involved in creating and implementing an outcome measurement system are developed further in another article.

The Unit of Outcome Measurement

Outcomes are the results of care in terms of patients' health over time. They are distinct from care processes or interventions designed to achieve the results, and from biologic indicators that are predictors of results. However, discomfort, timelines, and complications of care are outcomes, not process measures, because they relate directly to the health status of the patient (1). Patient satisfaction with care is a process measure, not an outcome. Patient satisfaction with health is an outcome measure.

Principles of Outcome Measurement

In any field, quality should be measured from the customer's perspective, not the supplier's. In health care, outcomes should be centered on the patient, not the individual units or specialties involved in care. For specialty care, outcomes should be measured for each medical condition or set of interrelated patient medical circumstances, such as asthma, diabetes, congestive heart failure, or breast cancer. A medical condition includes common complications, coexisting conditions, or co-occurring conditions. Each medical condition will have a different set of outcomes. For primary and preventive care, outcomes should be measured for defined patient populations with similar health circumstances, such as healthy adults, disabled elderly people, or adults with defined sets of chronic conditions.

Outcomes should be measured for each medical condition covering the full cycle of care, including acute care, related complications, rehabilitation, and reoccurrences. It is the overall results that matter, not the outcome of an individual intervention or specialty (too narrow), or a single visit or care episode (too short). If a surgical procedure is performed perfectly but a patient's subsequent rehabilitation fails, for example, the outcome is poor. For chronic conditions and primary and preventive care, outcomes should be measured for periods long enough to reveal the sustainability of health and the incidence of complications and need for additional care.

Generalized outcomes, such as overall hospital or departmental infection rates, mortality rates, medication errors, or surgical complications, are too broad to permit proper evaluation of a provider's care in a way that is relevant to patients. Such generalized outcomes also obscure the causal connections between specific care processes and outcomes, since results are heavily influenced by many different actors and the specific mix of medical conditions for which care is provided.

Health care's current organizational structure and information systems make it challenging to properly measure outcomes. Thus, most providers fail to do so. Providers tend to measure only what they directly control in a particular intervention and what is easily measured, rather than what matters for outcomes. Providers also measure outcomes for the interventions and treatment they bill for, rather than outcomes relevant for the patient. Outcomes are measured for departments or billing units, rather than for the full care cycle over which value is determined. Much outcome work is currently driven by medical specialty expert or consensus panels, not by multidisciplinary groups for medical conditions. Faulty organizational structure also helps explain why physicians fail to accept joint responsibility for outcomes, defending this by their lack of control over "outside" actors involved in care (even those in the same hospital) as well as over patient compliance.

The first step in outcome measurement is to define and delineate the set of medical conditions to be examined (or the patient populations in primary care settings). Setting medical condition boundaries requires specifying the range of related diseases, coexisting conditions, and associated complications included, as well as the beginning and end of the care cycle.

For any medical condition (or patient population in primary care), defining the relevant outcomes to measure should follow several principles:

1. First, outcomes should involve the health circumstances most relevant to patients.
2. Second, the set of outcomes should cover both near-term and longer-term patient health, addressing a period long enough to encompass the ultimate results of care. For chronic conditions, ongoing and sustained measurement is necessary.
3. Third, outcomes should cover the full range of services (and providers) that jointly determine the patient's results.
4. Finally, outcome measurement should include sufficient measurement of risk factors or initial conditions to allow risk adjustment (see below).

The Outcome Measures Hierarchy

There are always multiple dimensions of quality for any product or service, and health care is no exception. For any medical condition or patient population, multiple outcomes collectively define success. The set of outcomes is invariably broad, ranging from immediate procedural outcomes, to longer-term functional status, to recovery time, to complications and recurrences. Survival is just one outcome, albeit an important one, as is the incidence of particular complications or medical errors. Medicine's complexity means that competing outcomes (e.g., near-term safety and long-term functionality) must often be weighed against each other. The full set of outcomes for any medical condition can be arrayed in a three-tiered hierarchy (see Figure D3.1). The top tier of outcomes is generally the most important, with lower-tier outcomes reflecting a progression of results contingent on success at higher tiers.

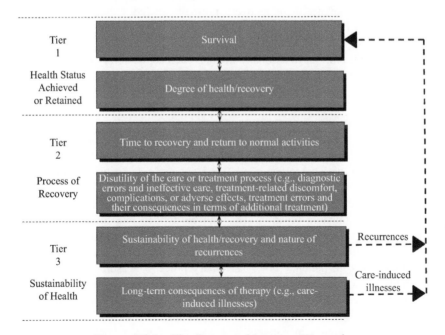

Figure D3.1   The Outcome Measures Hierarchy

Each tier of the hierarchy contains two broad levels, each of which involves one or more distinct outcome *dimensions*. Outcome dimensions capture specific aspects of patient health. These outcome dimensions are the critical dimensions of quality in health care. For each dimension, success is measured with one or more specific *measures* or *metrics*. Finally, for each measure there are often several choices in terms of the *timing* and *frequency* of when to measure it.

Tier 1

Tier 1 of the hierarchy is *patient health status achieved*, or for patients with some degenerative conditions, health status retained. The first level, survival, is of over-riding importance to most patients. Survival (or mortality) can be measured over a range of periods appropriate to the medical condition. For cancer, 1-year and 5-year survival are common metrics. Maximizing the duration of survival may not always be the most important outcome, however, especially for older patients who may weight other outcomes more heavily. I discuss the weighting of outcomes below.

Effective outcome-measurement systems must move well beyond survival, because survival alone omits many factors of great significance to patients. (Note that survival is sometimes used as a proxy for the broader effectiveness of care.) Measuring the full set of outcomes is also essential in order to reveal the connections between care processes or pathways and patient results.

The second level in Tier 1 is the *degree of health or recovery* achieved or retained. Regaining or preserving health is the ultimate purpose of most health care, with the exception of end-of-life or palliative services. Level two should capture the *peak* or *best steady-state* level of health achieved, defined according to the condition. Degree of health or recovery normally includes multiple dimensions such as freedom from disease and relevant aspects of functional status. For head and neck cancer, for example, level two outcomes include not only whether remission is achieved, but functional outcomes such as the ability to eat and speak normally, maintain appearance, and avoid depression (7).

Tier 2

Tier 2 of the outcomes hierarchy is the *process of recovery*. Recovery, or the process of achieving the best steady-state level of health attainable, can be protracted and arduous. Reducing the duration, complexity, and discomfort of recovery, in a manner consistent with achieving good Tier 1 outcomes, constitutes another group of important patient results.

The first level in Tier 2 is the *time* required to achieve recovery and return to normal or best attainable function. This can be divided into the time needed to complete various phases of care, such as time to diagnosis, time to treatment plan, time to care initiation, and duration of treatment. Cycle time is an outcome with major importance to patients, not a secondary process measure. Reducing cycle time yields direct benefits to the patient in terms of reducing the burden of recovery and can also affect health status achieved and its sustainability. For example, rapid initiation of therapy and avoidance of interruptions in therapy are often major influencers of prognosis in patients with cancer; after a myocardial infarction, faster time to reperfusion can improve function and reduce complications. The relationship between cycle time and health status achieved is just one of many instances in which outcomes at one level in the hierarchy can affect outcomes at other levels (see below).

The second level in Tier 2 is the *disutility of the care process* in terms of missed diagnosis, failed treatment, anxiety, discomfort, ability to work or function normally while undergoing treatment, short-term complications, retreatment, and errors, together with their consequences. This level can cover a wide range of dimensions depending on the condition. Ineffective or inappropriate treatments that fail to improve health will show up here, as will medical errors and treatment complications that lead to interruptions in care. Disutility of care will frequently affect the timeline of care.

Tier 3

Tier 3 is the *sustainability of health*. Sustainability measures the degree of health maintained as well as the extent and timing of related recurrences and consequences. The first level in Tier 3 is recurrences of the original disease or associated longer-term complications. Measures of time to recurrence and the seriousness of recurrence would fall here. The second level in Tier 3 captures *new* health problems created as a *consequence of the treatment itself*, or care-induced illnesses. When recurrences or new illnesses occur, some higher-tier outcome dimensions such as survival, degree of recovery from the recurrence, and so on, will also apply to measuring the outcome of these recurrences or illnesses (see the dotted lines in Figure D3.1).

With some conditions, such as metastatic cancers, providers may have limited impact on survival or other Tier 1 outcomes, or survival rates may be uniformly high. In these cases, providers can differentiate themselves on Tiers 2 and 3 by making care more timely, reducing discomfort, or limiting recurrences.

Defining Specific Outcome Dimensions and Measures

Each medical condition (or population of primary care patients) will have its own unique set of outcome measures. The importance of each tier, level, and dimension of outcomes will vary according to medical condition and sometimes according to the subgroup of patients. For most conditions, there will be multiple outcome dimensions at each level (with the possible exception of care-induced illness). The number of dimensions at each level will depend on the range of complications, the variety of treatment options, the duration of care, and so on. Broadly defined outcome concepts, such as functional status, must be subdivided into specific dimensions that are relevant to the condition. For example, rather than apply a generic activities of daily living assessment to all patients upon hospital discharge, the ability to eat and speak normally could be added to the measures tracked following head and neck cancer treatment.

Each outcome dimension may involve one or more specific measures and multiple periods. Survival is a single dimension, for example, but can be measured in a variety of ways and for several relevant periods. These choices will depend on the medical condition or patient population.

Selecting Outcome Dimensions

Figure D3.2 provides illustrative sets of outcome dimensions for breast cancer and acute knee osteoarthritis requiring replacement. These examples are not meant to be exhaustive, but to illustrate the structure for the comprehensive sets of outcome dimensions that are needed to fully describe patients' results — which most current measurement efforts fail to capture. No known organization systematically measures the entire outcome hierarchy for the medical conditions it addresses, though some are making good progress.

There are inevitably choices involved in selecting the set of outcome dimensions to measure. The most important criteria in making these choices should be importance to the patient, variability, frequency, and practicality. The outcome dimensions chosen should be important to the patient. Engaging patients and their families in defining this importance is an invaluable step, through focus groups, patient advisory councils, or other means. Outcome dimensions should be variable enough to require focus and improvement. Thus adverse outcomes chosen for measurement should occur often enough to justify the costs of measurement,

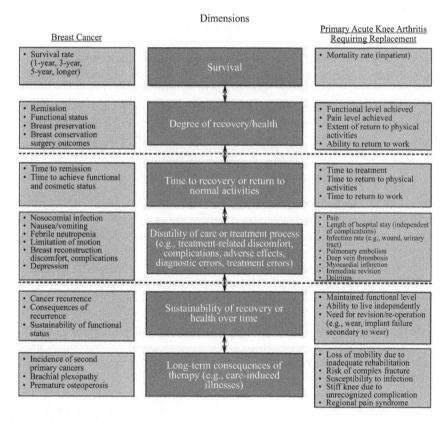

Figure D3.2   Illustrative Outcome Hierarchy for Breast Cancer and Knee Osteoarthritis

though very rare outcomes must be measured if they are very important to the patient. The practicality of accurate measurement must also play a role in determining what to measure, as noted above. Controllability, or the provider's current ability to affect the outcome, should be secondary because the key purpose of outcome measurement is to document problems that need to be studied and addressed.

At their outset, outcome-measurement efforts should include at least one outcome dimension at each tier of the hierarchy, and ideally one at each level. As experience and data infrastructure grow, the number of dimensions (and measures) can be expanded over time.

Relating Outcomes to Processes

To identify the set of outcome dimensions, a useful approach is to chart the cycle of care for the medical condition being examined. The care delivery value chain (CDVC), shown in Figure D3.3 for breast cancer, is a tool for mapping the full set of activities or processes involved in care (8). The CDVC highlights the full care cycle and all the involved entities or units. This full map of the care cycle allows a systematic identification of all the relevant outcome dimensions as well as when and where measurement should occur.

The CDVC not only helps to identify dimensions and measures, but also enables particular outcome dimensions to be linked to the specific processes of care from which they arise. The connections between the CDVC and outcomes, then, are important to guiding outcome improvement.

Selecting Particular Measures

To measure each outcome dimension, there are often a number of metrics or scales (e.g., the Medical Outcomes Study 36-Item Short-Form Health Survey [SF-36] or the Western Ontario and McMaster Universities Osteoarthritis Index [WOMAC]) that can be utilized. Some metrics, such as the EuroQol Group 5-Dimension Self-Report Questionnaire (EQ-5D) scale to measure health-related quality of life, are generic metrics that can be used for multiple medical conditions. Other measures or scales are tailored to disease classes (e.g., SF-36 for orthopedics) or to individual medical conditions.

The particular measures chosen for each outcome dimension should reflect a number of considerations:

1. First, measures should be selected that best capture the particular outcome from the perspective of the patient and medical science. Getting the measure right can have consequences. In in vitro fertilization (IVF), initial measurement focused on birth rates per IVF cycle, but this practice led to the implantation of numerous embryos and to a high number of multiple births (with a higher probability of complications). Over time, focus has shifted to birth rates per embryo implanted, and multiple births (especially triplet rates) have become a prominent outcome as well. The focus on measurement has played a major role in reducing triplet rates from 7 to 8% historically to less than 2%.

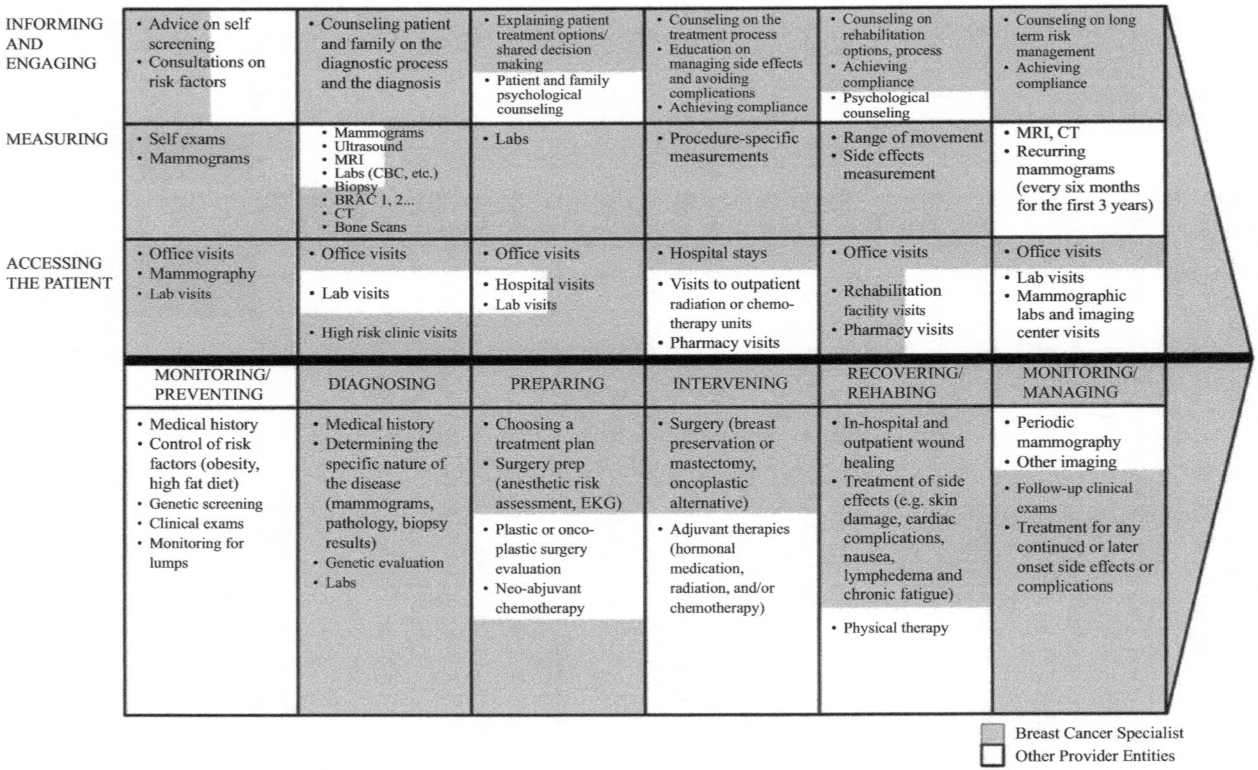

	MONITORING/ PREVENTING	DIAGNOSING	PREPARING	INTERVENING	RECOVERING/ REHABING	MONITORING/ MANAGING
INFORMING AND ENGAGING	• Advice on self screening • Consultations on risk factors	• Counseling patient and family on the diagnostic process and the diagnosis	• Explaining patient treatment options/ shared decision making • Patient and family psychological counseling	• Counseling on the treatment process • Education on managing side effects and avoiding complications • Achieving compliance	• Counseling on rehabilitation options: process • Achieving compliance • Psychological counseling	• Counseling on long term risk management • Achieving compliance
MEASURING	• Self exams • Mammograms	• Mammograms • Ultrasound • MRI • Labs (CBC, etc.) • BRAC 1, 2… • CT • Bone Scans	• Labs	• Procedure-specific measurements	• Range of movement • Side effects measurement	• MRI, CT • Recurring mammograms (every six months for the first 3 years)
ACCESSING THE PATIENT	• Office visits • Mammography • Lab visits	• Office visits • Lab visits • High risk clinic visits	• Office visits • Hospital visits • Lab visits	• Hospital stays • Visits to outpatient radiation or chemo-therapy units • Pharmacy visits	• Office visits • Rehabilitation facility visits • Pharmacy visits	• Office visits • Lab visits • Mammographic labs and imaging center visits
	• Medical history • Control of risk factors (obesity, high fat diet) • Genetic screening • Clinical exams • Monitoring for lumps	• Medical history • Determining the specific nature of the disease (mammograms, pathology, biopsy results) • Genetic evaluation • Labs	• Choosing a treatment plan • Surgery prep (anesthetic risk assessment, EKG) • Plastic or onco-plastic surgery evaluation • Neo-adjuvant chemotherapy	• Surgery (breast preservation or mastectomy, oncoplastic alternative) • Adjuvant therapies (hormonal medication, radiation, and/or chemotherapy)	• In-hospital and outpatient wound healing • Treatment of side effects (e.g. skin damage, cardiac complications, nausea, lymphedema and chronic fatigue) • Physical therapy	• Periodic mammography • Other imaging • Follow-up clinical exams • Treatment for any continued or later onset side effects or complications

Legend:
■ Breast Cancer Specialist
□ Other Provider Entities

Figure D3.3   The Care Delivery Value Chain (Breast Cancer)

2. A second consideration in choosing measures is that, other things being equal, the selection of standard and tested measures will improve validity and enable comparison across providers.
3. Third, measures should minimize ambiguity and judgment in scoring or interpreting, to ensure accuracy and consistency.
4. Fourth, patient surveys should be utilized to measure outcomes such as functional status and discomfort that reflect patients' realities and are difficult for outside parties to measure. Here, standardized scales such as the SF-36 or the Beck Depression Index are preferable when available. Compromises will often be necessary in measure selection, but the measures chosen can be improved over time.

Many outcome measures can be tracked at various times in the cycle of care or cover periods of varying durations. For example, as noted above, the time to recovery can be disaggregated into the time to diagnosis and treatment plan, the time between diagnosis and treatment, and the elapsed time during treatment itself. Timing and duration should reflect relevance to patients as well as periods long enough to reveal results.

Practical considerations, such as the availability of data and cost of information gathering, will also play a role in the measures selected. For example, billing data are often more easily accessible than data from chart reviews or new data entry, and measures calculated from billing data can be the place to start as information systems are improved. Practical considerations may also influence the number and duration of measurement periods chosen. For most conditions, immediate complications are far easier to track than longer-term measures that require patient follow-up. Overall, however, the orientation should be on reducing the cost of capturing the right measures rather than limiting measures to those that are easy to obtain.

Developments in electronic medical records are already making outcomes far less costly to measure. Information technology infrastructure should be designed to facilitate the extraction of clinical data for measurement purposes, in addition to supporting the care delivery process.

Relationships among Outcome Dimensions

The relative importance of particular outcome dimensions can vary according to individual patient preferences, as noted above. For example, the ability to restore full physical activity may be especially important to an avid athlete or to someone whose employment involves physical labor.

Measurement of the hierarchy can reveal that levels are mutually dependent, as represented in the figures by the bidirectional arrows between levels. Progress at one level sometimes positively affects other levels, reflecting complementarities among outcome dimensions. For example, reducing complications or eliminating errors will not only reduce the disutility of care but speed up recovery.

Such complementarities among outcome dimensions reveal important leverage points for care improvement. For example, error reduction can have special significance beyond its direct Tier 2 benefits because errors may have cascading consequences for recovery, time, discomfort, and risk of recurrence. Error reduction, then, has been a strategic type of outcome improvement to focus on.

Cycle time is another particularly leveraged outcome dimension for value improvement. As discussed, cycle time is an outcome itself, reflecting the duration of anxiety, discomfort, and poor health for the patient. However, speeding up diagnosis and treatment (e.g., avoiding interruptions in care) and better managing complications and rehabilitation often have major benefits for the likelihood and degree of recovery as well as its sustainability, such as in cancer care. The value benefits (outcomes achieved per cost incurred) of cycle time are amplified by its impact on cost. Faster cycle time usually means that fewer resources are required to care for the patient. Cycle time, then, is an outcome dimension that every provider should measure and work to improve, though few have yet begun to do so. Avoidable complications are another important set of outcome dimensions with important complementarity and cost effects.

Measurement of the hierarchy can also make explicit the tradeoffs among outcome dimensions. For example, achieving more complete recovery may require more arduous or time-consuming treatment or confer a higher risk of complications. Mapping these outcome tradeoffs, and seeking ways to reduce them, is an essential part of the care innovation process.

In cases where there are tradeoffs among outcome dimensions, patients may place different weights on each level and dimension of the outcome hierarchy. The discomfort of treatment willingly endured may be affected, for example, by the degree of recovery possible. The long-term sustainability of recovery, such as 20-year implant survival for patients who undergo hip replacement, may matter less to older patients than the degree and speed of recovery. Or considerations of disfigurement may weigh heavily against the risk of recurrence — for example, when determining the amount of the breast to be resected from a patient with breast cancer.

Differences in the value patients place on individual outcome dimensions does not reduce the need to measure the full hierarchy but makes it more important to do so.

Patients, their families, and their physicians, armed with information on a full set of outcomes, will be in a position to gain access to the treatments and providers that are best equipped to meet their particular needs (9–10). This level of outcome information goes well beyond what is currently available or even contemplated by medical societies and health plans in terms of consumer engagement.

Adjusting for Risk

The outcomes that are achievable will depend to some degree on each patient's *initial conditions*, sometimes also termed *risk factors*. Measuring and adjusting for initial conditions is therefore a crucial step in interpreting, comparing, and improving outcomes. In the case of breast cancer, for example, relevant initial conditions include the stage of disease at the initiation of care, the type of cancer (e.g., tubular, medullary, lobular, etc.), estrogen and progesterone receptor status (positive or negative), sites of metastases, and psychological factors, among others. Patients' compliance with treatment can also be interpreted as a risk factor — another reason why measurement of patient compliance is essential (1).

☐	Stage of disease
☐	Type of cancer (infiltrating ductal carcinoma, tubular, medullary, lobular, etc.)
☐	Estrogen and progesterone receptor status (positive or negative)
☐	Sites of metastases
☐	Previous treatments
☐	Age
☐	Menopausal status
☐	General health, including co-morbidities
☐	Psychological and social factors

Figure D3.4   Illustrative Risk Factors for a Patient with Breast Cancer

Risk adjustment is a complex topic, but I offer a number of strategic principles here. An illustrative set of initial conditions for breast cancer is shown in Figure D3.4. Initial conditions can affect all levels of the outcome hierarchy. Different initial conditions will often affect different outcome dimensions.

In order to evaluate outcomes for a medical condition, and especially to compare sets of outcomes over time or across providers, outcomes must be risk-adjusted or stratified by patient population based on the salient initial conditions. If initial conditions are not adjusted for, misleading conclusions can be drawn about the effectiveness of a treatment or provider that could mitigate the very purpose of outcome measurement. (An example of the risks of using outcome data without appropriate risk adjustment occurred when the state of Maine began to require drug-rehabilitation clinics to publish their outcomes). Subsequent studies have shown that the improvement in outcomes achieved in the years following the legislation were almost entirely attributable to clinics' turning away patients deemed likely to be problematic in order to increase their success rates (11). Several efforts to gather and report outcomes have failed due to inadequate risk adjustment, which has led to resistance and rejection by the medical community (8). That said, there are a growing number of successful risk-adjustment approaches that confirm its feasibility and impact.

Adjusting for risk is not only necessary for measuring outcomes accurately, but also for improving them. Understanding the link between risk factors and specific patient health outcomes is critical for care decisions.

Finally, risk adjustment is not only important for making comparisons, but is also essential to mitigating the risk that providers or health plans will "cherry pick" healthier patients to improve measured outcomes. Inadequate risk-adjustment methods, as well as poor understanding of actual costs, are root causes of the underpayment of providers for handling patients with more complex conditions, both in the United States and elsewhere (1). Flawed reimbursement for complex cases has many adverse consequences for value, ranging from inadequate care to excessive fragmentation of services as every provider is motivated to seek out "profitable" service lines and patient groups. Rigorous risk adjustment, coupled with corresponding reimbursement reform, will enable a move away from the current system of "profitable" and "unprofitable" interventions and patient populations and toward a system that encourages providers and health plans to focus on their areas of excellence.

Adjusting for initial conditions or risk normally involves two principal approaches. One is to stratify patient groups on the basis of the most important risk factors to allow outcomes for similar patients to be compared. This method is used in the area of in vitro fertilization, for example, where the Center for Disease Control reports birth rates according to maternal age cohorts and use of fresh or frozen embryos.

The other approach to risk adjustment is to utilize regression analysis to calculate expected outcomes, controlling for important patient risk factors. This allows average outcomes from different providers and periods to be adjusted for the patient mix or to be compared to expected outcomes for their particular patient populations. This method is utilized for outcome reporting in U.S. organ transplantation and in the Helios/AOK methodology in Germany focused on expected mortality for a wide array of medical conditions (4).

Both stratification and risk adjustment depend on having sufficiently large patient populations to support statistically meaningful comparisons. To accumulate adequate numbers of patients, it may be necessary to aggregate patients over time or to examine outcomes for teams rather than for individual practitioners. In U.S. organ transplantation, for example, data are normally reported for 3-year periods. In in vitro fertilization, one of the weaknesses in the current reporting system is that results are reported only for patients in the most recent year, not over longer periods.

However, statistical power should not be the principal objective or driver of outcome measurement. The principal benefit of outcome measurement is to inform and stimulate practice improvement. The measurement and tracking of outcomes have major benefits even if the number of patients does not allow fine comparisons. In organ transplantation, for example, only a subset of centers has outcomes that are statistically better or worse than expected. However, all centers track their progress, and centers with weaker outcomes work actively to improve them. I will discuss the difference between outcome measurement and traditional clinical trials further below.

The challenge of risk measurement has often been used as an argument against outcome measurement. Although adjusting for risk is surely challenging in some cases and will never be perfect, there is ample evidence that doing so is feasible and that inappropriate comparisons among providers can be minimized (12). Proven and accepted risk-adjustment methods for complex fields already exist in the United States and several other countries. There is also no doubt that risk-stratification and adjustment methods will continue to improve with experience and that gaming of measurement will be mitigated over time (8).

Risk Adjustment and Delivery Improvement

Even in its current imperfect state, risk adjustment is an essential tool for improving care delivery. Understanding and measuring patients' relevant initial conditions and their relationship to outcomes is indispensable to revealing new knowledge about medical conditions and their care.

The influence of initial conditions is partly inevitable — for example, the age of the mother appears to be a fundamental biologic influence on outcomes for in vitro fertilization (13). However, the influence of patient circumstances is partly a reflection of the state of understanding of a medical condition and its treatment. As clinical knowledge improves, certain risk factors may no longer meaningfully affect the outcomes of care, even though they may continue to influence the care process.

In vitro fertilization illustrates this learning process. Here, the biologic influences of age have been shown to weigh more heavily on egg production than on the ability to have a successful pregnancy. Through the use of donor eggs and improved technology for freezing a woman's own eggs, for example, older mothers are increasingly able to give birth to healthy children. So the impact of a mother's age has changed in terms of risk adjustment for the medical condition of infertility.

As learning occurs, risk adjustment for some initial conditions will become less necessary or even unnecessary for outcome comparison as providers manage them better. At the same time, new risk factors can emerge as sophistication in understanding a disease and in care delivery increases. This process of understanding and dealing with risk factors, then, is fundamental to driving value improvement. Advances in knowledge will reveal new, and perhaps more fundamental, initial conditions, such as genetic makeup. Yet improvements in care delivery over time can transform even genetic makeup from a risk factor to be adjusted for in comparing outcomes to a patient attribute that determines the best approach to successful care. Without systematic measurement of outcomes and risk factors, however, outcome improvement is hit-or-miss. The process of outcome measurement and risk adjustment is not only or even principally about comparing providers, then, but about enabling innovation in care.

These considerations suggest that it is preferable to err on the side of measuring more initial conditions rather than less and to create an explicit process for gradually revising the set of initial conditions used for risk adjustment. Most of all, the number and breadth of risk-adjustment studies and associated data collection must expand in every area of medicine to accelerate the rate of learning about care delivery.

The Outcomes Hierarchy and the Process of Value Improvement

Value improvement starts with defining and measuring the total set of outcomes for a medical condition and determining the major risk factors. Innovation in care delivery comes not only from focusing on individual outcome dimensions, but harnessing complementarities among various aspects of quality and reducing tradeoffs among outcome dimensions.

In medicine, as in most fields, progress in improving outcomes and value will be iterative and evolving. The outcomes hierarchy emphasizes that the pace of progress can vary across levels, and also among outcomes at a given level. As survival rates get high, for example, attention can shift to the speed and discomfort of treatment. Once the degree of recovery reaches an acceptable level, focus can shift to reducing tradeoffs between recovery and the risk of complications or care-induced illness, as in cancer therapy. Measurement of the entire outcome hierarchy not only encourages such improvements, but makes them more systematic and transparent.

Measuring the full hierarchy not only highlights multiple quality dimensions for improvement, but also expands the areas in which providers can distinguish themselves. As noted earlier, providers may achieve parity on some dimensions and then have to look to other dimensions to distinguish themselves. Or providers can concentrate on certain outcome dimensions that are weighted heavily by particular groups of patients.

In order to drive innovations in care, outcomes should be measured continuously for every patient, not just retrospectively in the context of discrete studies or evaluations. Whenever possible, outcomes should be measured in the line of care and inform continuous learning. The current approach to outcome measurement is skewed toward retrospective clinical studies, usually focused on a single end point. This bias towards clinical study methods is one of the reasons that outcome measurement remains so limited, despite its overwhelming benefits.

Comprehensive outcome measurement will enable a new type of clinical research, which focuses on overall care instead of controlled experiments around single interventions. Patient care is inevitably multidimensional, and actual care requires simultaneous choices on multiple variables and among numerous options. Conventional statistical methods need to be supplemented by careful study by clinical teams of patientspecific successes and failures. This kind of analysis seeks to identify common problems that arise, to discern patterns, and to develop hypotheses that give rise to learning, innovation, and further study.

Outcome Improvement and Cost Reduction

A major challenge in any field is to improve efficiency, and this is especially urgent in health care. One of the most powerful tools for reducing costs is improving quality, and outcome measurement is fundamental to improving the efficiency of care. Measuring the full outcome hierarchy provides a powerful tool for cost improvement that has been all but absent in the field. Comprehensive measurement of outcomes provides the evidence that will finally permit evaluation of whether care is actually benefitting patients and which treatments are most effective for each medical condition.

Historically, the overwhelming attention in outcome measurement has been directed at Tier 1 (health status achieved), particularly survival or mortality rates. At Tier 1, achieving better outcomes may (though by no means always does) require higher expenditures, especially when a new and expensive treatment or technology represents the only effective therapy. Such cases have led many observers to claim that innovation and new technology drive up health care costs. However, broader measurement of Tier 1 outcomes, notably functional status, will often open up opportunities for cost reduction. Improving the ability to function independently or return to work has huge cost consequences for the system.

Moreover, improvements in Tier 2 (process of recovery) and Tier 3 (health sustainability) outcomes almost invariably lower cost. Faster cycle time, fewer complications, and fewer failed therapies, for example, will have huge costs consequences. Tier 2 and 3 improvements can also reduce the cost of improving Tier 1 outcomes, because of the complementarities previously noted. For example, speeding up cycle time can also lead to more complete recovery, as is the case in cancer. Opportunities for dramatic improvement in Tier 2 and 3 outcomes engender great optimism for future cost containment; these opportunities have been over-looked because outcomes at these levels have been largely unmeasured and ignored.

Over the past several decades, joint replacement, new cancer therapies, organ transplantation, and many other new therapies were developed. In parallel, advancements in testing and diagnostic methods have allowed previously hidden conditions to be discovered or revealed much earlier. This stage of innovation, involving the development of new therapies for previously untreatable conditions and the discovery of previously hidden conditions, will almost inevitably raise cost, at least initially.

Today, however, the opportunity is different. Advancements in medical science have led to therapies that address most medical conditions in some way, albeit imperfectly. There will continue to be new tests and therapies where there were none before. However, the more common opportunity will be to drive dramatic value improvement in existing diagnostics and therapies, as well as to develop new, higher value therapies that address diseases at earlier stages or more fundamental levels. A new era of rapid improvement in value in health care is possible. Comprehensive outcome and cost measurement, together with supporting changes in care organization, reimbursement, and market competition, will be needed to unlock and drive such value based innovation.

Improving Value versus Rationing Care

Measuring the outcome hierarchy for each medical condition (and patient population receiving primary and preventive care) is indispensable for informing outcome improvement, assessing the value of alternative treatment approaches, and finding ways to deliver better outcomes more efficiently. Comparative-effectiveness research, in its present form, is important but not sufficient. It focuses largely on single interventions in highly controlled settings and sometimes incorporates just a single outcome or narrow set of outcomes. The outcome hierarchy is an important foundation for broadening and enriching clinical and comparative-effectiveness research at the medical condition level, as I have discussed. There have been efforts to monetize outcomes for purposes of calculating a benefit–cost ratio for alternative treatments. However, many such efforts tend to focus only on survival, even though survival is always one of a broader set of outcomes that matter to patients. Even for survival, assigning a monetary value is fraught with complexity, not to mention ethical issues. Is job productivity or earning power really a sufficient way to compare the health benefits of care, for example? Monetizing other important outcomes in the hierarchy from a benefit standpoint is even more challenging. For example, how should we value restoring the appearance of a patient with cancer or preserving a patient's normal voice?

The use of quality-adjusted life-years (QALYs) or disability-adjusted life-years (DALYs) represents a broader approach to collapsing outcomes into a single measure. Such measures embody a weighting of life expectancy based on quality of life. Quality of life is collapsed into a single number, determined using a variety of methods, despite the fact that it is inherently multidimensional and the relevant dimensions vary by medical condition.

At the medical condition level, we believe that there is little justification for shortcuts in measuring outcomes in driving value improvement. The full hierarchy of important outcomes needs to be measured and compared to cost. In evaluating alternative care delivery approaches, the task is to examine how the set of outcomes improves, and how improvement in the set of outcomes relates to cost. If one or more outcomes in the hierarchy improve while others remain stable, the set of outcomes improves. Value improves if outcomes improve at equal or lower cost, or if outcomes are stable at meaningfully lower cost.

There is no benefit to collapsing or suppressing outcome dimensions in making this evaluation at the medical condition level — quite the contrary. All parts of the outcome hierarchy are important to patients, and progress on each dimension is beneficial. Examinations of Tier 2 and Tier 3 outcomes, which are rarely considered in comparative-effectiveness studies, are powerful tools not only for outcome improvement but also cost reduction. There are certainly cases of tradeoffs — in which better outcomes occur only at much higher costs. However, there are virtually unlimited opportunities for improvement in the outcome hierarchy that do not involve such tradeoffs, and this is where attention in care improvement should be focused.

Monetization of outcomes and QALYs or DALYs are often used to compare the value of care across medical conditions. We know that for each medical condition, the set of relevant outcomes will be different. QALYs and DALYs focus just on those outcomes that can be readily standardized — again, survival and certain generic aspects of quality of life. Once again, the validity and comparability across conditions of these measures is highly questionable.

This effort to standardize and collapse outcomes to a single measure also suffers from a deeper problem. The whole approach assumes that the value of care for each medical condition is fixed and that care must be rationed. Optimizing within fixed constraints comes naturally to some economists but has proven shortsighted time and time again. In a field where outcomes are all but unmeasured, and where cost is poorly understood, there are major opportunities to improve outcome and value in the care for every medical condition. This is where the field should focus. Setting policies to enable and incentivize innovation should be our approach, rather than assuming that the value is fixed and focusing on choosing which patients should receive care. Given the major improvements in outcomes and efficiency observed in areas where there has been rigorous outcome measurement, there is every reason to hope that rationing will not be necessary except in extreme cases.

Health care is on a dangerous path if the primary rationale for outcome measurement is rationing of care rather than outcome and value improvement. Standardized outcome-measurement approaches will not well serve the needs of improving clinical practice, and they will disenfranchise providers. Turning to rationing without taking aggressive steps toward improving outcome and efficiency is a failure of policy — and will also prove unacceptable to patients and their families. Moreover, such policy will fail to be implemented when political realities intrude.

Conclusion

Outcome measurement is the single most important tool to drive innovation in health care delivery. The feasibility, practicality, and impact of outcome measurement have been conclusively demonstrated. Every provider can begin to measure the outcomes hierarchy in the medical conditions it serves, and track its progress versus past performance. Outcome measurement can begin for a subset of medical conditions and expand over time as infrastructure and experience grow.

This article provides a framework for systematically identifying the full set of outcomes for each medical condition, exploring the relationships among them, and revealing risk factors. Today, numerous voluntary and mandatory programs track different measures for subsets of providers, payers, and patient populations. The challenge is to make outcome measurement ubiquitous and an integral part of health care delivery.

Over time, the goal should be to establish uniform national and international outcome-measurement standards and methods. The feasibility of such standards has been conclusively demonstrated. Rather than resting with today's consensus organizations or government entities that are caught up in politics, responsibility for outcome measurement standards should be delegated to a respected independent organization, such as a new affiliate of the Institute of Medicine. Measurement and reporting of outcomes should eventually become mandatory for every provider and health plan. Reporting by health plans of health outcomes for its members, according to medical condition and patient population, using data drawn from providers' reporting, will help to shift health plans' focus from short-term cost reduction to value improvement.

As comprehensive outcome measurement is being phased in, every provider should report experience (i.e., the volume of patients treated for each medical condition), along with the procedures and treatment approaches utilized. Experience reporting will begin to help patients, their doctors, and health plans find the providers with the expertise that meets their needs. It will also highlight the fragmentation of care across facilities and providers and inform a rationalization of service lines. The most important users of outcome measurement are providers, for whom comprehensive measurement will lead to substantial improvement (5). The most important purpose of outcome measurement is improvement in care, not keeping score. Outcome measurement is also a powerful vehicle for bringing teams together and improving collaboration in a fragmented field. There is much evidence that the very act of measuring outcomes leads to substantial improvement. Public reporting of outcomes is not necessary in order to reap important benefits, and studies have revealed that confidential, internal reviews can motivate providers to improve their performance (14). Public reporting must be phased in carefully to win provider confidence. However, eventual progression to public reporting will accelerate innovation by further motivating providers to improve and permitting all stakeholders to benefit fully from outcome information.

From Harvard Business School, Boston.

References

[1]   Porter M.E. What is Value in Health Care? *The New England Journal of Medicine*. 2010; 363:2477–81 (10.1056/NEJMp1011024)

[2]   Porter M.E. A Strategy for Health Care Reform—Toward a Value-Based System. *The New England Journal of Medicine*. 2009; 361(2):109–112

[3]   Porter M.E. Defining and Introducing Value in Health Care. Evidence-Based Medicine and the Changing Nature of Health Care. 2007 IOM Annual Meeting Summary, 2008. pp. 161–72

[4]   Porter M.E., Molander R. Outcomes Measurement: *Learning From International Experiences*. Institute for Strategy and Competitiveness, Harvard University Working Paper, 2010. Available at http://www.hbs.edu/rhc/prior.html [Accessed on 29 Oct 2008]

[5]   Porter M.E., Baron J.F., Chacko J.M., and Tang R. *The UCLA Medical Center: Kidney Transplantation*. Harvard Business School Case 711-410, Boston: Harvard Business School Publishing, 2010

[6]   Porter M.E., Rahim S., and Tsai B. *In-Vitro Fertilization: Outcomes Measurement*. Harvard Business School Case 709-403, Boston: Harvard Business School Publishing, 2008

[7]   Feely T.W., Fly H.S., Albright H., Walters R., and Burke T.W. A Method for Defining Value in Healthcare Using Cancer Care as a Model. *Journal of Healthcare Management*. 2010; 55:399–412

[8]   Porter M.E., Teisberg E.O. *Redefining Health Care*. Boston: Harvard Business School Press, 2006

[9]   Wennberg J.E. Dealing with Medical Practice Variations: a Proposal for Action. *Health Affairs*. 1984; 3(2):6–32

[10] Wennberg J.E. Improving the Medical Decision-Making Process. *Health Affairs*.1988; 7(1):99–106

[11] Shen Y. Selection Incentives in a Performance-Based Contracting System. Health Services Research, April 1, 2003

[12] Boyce N. Using Outcome Data to Measure Quality in Health Care. *International Journal for Quality in Health Care*. 1996; 8(2):101–104

[13] Centers for Disease Control and Prevention. 2004 Assisted Reproductive Technology Success Rates: National Aummary and Fertility Clinic Report. Available at http://www.cdc.gov/ART/ART2004/index.htm [Accessed on 29 Oct 2008]

[14] Antonacci A.A., Lam S., Lavarias V., Homel P., and Eavey R.A. A Report Card System Using Error Profile Analysis and Concurrent Morbidity and Mortality Review: Surgical Outcome Analysis, Part II. *Journal of Surgical Research*. 2009; 153(1):95–104

[15] Porter M.E., Larsson S., and Lee T.H. Standardizing Patient Outcomes Measurement. *The New England Journal of Medicine*. 2016; 374(6): 504–506

Appendix D4

D4.1 Generic framework to Multi-Perspective Modeling and Holistic Simulation

As demonstrated in this book, a complete application of the Multi-Perspective Modeling and Holistic Simulation (MPM&HS) approach has been realized for healthcare systems analysis and design, leading to a framework under which healthcare systems simulation models develop facets related to Population dynamics, Individual behavior, Resource allocation and Health diffusion, either, one at a time, or by combining two or more of them. These facets cover the full set of healthcare M&S concerns as revealed by literature review, which, though interrelated, are often treated separately and the impact of other concerns on any one of them being approximated by parameters. However, a legitimate question is: "how do we identify the facets/perspectives in a general context and in a systematic way?"

We derive here a disciplined approach to building an ontology for the M&S of a domain of interest, as detailed hereafter.

The domain analysis ontology must provide a formal way to capture all the knowledge that might be in the range of M&S of the domain for which it is likely to be used. Therefore, it must capitalize on the abstractions used for the simulation of the entire targeted domain, beyond aspect specific modeling. Thus, the generic approach to domain analysis in MPM&HS is a 4-layered ontology which highlights at each layer a key generic characteristic.

As depicted by the SES presented in Figure D4.1, the following layers are defined:

Level 0. **System level:** This level recognizes the whole complex system as a juxtaposition of multiple facets, while various specializations can be identified as possible instances of the same integrated set of facets in various specific contexts. Examples in healthcare systems are specializations into primary, secondary, ternary and home care (Chapter 1). Similar examples in transportation/military/ ... systems are air, ground, marine ... specializations.

Level 1. **Facet level:** This level establishes three generic facets, i.e., "production facet," "consumption facet," and "coordination facet." In other words, a complex system is a whole made up of one or various facets, each of which being a production system (hence, leading to a ProF model), a consumption system (that gives a ConF model), or a coordinating system between production and consumption (giving a CooF model). These patterns encompass the

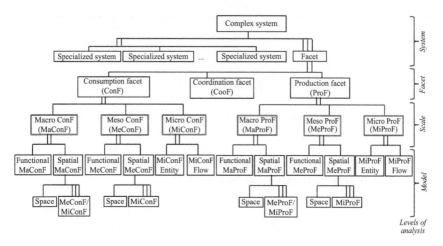

Figure D4.1   Generic SES for MPM&HS

traditional supply-demand duality that often characterizes complex systems (Chapter 2). The notion of "Production" encompasses the notion of "Supply" in that it involves not only the intentional supply of services needed, but all phenomena that produce positive and negative impacts on the system's stakeholders (such as vaccination and information diffusion to produce ease in healthcare, and contamination and epidemics to produce disease). Similarly, the notion of "Consumption" encompasses the notion of "Demand," as consumers may not necessarily be intentionally in demand.

Level 2. **Scale level:** This level emphasizes on that a characteristic feature of complex systems is the occurrence of interactions between heterogeneous components at different spatio-temporal scales with various interpretations of the notion of scale, and a major concern about scale transfer processes where inter-scale interactions must be properly described, as emphasized in Chapter 4. Thresholds between scales are critical points along the scale continuum where a shift in the importance of variables influencing a process occurs. As a result, the generic ontology proposed exhibits macro, meso, and micro levels of abstraction both within the consumption and the production facets, leading, respectively, to the generic MaConF, MeConF, MiConF, MaProF, MeProF, and MiProF models.

Level 3. **Model level:** This level identifies conventional models often originating from decades of theoretical findings as reusable artifacts to be selected and integrated in new M&S studies of complex systems. It defines the abstractions that can be directly simulated, by distinguishing four generic types of model, i.e., entity models, flow models, functional models and spatial models. While Entity models describe autonomous individuals with specific attributes and with or without goal-driven behavior, Functional models are formulated as mathematical equations, Spatial models are composed of individuals geographically located in a space model, and Flow models capture scenarios an individual can undergo. Consequently, the generic ontology has in each facet, entity and flow models at the micro level of abstraction, and functional and spatial models at the macro and meso levels of abstraction. Noteworthy is the fact that a spatial model at any macro level involves a space model where are located abstractions described at lower levels (i.e., meso and micro), and that similarly, a spatial model at any meso level involves a space model where are located abstractions described at the immediate lower levels (i.e., micro).

The generic ontology is meant to be instantiated in the analysis of any new domain of interest in view of its M&S. Such an instantiation provides the domain-specific ontology that will drive the MPM&HS process of the targeted domain.

Index